394e

NATURE
NURTURE
& PSYCHOLOGY

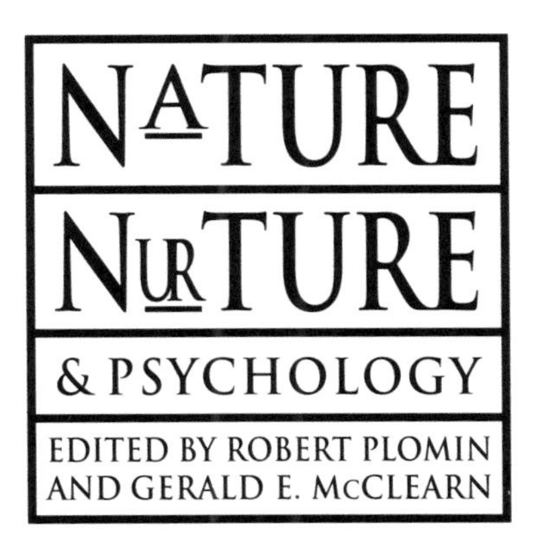

AMERICAN PSYCHOLOGICAL ASSOCIATION
WASHINGTON, DC

Published by
American Psychological Association
750 First Street, NE
Washington, DC 20002

First Printing July 1993
Second Printing March 1994

Copies may be ordered from
APA Order Department
P.O. Box 2710
Hyattsville, MD 20784

In the UK and Europe, copies may be ordered from
American Psychological Association
3 Henrietta Street
Covent Garden
London WCZE 8LU
England

Typeset in Century Book by Techna Type, Inc., York, PA

Printer: Quinn-Woodbine, Inc., Woodbine, NJ
Cover and Jacket Designer: Ethel Kessler Design, Inc., Bethesda, MD
Technical/Production Editor: Mark A. Meschter

Library of Congress Cataloging-in-Publication Data

Nature, nurture, and psychology / edited by Robert Plomin and Gerald E. McClearn.
p. cm.
Includes bibliographical references and index.
ISBN 1-55798-202-3 (acid-free paper)
1. Nature and nurture. 2. Individual differences. 3. Behavior genetics. 4. Genetic psychology. I. Plomin, Robert, 1948– . II. McClearn, G. E., 1927– .
BF341.N38 1993
155.7—dc20 93-9822
CIP

British Library Cataloguing-in-Publication Data

A CIP record is available from the British Library

Printed in the United States of America

Contents

Part Three: Nature–Nurture and the Development of Personality

Part Four: Psychopathology: Genetic and Experiential Factors

Contributors

Duane Alexander, National Institute for Child Health and Human Development, Bethesda, Maryland

Doreen Arcus, Department of Psychology, Harvard University

Anthony Bailey, Institute of Psychiatry, London

Patrick Bolton, Institute of Psychiatry, London

Thomas J. Bouchard, Jr., Department of Psychology, University of Minnesota

Nathan Brody, Department of Psychology, Wesleyan University

Urie Bronfenbrenner, Department of Human Development and Family Studies, Cornell University

Lon R. Cardon, Department of Mathematics, Stanford University

Stephen J. Ceci, Department of Human Development and Family Studies, Cornell University

S. S. Cherny, Institute for Behavioral Genetics, University of Colorado

John C. DeFries, Institute for Behavioral Genetics, University of Colorado

Lindon Eaves, Department of Human Genetics, Virginia Commonwealth University

David W. Fulker, Institute for Behavioral Genetics, University of Colorado

Jacquelyn J. Gillis, Institute for Behavioral Genetics, University of Colorado

H. H. Goldsmith, Department of Psychology, University of Wisconsin

Irving I. Gottesman, Department of Psychology, University of Virginia

John K. Hewitt, Institute for Behavioral Genetics, University of Colorado

Frances Degen Horowitz, Graduate School and University Center, City University of New York

William G. Iacono, Department of Psychology, University of Minnesota

Jerome Kagan, Department of Psychology, Harvard University

Randy Katz, Department of Psychology, University of Toronto

CONTRIBUTORS

Gregory A. Kimble, Department of Psychology, Duke University
Anne Le Couteur, Child Guidance Clinic, London
David T. Lykken, Department of Psychology, University of Minnesota
Gerald E. McClearn, College of Health and Human Development, Pennsylvania State University
Matt McGue, Department of Psychology, University of Minnesota
Peter McGuffin, Department of Psychological Medicine, University of Wales, Cardiff, United Kingdom
Joanne Meyer, Department of Human Genetics, Virginia Commonwealth University
Michael Neale, Department of Human Genetics, Virginia Commonwealth University
Andrew Pickles, Institute of Psychiatry, London
Robert Plomin, Center for Developmental and Health Genetics, Pennsylvania State University
David Reiss, Department of Psychiatry, George Washington University
David C. Rowe, Department of Psychology, University of Arizona
Michael Rutter, Institute of Psychiatry, London
Judy Silberg, Department of Human Genetics, Virginia Commonwealth University
Emily Simonoff, Institute of Psychiatry, London
Nancy Snidman, Department of Psychology, Harvard University
Theodore D. Wachs, Department of Psychological Sciences, Purdue University
Irwin D. Waldman, Department of Psychology, Emory University

Preface

The modern origins of genetic research in psychology began over 125 years ago with the work of Francis Galton, who coined the phrase "nature and nurture" to refer to the two major sources of individual differences—genetics and environment. Although progress in research was slow in the century following Galton, during the past decade, the pace of advances has accelerated dramatically as genetics has begun to flow into the mainstream of psychological research. This is a propitious time for a book that introduces psychologists to genetic research because researchers are witnessing the culmination of a century of slow progress toward gaining acceptance.

That acceptance has been hastened by several critical discoveries made recently by genetic researchers about the origins of individual differences, including differences in cognitive abilities and disabilities, differences in personality, and differences in psychopathology. But the reader need not take our word on this. The chapters in this book provide summaries of current findings written by some of the top researchers in these fields.

As researchers begin looking toward the new century, they can better set their course by taking stock of past efforts and considering future challenges. With that goal in mind, we have organized this book to provide an overview of the past, a summary of the present, and a glimpse of the future of genetic research in psychology. This information is offered in seven parts, with the first recounting the history of the field and the last examining future options. Of the five parts that lay in between we will say little because they are ably introduced by section editors who discuss both the conquests and the controversies that are presented therein.

The Past

This book begins in Part One with a two-chapter matched set of historical inquiries, the first by Gregory A. Kimble, a psychologist, and the second by Gerald E. McClearn, a behavioral geneticist.

Behavioral genetics marks its modern origins with Galton's research on hereditary genius nearly 125 years ago. Although there has been a steady stream of genetic research in psychology since that time, behavioral genetics is scarcely mentioned in histories of psychology (e.g., Boring, 1950).[1] In part, this neglect stems from a preoccupation with traditional experimental psychology that, beginning with William James,[2] ignored research on individual differences. Another factor was the environmentalistic hegemony that dominated psychology until the 1960s.

Nonetheless, nature and nurture has been one of the most enduring interests of psychologists and has attracted some of psychology's top scientists. For example, Galton and his student Karl Pearson invented the fundamental statistics of regression and correlation to solve the genetic problem of representing resemblance between family members. After studying with Wilhelm Wundt, James McKeen Cattell insisted on studying individual differences and left Leipzig to work for a while with Galton, which inspired some of the earliest experimental work in behavioral genetics. Cattell was a member of the organizing committee and attended the first American Psychological Association (APA) convention in 1892, and he hosted the second APA conference in 1893. One of the first twin studies was conducted by Edward L. Thorndike.[3] Ivan Pavlov's research on individual differences in dogs early in the century led him to the concept of temperament, by which he meant the inborn strength, equilibrium, and lability of nervous system processes. Gordon Allport also considered temperament as "largely hereditary in origin."[4] The letter *H* in Edward C. Tolman's HATE acronym of individual differences, which was

[1]Boring, E. G. (1950). *A history of experimental psychology*. New York: Appleton-Century-Crofts.
[2]James, W. (1890). *The principles of psychology*. New York: Holt.
[3]Thorndike, E. L. (1905). Measurement of twins. *Archives of Philosophy, Psychology, and Scientific Methods, 1*, 1–64.
[4]Allport, G. W. (1937, p. 54). *Personality: A psychological interpretation*. London: Constable.

presented in his APA presidential address, stands for heredity. His student, Robert C. Tryon,[5] who must have been the first psychologist to obtain a joint PhD in genetics, carried forward Tolman's[6] early efforts to select rats for maze performance. An important milestone was Calvin Hall's chapter[7] in the 1951 edition of the *Handbook of Experimental Psychology*. Another was the 1960 publication of *Behavior Genetics* by John Fuller and W. Robert Thompson.[8] Genetic research in the human species was slower to follow Galton's lead, but research projects on cognitive development spanning several decades were conducted by well-known psychologists such as Cyril Burt[9] and Marie Skodak and Harold M. Skeels.[10] Genetic research on psychopathology also has a long lineage (see Gottesman, 1982).[11]

The Present

When attention is turned to the front lines of genetics research, researchers must ask themselves to identify the areas that are yielding the most promising results. As noted earlier, many exciting discoveries are being made in several domains that will ultimately provide a clearer understanding of the origins of individual differences. Three sections in this book are focused on those domains.

The chapters in Part Two, which was organized by John C. DeFries, focus on the genetics of cognitive abilities and disabilities. Chapter 3 (McGue et al.) provides an overview of research throughout the life span; chapter 4 (Fulker, Cherny, & Cardon) discusses developmental genetic

[5]Tryon, R. C. (1940). Genetic differences in maze-learning in rats. *Yearbook of the National Society for the Study of Education, 39*, 111–119.

[6]Tolman, E. C. (1924). The inheritance of maze-learning ability in rats. *Journal of Comparative Psychology, 4*, 1–18.

[7]Hall, C. S. (1951). The genetics of behavior. In S. S. Stevens (Ed.), *Handbook of experimental psychology* (pp. 304–329). New York: Wiley.

[8]Fuller, J. L., & Thompson, W. R. (1960). *Behavior genetics*. New York: Wiley.

[9]Burt, C. (1966). The genetic determination of differences in intelligence: A study of monozygotic twins reared together and apart. *British Journal of Psychology, 57*, 137–153.

[10]Skodak, M., & Skeels, H. M. (1949). A final follow-up of one hundred adopted children. *Journal of Genetic Psychology, 75*, 85–125.

[11]Gottesman, I. I. (1982). *Schizophrenia: The epigenetic puzzle*. Cambridge, UK: Cambridge University Press.

analyses of general cognitive ability; chapter 5 (Cardon & Fulker) presents multivariate genetic analyses of specific cognitive abilities; and chapter 6 (DeFries & Gillis) explores the research on learning disability. In chapter 7, Duane Alexander, the director of the National Institute of Child Health and Human Development, contributes an interesting discussion of these advances.

In Part Three, the authors explore nature–nurture and the development of personality. H. H. Goldsmith organized this section, which includes his own concise summary of recent findings in the field as well as contributions on intelligence (Brody in chapter 8), genetics and personality (Rowe in chapter 9), and temperament (Kagan, Arcus, & Snidman in chapter 10).

The genetic and experiential factors contributing to psychopathology are discussed in Part Four. Given the depth and breadth of research in this area, it was difficult for the section organizer, Irving I. Gottesman, to choose five contributions that would adequately represent the field. To exemplify major advances, he chose to include research on mood disorders (McGuffin & Katz in chapter 11), schizophrenia (Gottesman in chapter 12), alcohol dependence (McGue in chapter 13), and autism (Rutter et al. in chapter 14) as well as an analysis of liability (Eaves et al. in chapter 15).

The Future

In addition to providing an overview of the present state of knowledge, we hope that these chapters also convey the sense of excitement about the future that pervades behavioral genetics. In part, this excitement springs from the synergism of applying genetic theory and research to issues of mainstream psychology, which will help researchers to look at old issues in new ways. Although most of the chapters in this book are forward-looking in this sense, the focus of the last two parts is the future. These parts evolved from roundtable discussions between behavioral geneticists and other eminent psychologists concerning the future of nature–nurture research. These discussions were conducted during a 10-hour series of symposia on behavioral genetics, which were held at the 1992

APA Centennial Convention in Washington, DC, and served as a springboard for this book.

One of the most promising areas for future progress lies in the phrase *nature–nurture*. This involves taking up the challenge posed by Anne Anastasi in her 1958 APA presidential address[12] to move beyond questions of how much variance is accounted for by genetic and environmental factors to the question of how genetic and environmental variables coact during development. Part Five, Bridging the Nature–Nurture Gap, includes chapters by three of the most eminent environmental researchers and theorists—Urie Bronfenbrenner, Frances Degen Horowitz, and Theodore D. Wachs, who organized the section. Each proposes an environmental theory that attempts to encompass genetic influence: a bioecological model (Bronfenbrenner & Ceci in chapter 16), a comprehensive new environmentalism (Horowitz in chapter 18), and a multidetermined probabilistic systems framework (Wachs in chapter 20). Also included in this section are two chapters by behavioral geneticists (Goldsmith, chapter 17, and Rowe & Waldman, chapter 19) who discuss their ideas for thinking about and exploring the vast hinterland of the gap between nature and nurture.

The final part of this book was organized by Michael Rutter to tackle the question, "Where do we go from here?" Rutter's section includes one chapter by a behavioral geneticist (John K. Hewitt in chapter 21). The three other chapters in this section were written by a psychologist and two psychiatrists who would not previously have been identified as behavioral geneticists: Jerome Kagan, David Reiss, and Michael Rutter. However, in recent years, each of these individuals has become actively involved in genetic research. Attracting such talented scientists to the field may be the best reason to be bullish about the future of behavioral genetics.

Several of the chapters hint at another source of great excitement in the field—the breathtaking pace of advances in molecular genetics. Researchers are at the dawn of a new era in which molecular genetics

[12]Anastasi, A. (1958). Heredity, environment, and the question "How?" *Psychological Review, 65*, 197–208.

techniques will revolutionize behavioral genetics. Just as important, the quantitative genetic perspective of behavioral genetics will transform the use of these techniques as researchers continue to explore the role of nature and nurture in the most complex of phenotypes, behavior.

Robert Plomin
Gerald E. McClearn

PART ONE

A Century of Nature and Nurture

CHAPTER 1

Evolution of the Nature–Nurture Issue in the History of Psychology

Gregory A. Kimble

One of the earliest "experiments" on the nature–nurture issue dates to the 13th century, when Frederick II, King of Germany,

> wanted to find out what kind of speech and what manner of speech children would have when they grew up if they spoke to no one beforehand. So he had foster mothers to suckle the children, to bathe and wash them, but in no way to prattle with them, or to speak with them, for he wanted to learn whether they would speak the Hebrew language, which was the oldest, or Greek, or Latin, or Arabic, or perhaps the language of their parents, of whom they had been born. But he labored in vain because the children all died. For they could not live without the petting and joyful faces and loving words of their foster mothers. (Stone & Church, 1973, p. 104)

Three centuries later, the Mogul emperor Akbar, a descendant of Tamerlane and Genghis Khan, reared children in isolation to discover whether their natural religion would be Hinduism, the Christian faith, or some other creed. But this experiment was also a failure: It only produced deaf mutes (Broadhurst, 1968). The implicit assumption in these so-called experiments was that environment contributes more than inheritance to human talents: Experience has the power to hide the inborn nature of an individual.

Early Dominance of Environmentalism

The environmental emphasis continued when the philosophical ancestors of psychology discussed the origins of knowledge. For example, the British empiricist John Locke (1690/1979) answered the nature–nurture question this way in a much-quoted passage from his *An Essay Concerning Human Understanding*:

> Let us suppose the mind to be, as we say, white paper, void of all characters, without any ideas:—How comes it to be furnished? Whence comes it by that vast store which the busy and boundless fancy of man has painted on it with an almost endless variety? Whence has it all the materials of reason and knowledge? To this I answer, in one word, from *experience.*

Later on, support for environmentalism came from science. For example, Helmholtz (see Southwell, 1924–1925) held that any aspect of perception that can be modified by experience must be the product of experience. He concluded that, by this criterion, only the quality and intensity of sensation are innate. Primitive consciousness is a pandemonium of sights, sounds, and all the other sensations, differentiated only by these fundamental properties. How does perception create an environment made up of objects with coherence? A reasoning-like process called *unconscious inference* constructs a perceptual world from the data of the senses.

Meanwhile, Bell (1811/1948) and Magendie (1822) had independently discovered the structural and functional independence of sensory and motor neurons in the spinal cord, thus setting the stage for an environmental interpretation of acts as well as percepts. In 1860 the Russian physiologist Sechenov (see 1935) stated the position forcefully: "The real cause of every human activity lies outside man" (p. 334), and "...999/1,000 of the contents of the mind depends on education in the broadest sense, and only 1/1,000 depends on individuality" (p. 335). But these statements were less colorful than that made by Watson (1924):

> Give me a dozen healthy infants and my own specified world to bring them up in, and I'll guarantee to take anyone at random and train him to become any kind of specialist I might select—doctor, lawyer, artist, merchant-chief, and yes, even beggar and thief, regardless of his talents, penchants, tendencies, abilities, vocations and race of his ancestors. (p. 104)

The Beginnings of Behavioral Genetics

More than half a century before Watson's (1924) proclamation, Charles Darwin (1859) and Gregor Mendel (1866) had sown the seeds of a genetic revolution. In his theory of evolution, Darwin described three requirements for the occurrence of that process: (a) *heredity* (offspring resemble their parents), (b) *variation* (although similar, offspring differ), and (c) *selection* (the fittest variants survive, the unfit perish in the struggle for existence).

Mendel's (1866) studies of genetics provided the details of the first of Darwin's three requirements. Mendel contributed the distinction between *phenotype* (observable characteristics of organisms) and *genotype* (underlying genetic makeup) and provided the laws of *segregation* (a genotype consists of pairs of genes, one allele from each parent) and *independent assortment* (the genes for traits are separate: inheriting wavy hair does not make it more likely that a person's hair will be blonde or red or black) and the concepts of *dominance* and *recessiveness* (dominant genes gain phenotypic expression whether an organism has one or two of them; recessive genes are expressed only if the individual receives one from each parent). Mendel's work was lost until the beginning of the 20th century and went largely unnoticed in psychology until later. Beginning with William James (1890), Darwin's theory became the model for the school of functionalism in psychology, which held that the adaptive value of responses preserves those that survive the slings and arrows of outrageous fortune, just as it preserves the fittest of a species.

Hereditary Genius

Darwin (1859) proposed that evolution selects for mental traits as well as anatomical structures, and Francis Galton (1869) argued for the inheritance of the traits that lead to human eminence. To obtain the evidence for that conclusion, Galton began by identifying eminent scientists, judges, authors, musicians, military leaders, statesmen, and (despite the popular opinion that "the children of religious parents frequently turn out badly" [1869, p. 258]) theologians. He then searched biographies and other records to obtain data like those summarized in Table 1. Using the most eminent person in each family as a referent, this table presents percentages of relatives, who were eminent themselves, in four clusters representing decreasing closeness of relationship. The corresponding decrease in percentages of eminent relatives makes the case for heritability.

Hereditary Defect

Studies of the genealogies of defective families carried the genetic thesis to the other end of the continuum of intellect. One of these was a study of the Jukes family (Dugdale, 1877), which came to the attention of New York state officials when they discovered that six of the Jukes were in prison at the same time in just one county. A detailed genealogical survey,

TABLE 1
Blood Relationship and Eminence

Nature of relationship	Eminent relatives (%)
Father	31
Brother	41
Son	48
Grandfather	17
Grandson	14
Uncle	18
Nephew	22
Great grandfather	3
Great grandson	3
Great uncle	5
Great nephew	10
First cousin	13
More remote relationships	0

which covered seven generations and about 750 Jukeses, revealed that the family had cost the state over $1.5 million through crime, pauperism, and vice. A second study was that of Martin Kallikak (Goddard, 1921), a revolutionary soldier who had fathered two lines of offspring, one that began with an illicit mating with a tavern maid who was thought to be retarded, and the other with a woman of at least average intelligence whom he married 2 years later. Over several generations, the first line produced many cases of apparent mental retardation; the second line had none.

Environmentalism Undaunted

The conclusion sometimes drawn from these data was that intelligence is inherited. Although these studies were suggestive, carried out without objective tests, the hypothesis that they were studies of intelligence, instead of motivation or personality, is risky. More important, in all of these studies, good or bad genes, good or bad environments, and good or bad outcomes went together. It is possible that the environment, not inheritance, causes eminence or incompetence. Because of its history, psychology had a preference for environmentalist interpretations, a preference that received support from other studies.

The Wild Boy of Aveyron

Among the earliest of these studies was that of the French physician Jean Itard (1807/1932), who undertook the task of training Victor, the so-called "wild boy of Aveyron." When captured in the woods, Victor was 12 or 13 years old. He could not speak and only shrieked and grunted. He crouched and trotted like a beast and, even when still, swayed back and forth incessantly. Dirty and naked, indifferent and inattentive, he subsisted on roots and raw potatoes. Itard believed that the boy had become retarded as a result of social isolation. By supplying Victor with the necessary experiences, he hoped to reverse the process. Itard did manage to teach Victor to discriminate objects by sight, touch, and taste; to connect objects to abstract symbols; to write a few words; and eventually to respond positively to affection. Although Itard was never able to educate Victor to the point of normalcy, the success he did achieve encouraged the conclusion that mental development depends on environmental opportunity.

Wolf Children

The Indian missionary J. A. L. Singh (Singh & Zingg, 1942) discovered a second case of wild children. In his missionary tours, Singh became intrigued with stories of a "man–ghost" living in a nearby jungle and went to locate him. The man–ghost habitat to which Singh was taken turned out to be an abandoned white-ant mound that wolves had turned into a den with numerous entrances and exits. The missionary had a shooting stand erected and watched the den with field glasses. On October 9, 1920, his patience was rewarded: Two female wolves and two wolf cubs came out of the den, followed by a hideous looking creature with a human body and a head so covered with a ball of hair that only the sharp contour of a human face was visible. Close behind this creature came a second smaller one.

Eight days later, the two man–ghosts were captured. They proved to be girls, one about 1.5 years old and the other about 8 years old, who had been nurtured by the wolves. They were reported as having sharp, piercing eyes and being more vicious than the wolf cubs with whom they were taken. Their captors named them Amala and Kamala. Amala, the younger child, died about a year later, having shown few signs of developing human capabilities. Kamala lived in the missionary orphanage for almost 9 years.

During this period, Kamala changed from something more animal than human into a recognizable child. Initially she ran on all fours, lapped her food rather than using her hands for eating, showed a preference for raw meat, and seemed more attracted to the farm animals than to the human inhabitants of the orphanage. Her only vocalizations were peculiar howling sounds that she often made at night, apparently to announce her presence in a certain location. However, Kamala gradually learned to eat using her hands, and she developed a taste for nutrients that she had previously rejected. She learned to walk, first on her knees and then eventually in an upright manner. Her savagery dissipated, and she developed an affection for certain people. She began to notice colors and seemed to prefer red. Her crowning achievement was the acquisition of a vocabulary of about 30 understandable words. Kamala died on November 11, 1929, after a long illness. What she had accomplished in her years of human contact and the fact that her improvement was accelerating at the time she died lent credence to an environmental interpretation of her initial deficits.

Joint Contribution

Again, although these studies were suggestive, they did not provide a convincing argument for the environmental determination of intellect because the defects of the feral children might have been genetic; they might have been abandoned in the first place because they were retarded. Facts that can have either an environmental or a hereditary interpretation obviously cannot decide between the two. Now about a century of more formal research has shown that both factors are important.

Four Faces of the Issue

The nature–nurture issue may be the most pervasive in psychology, but in different settings, it goes by different names. First, in the fields of sensation and perception it is the nativism–empiricism issue. Second, in developmental psychology it is maturation versus learning. Third, in learning and cognition it is a premise of environmental equipotentiality versus a principle of biological preparedness. Fourth, in differential psychology it is heredity versus environment as the determiner of human variation.

Perception: Nativism Versus Empiricism

All of the human senses are functional at birth. Soon thereafter, infants have the ability to perceive the psychologically primary colors (Borstein, Kessen, & Weiskopf, 1976), visual depth (Gibson & Walk, 1960), the location of auditory stimuli in space (Wertheimer, 1961), the contours of the human face (Fantz, 1961), emotional expression (Field, Woodson, Greenberg, & Cohen, 1982), and the human form in motion (Cutting, 1981)—all of which provide powerful arguments for a nativistic contribution.

But on the side of empiricism, William James (1890) knew that the senses are educable:

> [W]e have the well-known virtuosity displayed by the professional buyers and testers of various kinds of goods. One man will distinguish by taste between the upper and the lower half of a bottle of old Madeira. Another will recognize, by feeling the flour in a barrel, whether the wheat was grown in Iowa or Tennessee. (p. 509)

As might be expected from these positive contributions of experience, sensory deprivation has negative effects. Humans have occasionally lost their sight at an early age and, later on, regained it. One such individual, who was studied by Gregory and Wallace (1963), had color vision when his sight was restored, but his depth perception was very bad; he lacked most of the visual illusions and knew that a "blur" in front of him must be a human face only because it talked and because he knew that speech came from faces. Several decades of research have converged on the conclusion that a learned coordination of sight and bodily motion is critical to the development of visual perception (Held & Hein, 1963; Stratton, 1897).

Development: Maturation Versus Learning

The history of maturation versus learning is a history of the discovery that abilities that offhandedly seem to be learned depend on biological maturation. In a classical study, Coghill (1929) demonstrated that the development of locomotion in the salamander is a maturational process involving sequences (proximodistal, cephalocaudal, and mass action-differentiation) that depend on neural growth. Even earlier, Carmichael (1926) did an experiment to show that experience has little or nothing to do with the process. He kept salamander tadpoles in a chemical solution that prevented movement, during the period when swimming normally appears. When removed from the solution at the end of that period, they swam almost immediately, unaffected by the lack of practice.

Soon there were demonstrations that Coghill's (1929) sequences appear in the development of such human skills as walking and grasping (Shirley, 1931), and in a study similar to Carmichael's (1926), Dennis (1940) showed that confining Hopi children to a cradle-board for the first year or so of life, which thus deprives them of exercise, does not postpone the age of walking. It is apparent that what we informally call "learning to walk" is largely maturation. The results of other studies suggest that this conclusion applies in many contexts: emotional reactions (Bridges, 1932), smiling (Spitz & Wolf, 1946), and even language (Lenneberg, 1967). Other studies began to hint that critical periods are important to the interaction between maturation and learning. Imprinting in precocial birds is most rapid during a brief age span when the animal is biologically ready (Hess, 1958). Educational psychologists have known for years that a similar readiness is required for reading (Morphett & Washburn, 1931).

Learning and Cognition: Equipotentiality Versus Preparedness

Pavlov was an extreme environmentalist in the Sechenov tradition. He favored what Seligman and Hager (1972) later called the "premise of equipotentiality," the hypothesis that any response an organism can make can be conditioned to any stimulus it can detect and that all of these associations are formed with equal ease. Coupled with acceptance of Lamarck's theory of the inheritance of acquired characteristics, this outlook was the basis for the Soviet belief that an entire population could be trained to Marxist philosophy and that such an education could be passed along to future generations (Kimble, 1967). Lamarckian theory did not survive, but for a time, the premise of equipotentiality looked promising. The Russian laboratories produced many demonstrations of interoceptive conditioning, in which the conditioned and unconditioned stimuli or responses were internal to the learner (Razran, 1961). American experimenters later added their support when they found evidence for the learning of allergic reactions (Russell, Dark, Ellman, Callaway, & Peeke, 1984), resistance to the lethal effects of drug overdose (Siegel, 1984), and analgesia (Maier, 1986).

Doubts about the ubiquity of conditioning arose when research began to show that inborn sympathies between certain stimuli and responses modulate the process. Soon after Watson and Rayner (1920) demonstrated the conditionability of fear in their experiment with "Little Albert," Valentine (1930) proposed a role for biological preparedness when he discovered that such conditioning did not happen when the conditioned stimulus was a pair of opera glasses rather than a rat. Later on, Garcia and Koelling (1966) found, in rats, that aversions based on illness are easy to condition to tastes but difficult or impossible to condition to sights or sounds; they found that the reverse was true for shock-elicited avoidance reactions.

The principle of preparedness also applies to instrumental learning, a fact that is well known to animal trainers. For example, Breland and Breland (1961) described a "dance" performed by a chicken in response to the music from a juke box, which turned out to be little more than the natural scratching pattern that chickens use in their search for food, brought under stimulus control by reinforcement. Even language observes the principle of preparedness. Although the importance of experience is obvious from the fact that children learn their native tongues—instead of Arabic or Hebrew, as King Frederick thought they might—biological constraints are much in evidence. The neuroanatomy of aphasia proves that language has specific representation in the human brain. Babies come into the world with "wired-in" rudiments of linguistic competence. Their babblings are the same the whole world over, even if they are born to parents who are mute (Lenneberg, 1967). Furthermore, they discriminate phonemes categorically (Eimas & Tartter, 1979) just as they do primary colors.

Individual Differences: Heredity Versus Environment

The research that introduced psychology to the science of behavioral genetics was Tryon's (1940) selective breeding study of maze learning. Of all these studies of the nature–nurture issue, Tryon's was the only one that dealt with individual differences in behavior. Tryon mated "maze-bright" and "maze-dull" rats with others of the same kind for 21 generations. The members of the parent generation made about 14 errors in 19 trials on a complex maze. By the 7th generation, the progeny of the maze-bright matings were making an average of about 8 errors; those produced by the maze-dull matings were making approximately 18, which thereby suggests that maze-learning ability is genetic. Hall (1951) obtained similar evidence for the inheritance of fearfulness in rats, and Buss (1985) made a convincing case that something like selective breeding goes on with human beings because, contrary to a popular opinion, opposites do not attract. People choose mates like themselves and, to the extent that the similarities are genetic, pass them along to their children.

As happens in psychology, the behavioral genetic plot soon thickened. At the University of California, Berkeley, where Tryon did his pioneer investigation of the inheritance of maze talent, Krech, Rozenzweig, and Bennett (1962) obtained evidence for a surprising environmental influence: Rats reared in an "enriched" environment—in cages with other rats and objects to manipulate and crawl on—developed heavier brains and better problem-solving ability than did rats raised alone in an unfurnished, "impoverished" environment.

Further research with Tryon's rats exposed additional complexities. Searle (1949) found that the superiority of Tryon's maze-bright rats disappeared when they were tested in other mazes. Moreover, by comparison with the maze-dull rats, the maze-bright rats had a stronger hunger drive, a lower level of spontaneous activity, and greater running speed—any of which might have been responsible for their better performance on Tryon's maze.

There are similar problems at the human level. What we call intelligence is actually a collection of cognitive abilities that are genetic, but to varying degrees (Plomin, 1990). It seems certain that, if someone bred a strain of humans who excelled on one cognitive ability, they would not excel on others. The organization of human traits remains to be worked out, and until it is, determining the effects of nature and nurture on behavior will face the handicap that E. G. Boring once described—I can't remember where—when he said that psychology might be more advanced today if the giant in its history had been a Linnaeus instead of a Wundt.

The Issue at Mid-Century

When I was a student at Carleton College (1936–1940), psychology was roughly half-way through this history. It understood that the nature–nurture issue would not be settled in favor of one factor or the other and that asking whether individual differences in behavior are determined by heredity or environment is like asking whether the areas of rectangles are determined by their height or width. The accepted wisdom was that the right question was not "which?" but "how much?" But World War II began; for a decade, psychology was otherwise preoccupied, and there was little progress on the problem. When it returned to normal business at mid-century, it found learning theory in charge of scientific psychology. Operating in the Watsonian tradition, it played down the importance of genetic influences on behavior.

Common Themes and Variations

The basis for a constructive relationship between traditional psychology and the emerging science of behavioral genetics was actually already available because of the applicability to both disciplines of two powerful ways of thinking, the first of which was R. A. Fisher's method of the analysis of variance (ANOVA). The basic theorem of ANOVA is that the total variance of phenomena is the sum of all variances produced by the influences that control them (an "all-causes" variance) plus variance from two sources of error: accidents of sampling and unreliability of measurement. Thus, in the most general case:

$$S^2_{\text{Total}} = S^2_{\text{All Causes}} + S^2_{\text{Error}}, \qquad (1)$$

where S^2 is variance.

The second powerful way of thinking is one that permeates psychology: Behavior is the consequence of relatively enduring potentials for, and relatively more temporary instigations to, performance. A list of commonly contrasted potentials and performances, together with their instigators, in several different contexts makes this point (see Table 2).

The inclusion of genotype as a potential and phenotype as an outcomelike behavior in Table 2 underscores the fact that the table presents a way of thinking—a metaphor or a model. In terms that are more concrete, it is useful to think that genetics provides individuals with a "potential potential" and that the environment determines the degree of "realized potential." The potentials of traditional psychology are these realized potentials.

One implication of this model is that the all-causes variance can be divided into component variances that are assignable to potential and

TABLE 2
A Common Theme in Psychology

	General pattern		
Context:	Potential	Instigation	Behavior
	Several Examples		
Behavioral genetics:	Genotype	Environment	Phenotype
Cognitive psychology:	Knowledge	Situation	Performance
Psycholinguistics:	Deep structure	Communicative need	Surface structure
Nerve impulse:	Resting potential	Adequate stimulation	Action potential
Signal detection:	Sensitivity	Stimulus (above/below) criterion	"Hit"/"miss"
Classical learning theoretical:	Learning	Motivational minus inhibition	Performance
Skinnerian "nontheory":	Reflex reserve	Environmental occasion	Operant response
Intelligence:	Fluid	Environmental influence	Crystallized
Educational accomplishment:	Scholastic aptitude	Motivation/stimulation	School performance
Dream content:	Latent content	"Dream work"	Manifest content
Anxiety:	Trait anxiety	Threat	State anxiety
Ethology:	Action-specific energy	Releasing stimulus	Releasing reaction

instigation. Leaving aside for now the interactions among them, the following equation describes the simplest possible way in which these specific influences might work together:

$$S^2_{\text{Total}} = S^2_{\text{Potential}} + S^2_{\text{Instigation}} + S^2_{\text{Error}}. \quad (2)$$

Accounting for Variance in Behavioral Genetics

When the concepts of *polygenic inheritance* (many genes contribute to phenotypic variations in a trait) and *norm of reaction* (a genotype is a repertoire of different phenotypes that can appear in different environments) found their way into psychology, the stage was set for a parallel accounting from a genetic point of view. In this approach, total variance in behavior is phenotypic variance, and in the simplest case for which phenotypic variance is the sum of variances attributable to genes, environment, and error, the formula becomes:

$$S^2_{\text{Phenotypic}} = S^2_{\text{Genetic}} + S^2_{\text{Environmental}} + S^2_{\text{Error}}. \quad (3)$$

This equation also implies another, which defines a "coefficient of heritability" (h^2), the fraction of total variance that is genetic variance:

$$h^2 = S^2_{\text{Genetic}}/S^2_{\text{Phenotypic}}. \quad (4)$$

The similarities between Equations 2 and 3 suggest that it might be useful to look for other parallels between traditional experimental psychology and behavioral genetics.

The simplest experiments with outcomes that are treated to an ANOVA often manipulate conditions that have their hypothetical effect on either a potential or an instigator. An experiment that compares the recall of two groups of participants who learn materials with deep or superficial processing theoretically manipulates a potential, that is, the knowledge acquired by the participant (Table 2). An experiment on encoding specificity, which shows that participants can recall that knowledge better in the context in which they learned it than in a different one, manipulates one type of instigation. The analysis for these experiments begins by dividing total variance on recall into two components: between-groups variance, which is attributable to the effects of the independent variable, and within-groups variance, the variance of individuals in the same conditions. Because instigation does not vary in one experiment and because potential does not vary in the other, these variances drop out of the equations:

$$S^2_{\text{Recall}} = S^2_{\text{Level of Processing}} + S^2_{\text{Within-Groups}} \tag{5}$$

$$S^2_{\text{Total}} = S^2_{\text{Potential}} + S^2_{\text{Error}} \tag{6}$$

$$S^2_{\text{Recall}} = S^2_{\text{Context}} + S^2_{\text{Within-Groups}} \tag{7}$$

$$S^2_{\text{Total}} = S^2_{\text{Instigation}} + S^2_{\text{Error}}. \tag{8}$$

Now, moving to behavioral genetics, the beginning assumption of that science is that the total phenotypic variance of a trait is the sum of variances contributed by genes, environment, and error:

$$S^2_{\text{Phenotypic}} = S^2_{\text{Genetic}} + S^2_{\text{Environmental}} + S^2_{\text{Error}} \tag{9}$$

$$S^2_{\text{Total}} = S^2_{\text{Potential}} + S^2_{\text{Instigation}} + S^2_{\text{Error}}. \tag{10}$$

Behavioral genetics then proceeds to analyses that are parallel to those just described. It divides genetic variance into two components: *additive* polygenic variance and *nonadditive* variance, which is contributed by dominance and epistasis. As a reminder, dominance is a genetic effect that occurs at one genetic locus; epistasis refers to the fact that, in some cases, the products of genes at one locus depend on genes at others. Additive variance is like a main effect: The more genes an organism has for a trait, the greater the phenotypic value of that trait. Nonadditive variance is like within-groups variance (the error variance) in an ANOVA: Individuals with the same additive genotype may vary in nonadditive inheritance.

Also in step with traditional psychology, behavioral genetics divides environmental variance into *shared* and *nonshared* components. All children in a family share the same environment to the extent that, on the average, parenting in their family differs from that in other families. The result, across families, is a between-families variance like the between-groups variance in the results of an experiment. There is also a within-groups variance. Parents do not treat all their children in exactly the same way. The result is nonshared environmental variance, which accounts for more variation in behavior than does shared environmental variance. Two equations describe the relationships:

$$S^2_{\text{Total Genetic}} = S^2_{\text{Additive Genetic}} + S^2_{\text{Nonadditive Genetic}} \tag{11}$$

$$S^2_{\text{Total Environment}} = S^2_{\text{Shared Environmental}} + S^2_{\text{Nonshared Environmental}}. \tag{12}$$

As is well known, however, the variables that control behavior interact, and therefore, the equation must be longer:

$$S^2_{\text{Total}} = S^2_{\text{Potential}} + S^2_{\text{Instigation}} + S^2_{(\text{Potential} \times \text{Instigation})} + S^2_{\text{Within-Groups}}. \tag{13}$$

In the behavioral genetics version of this formula, the interaction is a genotype–environment interaction:

$$S^2_{\text{Phenotypic}} = S^2_{\text{Genetic}} + S^2_{\text{Environmental}} + S^2_{(\text{Genetic} \times \text{Environmental})} + S^2_{\text{Error}}. \tag{14}$$

The interaction term in this equation is a measure of the variance that occurs because the environment that is optimal for individuals with one genotype may not be optimal for others: The education that is best for a gifted child will not be the best for a child who is retarded. Experimental psychology acknowledges a similar interaction when it treats individual differences as *moderator variables* that interact with other independent variables. The best-known example is the history of research showing that, although arousal (sometimes anxiety) may facilitate performance on a simple task, it may impede performance on a complex one. Psychometric psychology calls its variation on that theme *trait–situation interaction*. In this case, individual differences are independent variables, and different environments are moderator variables.

Finally, to complete the catalog of parallels, behavioral geneticists have found that people tend to live in environments that match their inherited potential. The infant who has potential academic giftedness is apt to have parents who are bright themselves and who provide a stimulating environment. Such youngsters are often identified as gifted and offered special educational opportunities. If such opportunities are not offered, then the gifted child may seek them out. These *gene–environment correlations* make it impossible to assign causality for behavior to heredity or environment. When that happens in experimental psychology (i.e., when the effects of experimental treatments cannot be interpreted because subjects in experimental conditions differ systematically), the effect is called *confounding*.

Quantitative Behavioral Genetics

When Galton was assembling his data on hereditary genius, the idea of correlation occurred to him. He passed the thought along to his follower and biographer, Karl Pearson, who invented the product–moment correlation coefficient (r). Applied to data for pairs of people who differed in blood relationship, this correlation coefficient became a standard tool in the science of behavioral genetics. Analyses of such data soon revealed that environment and heredity both contribute to these correlations. In the case of IQ, for example (see chapter 3, this book), their values parallel the closeness of relationship of individuals whose scores are correlated, which is an argument for genetics. However, the following argue for environment: (a) When similarly related people who live together are compared with those who live apart, the correlations for the people who live apart are lower than those for the people who live together, and (b) when unrelated people live together, their IQs are positively correlated. Behavioral geneticists have developed quantitative measures of these influences.

Estimating the Contributions of Heredity and Environment

These measures come from a variety of behavioral genetics methods, of which I shall describe just one—the twin method. This method compares the correlations of measures obtained on identical (monozygotic [MZ]) twins, who have 100% of their genes in common, and fraternal (dizygotic [DZ]) twins, who share only 50% of their genes.

Data obtained with the twin method yield one formula for h^2:

$$h^2 = 2(r_{\text{MZ Twins}} - r_{\text{DZ Twins}}). \qquad (15)$$

For example, if the correlations were 0.85 for MZ twins and 0.60 for DZ twins, and both groups of twins were reared together, the formula becomes:

$$h^2 = 2(0.85 - 0.60) = 2(0.25) = 0.50. \quad (16)$$

This formula requires four comments. First, although the fraction of variance accounted for by a correlation coefficient is usually r^2, in the case of kinship correlations, it is just r (Jensen, 1971). Thus, correlations of 0.85 for MZ twins and 0.60 for DZ twins would mean that those percentages of the variance in their IQs are accounted for by the joint environmental and genetic influences that come together in those relationships. Second, on the practical assumption that the environments of MZ and DZ twins are similar, the difference between the correlations reflects the genetic contribution. Third, the doubling of that difference is necessary to assess the magnitude of the genetic contribution because the covariance of DZ twins, with 50% of their genes in common, includes only half of the genetic influence. Fourth, it is important to recognize that coefficients of heritability, as is true of every statistic, are subject to sampling error; the values of h^2 obtained in different studies are varied.

Implications

Behavioral genetic research has now demonstrated an influence of heritability in many areas of psychology, including intelligence and its components, school achievement, organic dementias (Alzheimer's disease, Huntington's chorea, alcoholism), what once was called "functional" psychopathology (schizophrenia, the affective disorders, anorexia nervosa, panic disorder), traits of personality (extraversion–introversion, emotionality–neuroticism), attitudes (conservatism–traditionalism), delinquency–criminality, and vocational interests (Plomin, 1990). These discoveries have powerful implications, not just for the nature–nurture problem, but for social policy and our attitudes toward social issues.

Environment Accounts for the Majority of Variance

Estimates of the magnitude of the influence of genetics suggest that inheritance accounts for something like 35% of the variance in this broad range of human traits and tendencies, more for cognitive abilities and less for traits of personality. These data mean that, on one hand, an influence of this magnitude cannot be ignored—in psychology it is rare for an experimentally manipulated variable to account for that much variation—but, on the other hand, the answer to the nature–nurture question appears to be that environment is more important.

High Heritability Does Not Mean That Environment Has No Effect

Even if the coefficient of heritability for a trait is 1.0, environment can have a large effect; but it will be the same for everyone. There can be no gene–environment interaction because genetics accounts for all the variance. Over the centuries, something of that sort seems to have happened to human height, for which the coefficient of heritability is about 0.90. Better child care and nutrition have produced a substantial, but very general, increase in the height of the population. As exhibits in American museums often demonstrate, the clothing and beds used by early English colonists would be much too small for Americans today.

Heritability Within Groups Is Not the Same as Heritability Between Groups

Coefficients of heritability are extremely sensitive to the populations on which they are based. Measures obtained on one population (e.g., race) cannot be generalized to other populations or to differences between populations. Such differences are phenotypic differences. There may or not be genotypic differences, and if there are, they may favor either group. Psychology and the world at large could have been spared much bitter controversy if they had understood that it is important to distinguish between within-groups and between-groups heritability. The coefficients of heritability discussed in this chapter are fractions of the variance within populations that are genetic. Although there is a coefficient of between-groups heritability, which potentially provides a measure of the fraction of differences between the means of groups that is genetic, it presently is of no use because its calculation requires measures that, so far, are not available (McClearn & DeFries, 1973).

Conclusion

The nature–nurture issue in psychology is older than psychology itself. A decade or so before Wilhelm Wundt set up his laboratory in Leipzig in

1879, Sechenov (1935) and Galton (1869), respectively, had already pronounced the environmentalist and hereditary agendas. In the century and more since then, events have conspired to hide the fact that, going their separate ways, the sciences of Sechenov and Galton have arrived at similar destinations. The scope of psychology expanded. Modern Galtonism deals with topics that range from artistic aptitude to zoophilia; modern environmentalism, in the Sechenov tradition, studies everything from absolute thresholds to zero-sum games. Obsession with the details of such phenomena makes the common content of psychology hard to detect. Instead, war is waged over the content and method of psychology.

Perhaps the time has come to declare a cease-fire in the battles, to look for commonalities in psychology, and to explore the possibility that peace may be inherent in the consensus that we find. In the area of content, most psychologists agree that psychology is the science of behavior, before some of them go on to say that it also is the science of something more, like "mind," "brain," "mental processes," "human potential," or "experience." In the area of method, the question is that of what it means to be a science. Of the many answers that are available, the one that seems best to embrace the diversity of psychology comes from behavioral genetics. The several sciences of psychology are all in the business of accounting for the variance in behavior. The following two formulae, which were presented earlier, will serve as reminders of the argument:

$$S^2_{\text{Total}} = S^2_{\text{Potential}} + S^2_{\text{Instigation}} + S^2_{\text{Error}} \tag{17}$$

$$S^2_{\text{Phenotypic}} = S^2_{\text{Genetic}} + S^2_{\text{Environmental}} + S^2_{\text{Error}}. \tag{18}$$

In their prototypical versions, traditional psychology and behavioral genetics are identical in structure. A break-down of these general equations shows that the resemblance is more than just skin-deep. It would be easy to demonstrate that other psychological specialists accept these formulae. In their actual operations, they are more alike than is ordinarily understood.

References

Bell, C. (1948). Idea of a new anatomy of the brain: Submitted for the observation of his friends. Reprinted in W. Dennis (Ed.), *Readings in the history of psychology*. New York: Appleton-Century-Crofts. (Original work appeared in 1811)

Borstein, M. H., Kessen, W., & Weiskopf, S. (1976). The categories of hue in infancy. *Science, 191*, 201–202.

Breland, K., & Breland, M. (1961). The misbehavior of organisms. *American Psychologist, 16*, 681–684.

Bridges, K. M. B. (1932). Emotional development in early infancy. *Child Development, 3*, 324–344.

Broadhurst, P. L. (1968). Genetics. In D. L. Sills (Ed.), *International encyclopedia of the social sciences* (Vol. 6, pp. 96–102). New York: Macmillan.

Buss, D. M. (1985). Human mate selection. *American Scientist, 73*, 47–51.

Carmichael, L. (1926). The development of behavior in vertebrates experimentally removed from the influence of external stimulation. *Psychological Review, 33*, 51–58.

Coghill, G. E. (1929). *Anatomy and the problem of behavior*. New York: Macmillan.

Cutting, J. E. (1981). Coding theory adapted to gait perception. *Journal of Experimental Psychology: Human Perception and Performance, 7*, 71–87.

Darwin, C. (1859). *On the origin of species by means of natural selection or the preservation of favoured races in the struggle for life*. New York: Appleton.

Dennis, W. (1940). The effect of cradling practices upon the onset of walking in Hopi children. *Journal of Genetic Psychology, 56*, 77–86.

Dugdale, R. L. (1877). *The Jukes: A study in crime, pauperism, disease and heredity*. New York: Putnam.

Eimas, P., & Tartter, V. (1979). On the development of speech perception: Mechanisms and analogies. In H. W. Reese & L. Lipsett (Eds.), *Advances in child development and behavior* (Vol. 13). New York: Academic Press.

Fantz, R. L. (1961). The origins of form perception. *Scientific American, 204*, 66–72.

Field, T. M., Woodson, R., Greenberg, R., & Cohen, D. (1982). Discrimination and imitation of facial expressions by neonates. *Science, 218*, 179–181.

Galton, F. (1869). *Hereditary genius: An inquiry into its laws and consequences*. London: Macmillan.

Garcia, J., & Koelling, R. A. (1966). Relation of cue to consequence in avoidance learning. *Psychonomic Science, 4*, 123–124.

Gibson, E. J., & Walk, R. D. (1950). The visual cliff. *Scientific American, 202*, 2–9.

Goddard, H. H. (1921). *The Kallikak family: A study in the heredity of feeblemindedness*. New York: Macmillan.

Gregory, R. L., & Wallace, J. G. (1963). Recovery from early blindness: A case study. *Experimental Psychological Society Monographs* (No. 2).

Hall, C. S. (1951). The genetics of behavior. In S. S. Stevens (Ed.), *Handbook of experimental psychology*. New York: Wiley.

Held, R., & Hein, A. (1963). Movement-produced stimulation in the development of visually guided behavior. *Journal of Comparative and Physiological Psychology, 56*, 872–876.

Itard, J. M. G. (1932). *The wild boy of Aveyron*. New York: Appleton-Century. (Original work published in 1807)

James, W. (1890). *Principles of psychology*. New York: Holt.

Jensen, A. R. (1971). Note on why genetic correlations are not squared. *Psychological Bulletin, 75*, 223–224.

Kimble, G. (1967). *Foundations of conditioning and learning*. New York: Appleton-Century-Crofts.

Krech, D., Rozenzweig, M. R., & Bennett, E. L. (1962). Relations between brain chemistry and problem-solving among rats raised in enriched and impoverished environments. *Journal of Comparative and Physiological Psychology, 55*, 801–807.

Lenneberg, E. (1967). *Biological foundations of language*. New York: Wiley.

Locke, J. (1979). *An essay concerning human understanding* (P. H. Nidditch, Ed.). New York: Oxford University Press. (Original published in 1690)

Magendie, F. (1822). Expériences sur fonctions des racines des nerfs rachidiens [Experiments on the functions of the roots of the spinal nerves] *Journal de Physiologie Expérimentale et Pathologique, 2*, 276–279.

Maier, S. F. (1986). Stressor controllability and stress-induced analgesia. *Annals of the New York Academy of Sciences, 467*, 55–72.

McClearn, G. E., & DeFries, J. C. (1973). *Introduction to behavioral genetics*. San Francisco: Freeman.

Mendel, G. J. (1866). Versuche uber Pflanzen-Hybriden [Research on hybrid plants]. *Verhandlungen des Naturforshunden in Breuen, 4*, 3–47.

Morphett, M. V., & Washburn, C. (1931). When should children begin to read? *Elementary School Journal, 31*, 496–503.

Plomin, R. (1990). *Nature and nurture: An introduction to behavioral genetics*. Belmont, CA: Brooks-Cole.

Razran, G. (1961). The observable unconscious and the inferable conscious in Soviet psychology: Introceptive conditioning, semantic conditioning, and the orienting reflex. *Psychological Review, 68*, 81–147.

Russell, M., Dark, K. A., Ellman, G., Callaway, E., & Peeke, H. V. (1984). Learned histamine release. *Science, 225*, 734–735.

Searle, L. V. (1949). The organization of hereditary maze brightness and maze dullness. *Genetic Psychology Monographs, 39*, 279–325.

Sechenov, I. M. (1935). *Collected works*. Moscow: State Publishing House.

Seligman, M. E. P., & Hager, J. L. (1972) *Biological boundaries of learning*. New York: Appleton-Century-Crofts.

Shirley, M. M. (1931). *The first two years of life: A study of twenty-five babies* (Vol. 1). Minneapolis: University of Minnesota Press.

Siegel, S. (1984). Pavlovian conditioning and heroin overdose: Reports by overdose victims. *Bulletin of the Psychonomic Society, 22*, 428–430.

Singh, J. A. L., & Zingg, R. M. (1942). *Wolfchildren and feral man.* New York: Harper & Row.

Southwell, J. P. C. (Ed.). (1924–1925). *Helmholtz's treatise on physiological optics* (Vols. 1–3). New York: Optical Society of America.

Spitz, R. A., & Wolf, K. M. (1946). The origin of the smiling response. *Genetic Psychology Monographs, 34,* 71–75.

Stone, L. J., & Church, J. (1973). *Childhood and adolescence.* New York: Random House.

Stratton, G. M. (1897). Vision without inversion of the retinal image. *Psychological Review, 4,* 341–360.

Tryon, R. C. (1940). Genetic differences in maze-learning in rats. *Yearbook of the National Society for the Study of Education, 39,* 111–119.

Valentine, C. W. (1930). The innate bases of fear. *Journal of Genetic Psychology, 37,* 394–420.

Watson, J. B. (1924). *Behaviorism.* New York: Norton.

Watson, J. B., & Rayner, R. (1920). Conditioned emotional reactions. *Journal of Experimental Psychology, 3,* 1–14.

Wertheimer, M. (1961). Psychomotor coordination of auditory and visual space at birth. *Science, 134,* 1692.

CHAPTER 2

Behavioral Genetics: The Last Century and the Next

Gerald E. McClearn

The purpose of this chapter is to sketch the history of the field of behavioral genetics in the century since the founding of the American Psychological Association (APA) and forecast the next century. In any field of science, the latter part of this assignment would be daunting. In genetics, given the explosive growth of knowledge the field has been experiencing, predicting even next week's main events is hazardous. Behavioral genetics, having drawn its theories and techniques from the parent discipline of genetics with some lag time, is not much more predictable than that. I think, however, that I can discern some current developments the general course of which can be conjectured. These guesses may constitute at least a peek through the curtain into the future.

In dealing with the past history of the field, there is little to be gained here by extensive citation of the evidence for genetic influence on this or that behavioral phenotype. Reviews and textbooks are readily available to provide details of this kind (e.g., Plomin, DeFries, & McClearn, 1990).

In this limited space, I present some generalities about the range of behavioral properties investigated from the genetics perspective, and I cite a few specific examples. In the main, however, I wish to trace the unfolding of the ideational base of the field and indicate the relevance of that base to the parent discipline of psychology.

A useful starting point might be to consider what the participants of the first APA meeting in 1892 might have talked about if the issue of heredity and behavior had come up for discussion. The specific body of thought on inheritance of behavioral attributes available to these conferees might have included some speculations from the ancient Greek philosophers, who recorded thoughts on this as on all other matters; a vague appreciation shared with the population at large that behavioral properties "run in families"; Dugdale's (1877) sociological study on the Jukes family; and a few other topics. But, the conversation almost certainly would have been dominated by the more recent work of Francis Galton. Twenty-three years before the first APA meeting, Galton (1869) published his volume, *Hereditary Genius*, in which he examined the relatives of eminent persons and found that these relatives included a greater number of individuals of high mental ability than could be accounted for by chance. Seventeen years before APA's first meeting, Galton (1876) published "The History of Twins as a Criterion of the Relative Powers of Nature and Nurture," which established an alliterative phrase that, with a very heavy burden of accumulated meaning and emotion, still exists. Just 9 years previously (Galton, 1883), in his *Inquiries Into Human Faculty and Its Development*, Galton concluded:

> There is no escape from the conclusion that nature prevails enormously over nurture when the differences of nurture do not exceed what is commonly to be found among persons of the same rank of society and in the same country. (p. 241)

Galton is well known to have been an inveterate counter and measurer, and throughout his career, he made many contributions to the definition, measurement, and evaluation of behavioral traits. Even so, the taxonomic status of psychology of his time was necessarily primitive by current standards. For example, the "ability" that Galton evaluated was based largely on reputation and unsystematic report, and the concept appears intolerably vague to the modern observer.

The concepts of inheritance were only slightly less vague. By "nature," Galton meant whatever was transmitted between generations, but it was far from clear what the nature of that hereditary material was or what the rules of its transmission were. The prevailing theory of heredity that was available as a thinking tool for Galton (and still, years later, for the hypothetical APA founders) was the pangenesis theory of Darwin (1859), which was featured in his theory of evolution. Briefly, this theory postulated that some representation of body parts coalesced in germ cells and formed the essence of the hereditary material. There were two aspects to this theory (and other similar theories as well), which turned out to be wrong. The first was that the hereditary influences received from parents merged or blended in the offspring. The result of this blending would be that the hereditary material passed on by an individual would be different from that which contributed to that individual. Such blending should constitute a great leveling, with everyone's genetic make-up ultimately approaching a common condition. An abiding and unresolved puzzlement was the maintenance of population variability, which remained apparently unchanged over many generations. The second incorrect aspect of the theory was the presumption that environmental influences on the body parts would change their representation in the germ cells and would thus be transmissible to offspring.

Much of the available scientific literature comprised listings of outcomes of particular crosses, without generality. The results were complex, and as Lush (1951) described the situation, the students of heredity at the time APA was founded, and for quite some time to come, operated with two vague rules: (a) like begets like, and (b) like does not always beget like.

Mendelian Genetics

A new paradigm, destined to transform the life sciences, had actually been published by a monk named Gregor Mendel (1866) 27 years before the founding of APA. Mendel had examined a number of dichotomous traits in the pea plant. The results of various matings were explained by hypothesizing discrete hereditary elements, which occurred in pairs, with one member of each pair having been provided to an organism by each of its parents. These elements (later to be named *genes*) could exist in alternate states (later to be named *alleles*). There was no "blending" (the elements retained their characteristic nature regardless of what their partner element was). Furthermore, evidence mounted that the elements were buffered from the vagaries of specific environmental circumstances to which the organism was exposed. Thus, an individual would pass on to a particular offspring either one of the two alleles it possessed for each gene, and the transmitted allele would be in the same state when it "left" the individual as it was when it "entered" it. The maintenance of variability in succeeding generations was no longer an enigma.

Figure 1 represents some principal features of the Mendelian schema. Of fundamental importance is the distinction made between the hereditary material and the observable characteristics. Note that the genotype is on the X axis and that the phenotype is on the Y axis (these terms were introduced many years after Mendel's original proposal). Assuming the simplest case of two allelic forms for a gene, the possible genotypes can be represented by *aa*, *Aa*, or *AA*, for example. These possibilities can be quantified by considering the number of *A* alleles in the phenotype—0, 1, or 2. The phenotype can take one of two values, for example small and large. The classic relationship shown in Figure 1 between genotype and phenotype is a nonlinear, step-function relationship in which possession of either one or two *A* alleles has the same phenotypic consequence. This relationship is described as one of dominance: The *A* allele is dominant to the recessive *a* allele. Later research demonstrated that some genes could display other types of relationships, including the additive one shown in Figure 2. In this relationship, the possessor of one each of the allelic types is intermediate in phenotype, which imposes a requirement for trichotomous rather than dichotomous classification. This 3-point scale presages an important later development—a system for dealing with phenotypes represented on a continuous scale (discussed later).

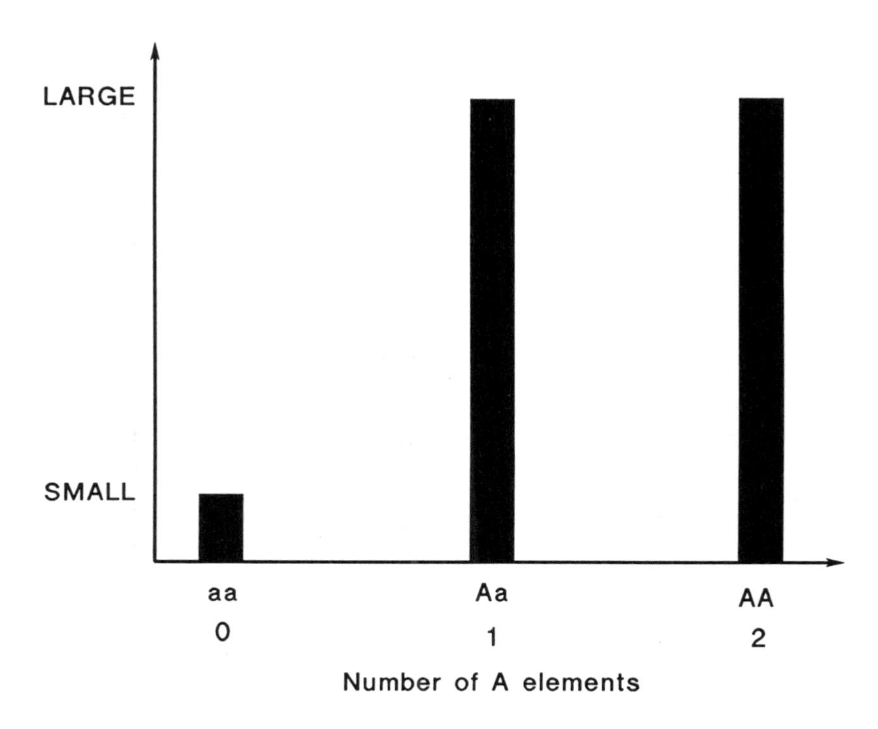

FIGURE 1. Relationship between three genotypes and two discrete levels of phenotypic expression in a hypothetical additive Mendelian condition. (A = dominant allele; a = recessive allele.)

The Mendelian model powerfully improved on its predecessors in giving clear predictions of the consequences of different matings. Regrettably, however, Mendel's results and the interpretations of them remained largely unnoticed by the biological community until 1900, when they were "rediscovered" by several investigators nearly simultaneously. The intellectual climate was now receptive, and an explosion of research effort ensued. The Mendelian model was found to apply to a large number of diverse phenotypes. In the enthusiasm of the time, the fit of model to data was sometimes forced, and some extravagant claims were made. Behavioral phenotypes were among those targeted in the early days, and Mendelian fits were attempted (and claimed) for nomadism, pauperism,

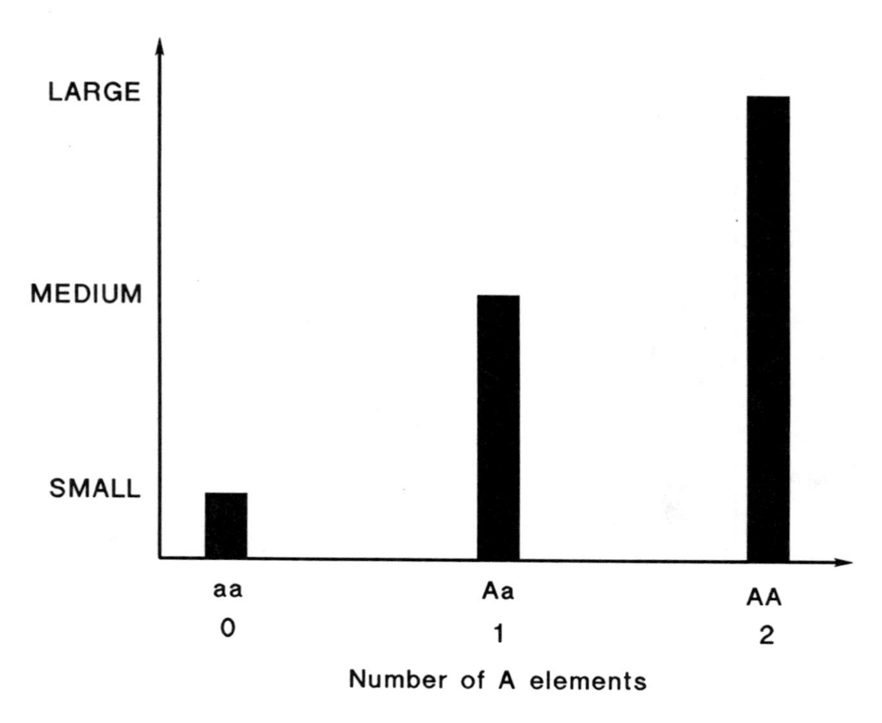

FIGURE 2. Relationship between three genotypes and three discrete levels of phenotypic expression in a hypothetical dominant/recessive Mendelian condition. (A = dominant allele; a = recessive allele.)

licentiousness, prostitution, alcoholism, and other characteristics that were of top concern to the reformists of the day.

As research progressed, rigor increased in conceptualization and measurement of phenotypes and in statistical methods and standards. Many of the early claims are now just historical curiosities. Perhaps the most prognosticative outcomes were in the general area of mental retardation, then called feeble-mindedness. The results of early studies that dealt with a conglomerate of conditions were unclear. Differential diagnosis was foreshadowed in the mid-1930s when Følling and colleagues (see Følling, Mohr, & Ruud, 1945) identified a biochemical anomaly in one type of retardation, and it was soon shown that this disease—phenylketonuria (PKU)—obeyed Mendelian rules.

Mendelian genetics has continued to the present as a powerful research interest. Many single-gene conditions have been identified in humankind, which demonstrates clearly that human beings, as well as plants and experimental and domesticated animals, are subject to Mendelian inheritance. A recent compilation (McKusick, 1992) describes over 2,300 single genes that have been assigned to specific chromosomes and at least as many that have been tentatively identified. Many of these conditions involve anomalies of one kind or another, and over 100 of them involve some degree of mental retardation. Other supposed single-gene conditions of interest to behavioral scientists include some sensory anomalies such as color blindness, deafness, and ability to taste certain substances. A particularly dramatic condition is Huntington's disease in which symptoms of neurological dysfunction, depression, and cognitive impairment first appear typically in the 3rd to 5th decades of life.

One avenue of modern research has focused strongly on attempts by sophisticated segregation analysis to demonstrate that a Mendelian gene underlies some condition and through linkage analysis to find the location of that gene on the chromosomes. Some spectacular successes have been recently attained, as in the case of Huntington's disease (Gusella et al., 1983) and muscular dystrophy (Davies et al., 1983). There have been disappointments in the attempts to show Mendelian laws at work in manic–depressive illness and in schizophrenia, however (McGuffin, Owen, & Gill, 1992; see also Gottesman, chapter 12). To understand one possible reason for these failures requires a return to the early days of the century and an entirely different tradition of studying inheritance.

Biometrical Investigation of Inheritance

Regression and correlational analyses, now ubiquitous and commonplace tools in psychological investigation, were invented in the context of research on inheritance by Galton, Karl Pearson, and their colleagues. The phenotypes in which this school was interested generally were continuously distributed. "Ability" and "temperament" were seen as existing in differing degree, not as the dichotomies for which Mendelian analysis was appropriate. Regression and correlation were, of course, admirably suited for describing the resemblance of relatives on these continuously distributed traits. In general, it was observed for a variety of phenotypes that resemblance increased as a function of the closeness of the biological relationship—parents resembled offspring more than aunts and uncles resembled nephews and nieces, and so on. Of particular relevance to behavioral genetics was a study of siblings by Pearson (1904) that compared the inheritance of behavioral ("psychical") properties (e.g., teachers' ratings of vivacity, assertiveness, introspection, popularity, conscientiousness, temper, ability) and physical ones (e.g., health, eye color, cephalic index). The average sibling correlation was slightly more than 0.50 for each class of traits, and Pearson concluded:

> We are forced, I think literally forced, to the general conclusion that the physical and psychical characters in man are inherited within broad lines in the same manner, and with the same intensity. (p. 156)

A dispute, one of the most animated in the history of science, ensued between the biometricians and the Mendelians. The Mendelians were demonstrating, at least to their own satisfaction, the general applicability of the Mendelian rules; the biometricians dismissed the Mendelian outcomes as special cases applicable only to abnormalities. A landmark contribution by Ronald Fisher (1918) resolved the issue. Briefly, Fisher proposed a model in which a large number of genes, each acting in a Mendelian manner, had an individually small effect on a common phenotype. The collective influence of these myriad "polygenes" (a term coined much later) would be to generate an essentially continuous distribution, susceptible to the statistical analyses of the biometricians. The field of *quantitative genetics* emerged from this fusion of Mendelian and biometrical genetics. Much of the subsequent elaboration and expansion of the basic model occurred in agriculture, with an enormous beneficial effect on the world's capacity to produce food. Because of the general predilection of behavioral geneticists for continuously distributed phenotypes, quantitative genetics gradually came to occupy a central role in the conceptualization of the field.

The Differential Model

The essence of the quantitative genetic approach is the decomposition of the measured variance of a phenotype into components representing genetic and environmental sources of influence. The fundamental expression of this perspective is given by the relationship

$$V_P = V_G + V_E,$$

in which the measured variance of the phenotype (V_P) is considered to be a linear composite of terms representing genetic variance (V_G) and environmental variance (V_E). These principal categories are capable of further subdivision, and interaction and covariance terms make for more complicated expressions in many applications of the analytic model. Those interested should consult Falconer (1989). The inclusion of environmental influence is of critical importance. Indeed, because environmental and genetic influences are given equal billing in the model, the conventional term quantitative genetics seems inappropriate and, indeed, misleading. For some time, and with practically no success, I have encouraged the term *differential model* as being more appropriate. This differential model is so different from the "hereditarian" view so often mistakenly attributed to behavioral genetics that it is worth some explication.

First, note that the model applies to genes that are "segregating" in the population. That is to say, it applies to those genes that contribute to individual differences among individuals within a species, not those that contribute to differences between species. There are undoubtedly many genes that all human beings have in common in an invariant allelic state. A subset of these will also be shared with other primates, and a subset of these with other mammals, and so on. Those genes for which alternative allelic states are compatible with life provide the genetic basis for individuality. Estimates of the number of these segregating genes in human beings typically range from 50,000 to 100,000.

An extremely important aspect of the model is the great power of a polygenic system to generate variability. With several tens of thousands of segregating genes, there is a truly awesome capability of sexual reproduction to generate individuality at the genetic level. As applied to human beings, all who are not members of an identical twin pair or similar multiple birth will have a unique genotype, never assembled before in the entire history of our species.

Just as each individual has a unique genotype, each individual also experiences a unique environmental history. It is important to note the breadth of definition of environment in this context. The term embraces everything not encoded in the genetic material and ranges from intimate intracellular chemical milieus to peer group influences.

Different study designs allow various compartments of environmental and genetic influence to be estimated. One frequently presented statistic is *heritability*, which is the proportion of the measured variance due to genetic differences among individuals in a studied population. In keeping with the even-handedness of the differential model, many investigators also, or alternatively, report *environmentality*, which is 1 minus heritability. Each of these measures is merely a descriptive statistic, not a statement of eternal verities. They estimate the state of affairs of a particular population with a particular gene pool in a particular array of environments. The answer could well be different in a different population or in different environmental circumstances. Note particularly that neither the genetic influence nor the environmental influence can be estimated without the other.

The principal tactics used to generate the data for such analyses include the measurement of resemblance of family members, twins, and adoptees in human studies and the use of inbred strains and selective breeding in experimental animals. The logic involved is fairly straightforward, although the detailed application can be algebraically complex. For example, any departure of the resemblance of identical twins from perfect similarity must be due to environmental sources; the excess similarity of identical twins over fraternal twins can, with certain assumptions, be attributed to the fact that identical twins are genotypically identical and that fraternal twins share only half of their segregating genes, like ordinary siblings; the resemblance of an adoptive child and an adoptive parent should reflect effects of rearing environment, as should resemblance between children separately adopted into the same home; and so on.

In the case of animals, inbreeding generates genetic homogeneity within each strain. Each animal within an inbred strain is essentially genetically identical to each other within the strain, and strains differ from each other. Thus, within-strains variance is due to environment, and between-strains variance is due to genetic factors. Generations derived by crossing unlike inbred strains permit further detailed estimates of variance components.

Selective breeding is an unambiguous demonstration of the efficacy of genes. Animals with like phenotype are mated together. For example, those with high values are mated together to begin a "high" line, and a contrasting "low" line is generated by mating of animals with low values. If genotype is totally unrelated to phenotype, then the selection will be ineffective. If the means of the lines diverge over repeated generations of selection, then it is proof positive of a genotypic influence over the phenotype, and the rate of divergence can be interpreted to estimate the relative contributions of genes and environment.

With Fisher's insight and the subsequent development of the quantitative genetic (i.e., "differential") theory, all of the ingredients were present for a declaration of peace in the intellectual war of the nature–nurture controversy. Some scholars did in fact declare the war over, but for a variety of reasons—detailed discussion of which would require much more space than available here—many segments of the behavioral and social sciences were reluctant to accept genes into the realm of their explanatory principles. To some extent, at least, this reluctance was based on an assumption that things were either "genetic" or "environmental" (remember nature versus nurture!). Furthermore, if something were genetic, then it was assumed to be unalterable. Thus, evidence of change in a behavioral phenotype under altered environmental conditions, such as training, for example, was often taken as proof that the phenotype could not be genetic. The differential model makes it clear that it is not a case of nature or nurture. Genes and environment coact. It is not necessary to choose sides. As for genetics representing unalterable fate, the example of PKU is particularly instructive. In this classical Mendelian condition, reseachers devised an environment (a nutritional environment, it happens) that ameliorates substantially the mental retardation.

The differential model has been applied with vigor to a great variety of behavioral phenotypes. The index of Plomin et al. (1990) includes ability, activity, aggressive behavior, aggressive conduct disorder, alcoholism, Alzheimer's disease, amaurotic idiocy, anorexia nervosa, antisocial personality disorder, anxiety neurosis, attention deficit disorder, attitudes, avoidance behavior, and avoidance learning—just in the As. At one time, the demonstration of a genetic influence in a behavioral phenotype was noteworthy (and, per se, publication-worthy). Now, it is almost a foregone conclusion that some degree of heritability will be found; the issue is its magnitude. It is similarly expected that there will be environmental influence. Indeed, it is becoming clear that a very powerful strategy for investigating environmental factors is to use a genetic (differential) design of some sort (e.g., Plomin et al., 1990). An important guide to search for specific environmental factors can be derived from the estimation of the relative importance of the anonymous categories of shared, nonshared, and prenatal environments, for example.

Mechanisms of Heredity

Mendel's elements were clearly hypothetical variables, but before long, the early genetics research began to identify physical aspects of their existence. First, it was observed that the Mendelian processes of genes paralleled the processes of intranuclear structures called chromosomes. Evidence rapidly accumulated that genes resided in particular places on these chromosomes, and detailed microscopic study gave some intimations about what they were like (e.g., their average size), and the basis was laid for the construction of the linear maps that permit the location of particular genes. Just as there were these hints as to what genes were, there were also some hints as to what they did. It was shown that at least some genes seemed to be involved in the generation of enzymes, and the conviction gradually grew that maybe that is what all genes did. Thus, halfway through APA's 1st century, there was some very hard-won empirical information and theory about gene structure and function, but the prospect of clear understanding of the nature of the gene seemed very remote, indeed.

In midcentury, an intellectual bombshell exploded, with the theory of Watson and Crick (1953a, 1953b). The result was a breathtaking series of major discoveries that have not only transformed biology but, with an impact comparable to that of the Copernican and Darwinian revolutions, also altered humankind's view of itself and its place in the scheme of things. So well have the basic terms suffused society that public media often no longer feel it to be necessary even to define the abbreviations DNA and RNA.

The basic dogma is that DNA is the repository of the genetic information. That information is coded as a series of bases. The hereditary information in the DNA is copied onto RNA by a complementation process. The RNA participates with certain intracellular structures to produce polypeptides, which compose proteins; these proteins include structural, transport, and catalytic proteins (enzymes). Enzymes are specially critical in facilitating the chemical reactions of life. It became clear that different allelic forms of a gene were due to different sequences of the bases in the DNA, which result in different sequences of amino acids in the protein, which might, as a consequence, alter functional properties.

Until quite recently, behavioral genetics has been an onlooker and cheerleader for molecular genetics. The principal consequence of the molecular revolution for behavioral genetics has been that we now have a plausible answer to the question of how genes can possibly influence labile behavior. It is a reductionist explanation: Genes influence the proteins that are critical to the functioning of the organ systems that determine behavior. Thus, two sets of beliefs are involved. First is that the nervous system, endocrine system, sensory systems, musculature, and so on each separately, and collectively, influence behavior. This is the stuff of biopsychology, the interface between physiology, biochemistry, and behavior. Second is the belief that allelic differences of genes influence these anatomical systems and their function. This is the area of molecular and physiological genetics. If the theories and databases of these domains are accepted, then it is not surprising that genes can influence behavior. It would be absolutely astonishing if they could not!

Simple graphic representations are useful in illustrating some important principles. In Figure 3 is displayed a network of organ systems and processes leading to two phenotypes, P1 and P2. The Gs indicate a genetic influence on each element in the network. Two considerations merit attention. First, each of the phenotypes is influenced by multiple genes. This, of course, is the basic differential model. Second, through branching pathways, particular genes may influence more than one phenotype. This phenomenon is named *pleiotropy* and translates into genetic correlation and comorbidity. The differential model is such that neither genetic nor environmental factors alone suffice for explanation. Figure 4 makes this point by indicating that each process is also susceptible to

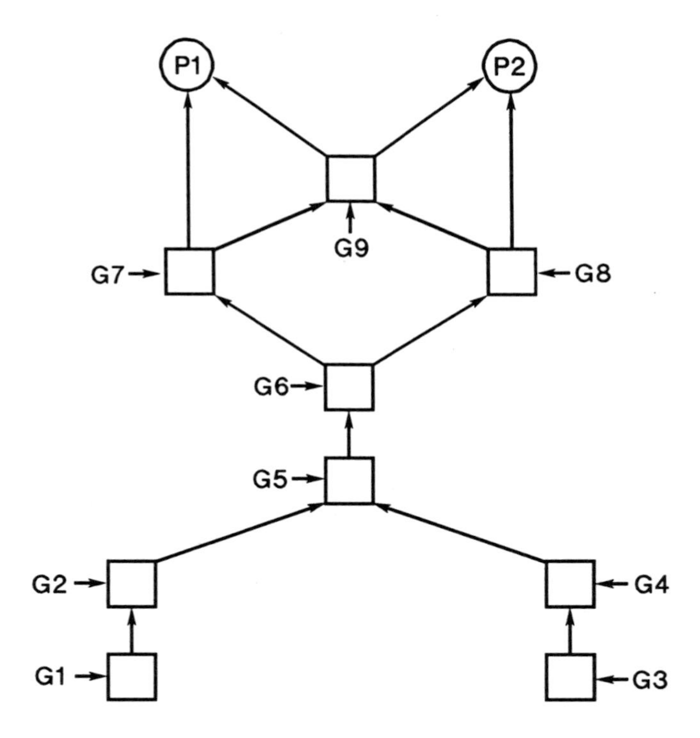

FIGURE 3. A representation of genetic (G) influence on elements in a causal network. (P = phenotype.)

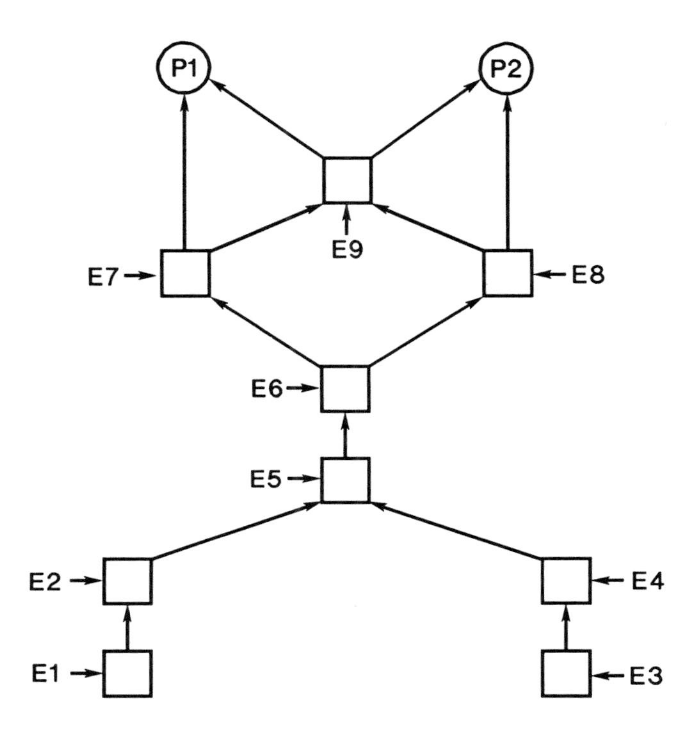

FIGURE 4. A representation of environmental (E) influence on elements in a causal network. (P = phenotype.)

environmental influence, and environmental analogs to polygeny and pleiotropy are clearly implied: multiple influences and manifold effects. The network of influences (causal field or causal matrix, if you prefer) shown also suggests a rationalization of single-gene influence. In the differential model, the Mendelian outcome is a limiting case in which the allelic configuration of a single gene accounts for such a large part of the variance that other genes and environment can, for most practical purposes, be ignored. In the network model presented in Figure 4, that outcome can be understood as a gene participating in a bottleneck in the network.

The model shown is both ridiculously symmetric and pathetically simplified. The true complexity of, say, biochemical processes is enormously greater. A gesture to the true world is made in Figure 5, which

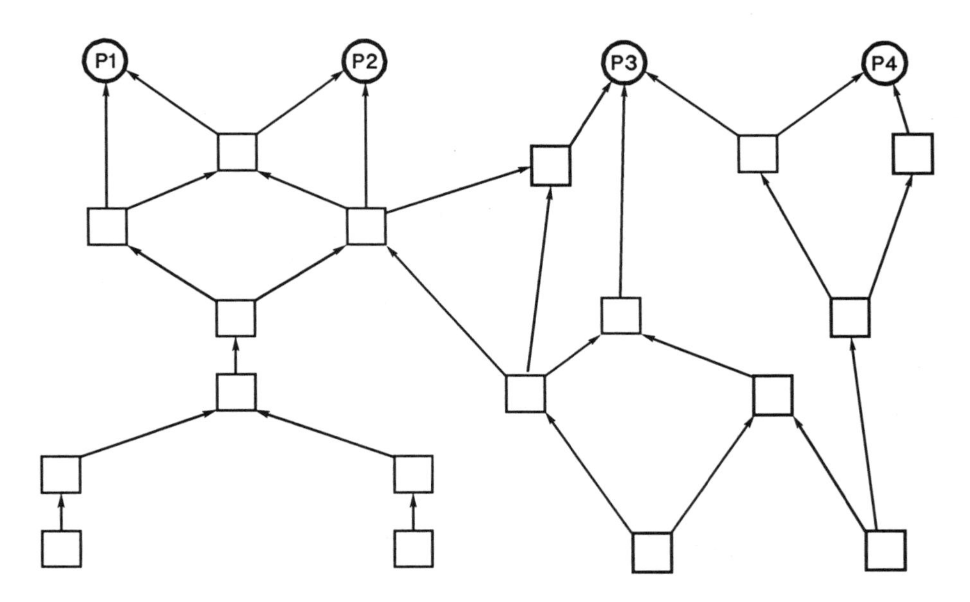

FIGURE 5. Representation of a network of influences on several phenotypes (P) illustrating pleiotropy, polygeny, and other bases for correlations involving phenotypes.

expands the causal field to embrace two more phenotypes and many more elements in the matrix (with genetic and environmental factors understood to be operative on all elements). The point of this figure is to illustrate better the implications of multiple pathways leading to phenotypic expression. These multiple pathways may be the explanation of the rarity of single-gene effects in the complex behaviors that are the focus of behavioral genetics interest.

One of the attributes of complex biological systems such as suggested here is that they are controlled or regulated. Although the arrows in the figures suggest simple unidirectional causality, real systems would be replete with feedback systems providing homeostatic regulation. Newtonian causality considerations become inadequate, and notions of circular or network causality will be required for more complete comprehension of the operation of any such network.

A major result from these considerations is that researchers are freed from the old notion that "genes are fate." If the environment is operating on the same causal field that mediates genetic influence, then compensation or remediation through environmental intervention is clearly reasonable to seek. The most vivid success story to date is that of PKU, and its example is encouraging even for polygenically influenced phenotypes. Enormous research efforts will obviously be required to identify the elements in any particular causal field and characterize their interrelationships. In many cases, the effort would be more than justified, and the dramatic recent advances in structural modeling applied to quantitatively distributed phenotypes has given us a grand strategy for the analysis.

Developmental Behavioral Genetics

It is perfectly apparent that a newborn resembles its parents less than it will when it is grown. Something about genetic influence happens developmentally. In a prescient remark about genetic change, Galton (1876) noted:

> It must be borne in mind that the divergence of development, when it occurs, need not be ascribed to the effect of different natures, but it is quite possible that it may be due to the appearance of qualities inherited at birth, though dormant. (p. 402)

Indeed, Galton's first study of twins investigated the extent to which twin resemblance changed during development (Galton, 1876). This phenomenon has been carefully described for intellectual function. For example, one study (Honzik, 1957) showed a dramatic increase in the correlation between a child's IQ and maternal education from about 0.05 at 2 years of age to about 0.35 at 6 years of age. The finding of increased heritability for cognitive ability has consistently been found in childhood (Fulker, DeFries, & Plomin, 1988) and, recently, in the last half of the life span in which heritability reaches 80% (Pedersen, Plomin, Nesselroade, & McClearn, 1992; see also McGue et al., chapter 3). As only one example in the domain of temperament, the correlation of identical twins with respect to activity level was found to increase from about 0.30 to about 0.58 between 3 and 24 months of age (Matheny, 1983; see also Plomin & Nesselroade, 1990).

Another indication that there are developmental timing aspects to genetic effects is provided by genetic diseases with late onset. The classic case of this sort is the condition of Huntington's disease, in which the onset is typically in young or middle adulthood.

There was a flowering of developmental genetics in the 1950s, featuring both single genes, particularly genes with lethal effects (e.g., Hadorn, 1955), and polygenic systems (e.g., Waddington, 1957). This line of research was overtaken by the excitement of molecular genetics discoveries, and the molecular perspective on development has become a torrid research area today.

A major event in the molecular approach was the demonstration in microorganisms that genes could be turned on and turned off. In fruit fly larvae, where chromosomes are particularly advantageous for microscopic study, it was subsequently discovered that in a critical developmental period there were visible manifestations of sequencing of activity of different genes on different chromosomes. These and similar observations have given rise to a field of study of regulatory genes—genes that control other genes (see MacLean, 1989, for a particularly accessible treatment of gene regulation).

This field of investigation greatly enriches the perception of genetic influence. Not only may researchers contemplate the consequences of the differences among individuals in genotype, but they may also consider intraindividual genetic differences. Although it is clear that individuals get all of their genetic information at conception, only parts of the total set are used by different organ systems, and different parts are called into play at different times. From this perspective, development involves programmatic change in what might be called the *effective genotype.* Whereas many genes are undoubtedly operating from the beginning and many others function for a while and then fade away, others heed their call to perform at some specific life stage, are locked on, and function from then on to the end; and still others put in brief appearances at critical developmental periods. (An interesting conjecture is that senescing genes may exist, which get turned on late in life.)

The control and regulation of genes is far from completely understood. The corpus of knowledge grows at a prodigious rate, however, and it is easy to predict that profound new insights into development will be the outcome of the current high interest in the field.

Behavioral genetics is increasingly using developmental genetics perspectives, with special journal issues (Plomin, 1983) and monographs (Hahn, Hewitt, Henderson, & Benno, 1990; Plomin, 1986) testifying to the current prominence of the developmental orientation within behavioral genetics.

As a summary, if the human genome could be visualized, a picture would be presented of an organism proceeding through its lifetime with programmed species-typical changes in effective genotype—changes in the array of genes that are turned on at a particular time. These changes are slow in relation to the life span and are unlikely to proceed at a uniform rate. The rate may be relatively greater in infancy than in later childhood, for example, and adolescence must mark particularly rapid change in effective genotype. For these changes, the differential model is an analytic tool of choice: It can partition interindividual variance, most appropriately among individuals at the same general stage in their species-typical trajectory.

The Forecast

The easy part of peeking into the future is to forecast more of the same. More and more behavioral phenotypes will be described, and their components of variance will be estimated with increasing precision by increasingly sophisticated statistical models. It will increasingly be seen that genetic methods are invaluable for researchers who are interested in detecting, characterizing, and estimating the importance of effects of environment. For an expanding number of behavioral phenotypes, insights will be won into the physiological, anatomical, and biochemical mechanisms that mediate the genetic influence. Indeed, genetic methods will likely come to suffuse the reductionist study of mechanisms of behavior. Studies in developmental behavioral genetics, in the main, have been characterized by quantitative perspectives. Increasing emphasis on molecular developmental genetics perspectives in the study of behavioral development is easily predictable.

Application of the quantitative genetic model in behavioral genetics has principally attended to the proximal input variables (the hypothesized plural genes of small effect) and the distal output variables (the measured behavioral phenotypes), with relatively little concern about the intervening variables. It is possible to imagine that each of the polygenes makes its contribution to the final phenotype through the same simple causal path, but the more complex scenarios suggested earlier (Figure 5) are probably more veridical for complex behavioral attributes in general. That is to say, the intervening variables from genes to behavior are likely to constitute complex systems. The study of the properties of complex systems and application to various scientific disciplines (including behavior; see Ford, 1987) has expanded prodigiously in recent years. As Murphy and colleagues have pointed out in a series of incisive articles (e.g., Murphy & Berger, 1988; Murphy & Trojak, 1986), the principles of complex systems have important implications for quantitative genetic research. With considerable confidence about the event, but less about the rate at which it will occur, I predict that complex-systems thinking will become an integral part of the differential model (McClearn, 1993; McClearn & Plomin, in press).

In a distinctly different approach, the intermediates between genes and behavior will be opened to investigation with powerful techniques. Until recently, researchers have been able to deal with two types of genes: the sledge-hammer, Mendelian genes and the puny polygenes. It has long been recognized that in the multiple gene situation there may be a whole continuum of effect size. Some genes may exert enough influence to be detectable if rather heroic efforts are made to locate them. Because of the rapid increase in molecular markers of location on chromosomes, plant geneticists recently began development of techniques for efficient identification of chromosomal regions containing such *quantitative trait loci* (QTL), and variants of these methods have now been adapted to animal research (Plomin, 1990). The approach should work as well for behavioral phenotypes as for any other class of phenotypes. QTL have already been tentatively identified with respect to various aspects of alcohol-related behavior (McClearn, Plomin, Gora-Maslak, & Crabbe, 1991) and other behavioral phenotypes (Plomin, McClearn, Gora-Maslak,

& Neiderhiser, 1991). Prospects now seem bright of identifying, within a polygenic system, a number of specific genes that collectively account for a substantial part of the genetic variance component estimated by the differential model.

Lifting the veil of anonymity for at least some of the genes that influence the complex systems of behavioral phenotypes will offer grand opportunities for the merging of molecular and quantitative genetics theories and methods. The way will be opened for application of the whole molecular genetic armamentarium in the elucidation of mechanism in polygenic systems. Furthermore, an unparalleled bridge may be generated between research with animal models and that with human beings. The fact that, for segments of chromosomes, the order of genes is the same in man and other animals offers prospects that identification of QTL in laboratory animals will point to specific areas of the human genome for location of the homologous human gene.

Finally, I suggest that the attention of behavioral geneticists may be drawn increasingly to the notion of intraindividual changes in effective genotype. The idea of developmental changes in arrays of effective genes, discussed earlier, opens the door to consideration of other possible types of alteration of composition of the genetic team on the field of play.

A rapidly growing body of literature describes the effects of certain long-term environmental circumstances on the types and amounts of RNA present in certain organs. The nutritional environment provides many useful examples. With appropriate dietary manipulations in rats, for example, variations have been shown in the amount of liver RNA for malic enzyme (Katsurada, Iritani, Fukuda, Noguchi, & Tanaka, 1987), glucose-6-phosphate dehydrogenase (Fritz & Kletzien, 1987; Tomlinson, Nakayama, & Holten, 1987), aldolase (Munnich et al., 1985), apolipoprotein E (Kim et al., 1989), and L-type pyruvate kinase (Munnich et al., 1984; Weber et al., 1984). These observations relate, of course, to the well-known phenomenon of enzyme induction. The specific observations on RNA are particularly pertinent to the present topic.

Other recent evidence in rats relates repeated immobilization stress expression of genes for adrenal catecholamine biosynthetic enzymes: tyrosine hydroxylase and dopamine β-hydroxylase (McMahon et al., 1992).

Mild footshock stress in rats has been shown to increase expression of the c-*fos* gene in the brain (Smith, Banerjee, Gold, & Glowa, 1992). Of special relevance to behavioral science is that the conditioned stressor alone (tone previously associated with footshock) is also effective in increasing this gene expression in certain parts of the brain (Smith et al., 1992).

It is irresistible to contemplate the effects of short periods of stress. For example, a single immobilization stress episode has been shown to increase tyrosine hydroxylase (McMahon et al., 1992). It is unlikely that this represents a lower limit to the temporal grain in which gene–environment interactions take place. It may well be that such interactions occur on the temporal scale appropriate for the processes of neurotransmitter production, reuptake, and disposition. So my momentary effective genotype in certain brain regions may be fluctuating as I formulate these arguments, and so may the reader's. There may lie herein a promising research area bridging behavioral genetics and neuroscience.

Some Family Notes

Behavioral genetics, it may be said, has had an uneasy childhood. For much of its existence, its parent discipline of genetics hasn't paid any attention to it. If the geneticists who engaged in the furiously paced search for the molecular basis of heredity paused at all to notice behavioral scientists, then it was to express astonishment that anyone would be so unwise as to select such messy phenotypes for study. However, more complex systems have recently emerged within genetics as challenges for the future (Bodmer, 1986). It is not impossible that some behavioral phenotypes will be attractive targets in this new emphasis.

The other parent, psychology, has known of behavioral genetics but has often appeared to wish to disown its offspring. Recent recognition perhaps means that some merit is now seen in the child. Some of its cousins and stepsiblings have come to like it in recent years and now welcome it warmly. Some have even joined the child. Some remain suspicious, some say "Who?", and some pick up their plates and move away as the child approaches the picnic table. So it goes!

All of the relatives in psychology should be assured that the behavioral genetics child is as interested as they are in the family business—the understanding and explanation of behavior—and just as excited about what lies ahead in the next century.

References

Bodmer, W. F. (1986). Human genetics: The molecular challenge. *Cold Spring Harbor Symposia on Quantitative Biology, 51,* 1–14.

Darwin, C. (1859). *On the origin of species by means of natural selection or the preservation of favoured races in the struggle for life.* New York: Appleton.

Davies, K. E., Pearson, P. L., Harper, P. S., Murray, J. M., O'Brien, T., Sarfrazi, M., & Williamson, R. (1983). Linkage analysis of two cloned DNA sequences flanking the Duchenne muscular dystrophy locus on the short arm of the human X chromosome. *Nucleic Acids Research, 11,* 2303–2305.

Dugdale, R. L. (1877). *The Jukes: A study in crime, pauperism, disease and heredity.* New York: Putnam.

Falconer, D. S. (1989). *Introduction to quantitative genetics* (3rd ed.). Essex, UK: Longman Scientific & Technical.

Fisher, R. A. (1918). The correlation between relatives on the supposition of Mendelian inheritance. *Transactions of the Royal Society of Edinburgh, 52,* 399–433.

Følling, A., Mohr, O. L., & Ruud, L. (1945). *Oligophrenia phenylpyrouvica:* A recessive syndrome in man. *Norske Videnskaps/Akademi I Oslo, Matematisk-Naturvidenskapelig Klasse, 13,* 1–44.

Ford, D. L. (1987). *Humans as self-constructing living systems.* Hillsdale, NJ: Erlbaum.

Fritz, R. S., & Kletzien, R. F. (1987). Regulation of glucose-6-phosphate dehydrogenase by diet and thyroid hormone. *Molecular and Cellular Endocrinology, 51,* 13–17.

Fulker, D. W., DeFries, J. C., & Plomin, R. (1988). Genetic influence on general mental ability increases between infancy and middle childhood. *Nature, 336,* 767–769.

Galton, F. (1869). *Hereditary genius: An inquiry into its laws and consequences.* London: Macmillan.

Galton, F. (1876). The history of twins as a criterion of the relative powers of nature and nurture. *Royal Anthropological Institute of Great Britain and Ireland Journal, 6,* 391–406.

Galton, F. (1883). *Inquiries into human faculty and its development.* London: Macmillan.

Gusella, J. F., Wexler, N. S., Conneally, P. M., Naylor, S. L., Anderson, M. A., Tanzi, R. E., Watkins, P. C., & Ottina, K. (1983). Apolymorphic DNA marker genetically linked to Huntington's disease. *Nature, 306,* 234–238.

Hadorn, E. (1955). *Developmental genetics and lethal factors.* New York: Wiley.

Hahn, M. E., Hewitt, J. K., Henderson, N. D., & Benno, R. (Eds.). (1990). *Developmental behavioral genetics.* New York: Oxford University Press.

Honzik, M. P. (1957). Developmental studies of parent–child resemblance in intelligence. *Child Development, 28,* 215–228.

Katsurada, A., Iritani, N., Fukuda, H., Noguchi, T., & Tanaka, T. (1987). Influence of diet on the transcriptional and post-transcriptional regulation of malic enzyme induction in the rat liver. *European Journal of Biochemistry, 168,* 487–491.

Kim, M. H., Nakayama, R., Manos, P., Tomlinson, J. E., Choi, E., Ng, J. D., & Holten, D. (1989). Regulation of apolipoprotein E synthesis and mRNA by diet and hormones. *Journal of Lipid Research, 30,* 663–671.

Lush, J. L. (1951). Genetics and animal breeding. In L. C. Dunn (Ed.), *Genetics in the 20th century* (pp. 493–525). New York: Macmillan.

MacLean, N. (1989). *Genes and gene regulation.* London: Edward Arnold.

Matheny, A. P., Jr. (1983). A longitudinal twin study of stability of components from Bayley's infant behavior record. *Child Development, 54,* 356–360.

McClearn, G. E. (1993). Genetics, systems, and alcohol. *Behavior Genetics, 23,* 223–230.

McClearn, G. E., & Plomin, R. (in press). Strategies for the search for genetic influences in alcohol-related phenotypes. In H. Begleiter & B. Kissin (Eds.), *Alcohol and alcoholism.* London: Oxford University Press.

McClearn, G. E., Plomin, R., Gora-Maslak, G., & Crabbe, J. C. (1991). The gene chase in behavioral science. *Psychological Science, 2,* 222–229.

McGuffin, P., Owen, M., & Gill, M. (1992). Molecular genetics of schizophrenia. In J. Mendlewics & H. Hippius (Eds.), *Genetic research in psychiatry* (pp. 25–48). New York: Springer-Verlag.

McKusick, V. A. (1992). *Mendelian inheritance in man* (10th ed.). Baltimore: Johns Hopkins University Press.

McMahon, A., Kvetnansky, R., Fukuhara, K., Weise, V. K., Kopin, I. J., & Sabban, E. L. (1992). Regulation of tyrosine hydroxylase and dopamine β-hydroxylase mRNA levels in rat adrenals by a single and repeated immobilization stress. *Journal of Neurochemistry, 58,* 2124–2130.

Mendel, G. J. (1866). Versuche uber Pflanzen-Hybriden [Research on hybrid plants]. *Verhandlungen des Naturforshunden in Breuen, 4,* 3–47.

Munnich, A., Besmond, C., Darquy, S., Reach, G., Vaulont, S., Dreyfus, J-C., & Kahn, A. (1985). Dietary and hormonal regulation of adolase B gene expression. *Journal for Clinical Investigation, 75,* 1045–1052.

Munnich, A., Marie, J., Reach, G., Vaulont, S., Simon, M. P., & Kahn, A. (1984). *In vivo* hormonal control of L-type pyruvate kinase gene expression. *Journal of Biological Chemistry, 259,* 10228–10231.

Murphy, E. A., & Berger, K. R. (1988). An approach to the genetics of physiological homeostasis. In B. S. Weir, E. E. Eisen, M. M. Goodman, & G. Namkoong (Eds.),

Proceedings of the Second International Conference on Quantitative Genetics (pp. 283–296). Sunderland, MA: Sinauer Associates.

Murphy, E. A., & Trojak, J. L. (1986). The genetics of quantifiable homeostasis: I. The general issues. *American Journal of Medical Genetics, 24,* 159–169.

Pearson, K. (1904). On the laws of inheritance of man: II. On the inheritance of the mental and moral characters in man, and its comparison with the inheritance of the physical characters. *Biometrika, 3,* 131–190.

Pedersen, N. L., Plomin, R., Nesselroade, J. R., & McClearn, G. E. (1992). A quantitative genetic analysis of cognitive abilities during the second half of the life span. *Psychological Science, 3,* 346–353.

Plomin, R. (1983). Developmental behavioral genetics. *Child Development, 54,* 253–259.

Plomin, R. (1986). *Development, genetics, and psychology.* Hillsdale, NJ: Erlbaum.

Plomin, R. (1990). Role of inheritance in behavior. *Science, 248,* 183–188.

Plomin, R., DeFries, J. C., & McClearn, G. E. (1990). *Behavioral genetics: A primer* (2nd ed.). New York: Freeman.

Plomin, R., McClearn, G. E., Gora-Maslak, G., & Neiderhiser, J. (1991). Use of recombinant inbred strains to detect quantitative trait loci associated with behavior. *Behavior Genetics, 21,* 99–116.

Plomin, R., & Nesselroade, J. R. (1990). Behavioral genetics and personality change. *Journal of Personality, 58,* 191–220.

Smith, M. A., Banerjee, S., Gold, P. W., & Glowa, J. (1992). Induction of c-*fos* mRNA in rat brain by conditioned and unconditioned stressors. *Brain Research, 578,* 135–141.

Tomlinson, J. E., Nakayama, R., & Holten, D. (1987). Repression of pentose phosphate pathway dehydrogenase synthesis and mRNA by dietary fat in rats. *Journal of Nutrition, 118,* 408–415.

Waddington, C. H. (1957). *The strategy of the genes.* New York: Macmillan.

Watson, J. D., & Crick, F. H. C. (1953a). Genetical implications of the structure of deoxyribonucleic acid. *Nature, 171,* 964–967.

Watson, J. D., & Crick, F. H. C. (1953b). Molecular structure of nucleic acids: A structure for deoxyribose nucleic acids. *Nature, 171,* 737–738.

Weber, A., Marie, J., Cottreau, D., Simon, M.-P., Besmond, C., Dreyfus, J.-C., & Kahn, A. (1984). Dietary control of adolase B and L-type pyruvate kinase mRNAs in rat. *Journal of Biological Chemistry, 259,* 1798–1802.

PART TWO

The Genetics of Cognitive Abilities and Disabilities

PART TWO

Introduction

John C. DeFries

Three decades before the inception of the American Psychological Association, Francis Galton conducted the first systematic study of the inheritance of mental ability. Using biographical information, published accounts, and direct inquiry, Galton evaluated the accomplishments of eminent judges, statesmen, peers, military commanders, artists, scholars, musicians, religious leaders, and their relatives. From these data, Galton identified the most eminent man in each family and then tabulated other individuals who attained eminence in one profession or another according to the closeness of their relationship to these most eminent men. Although Galton's measure of mental ability was highly subjective, the results of his study were striking: The number of eminent men in these families was far greater than would be expected to occur by chance alone; moreover, the closer the relationship to the most eminent men, the higher the incidence of eminence.

The results of Galton's study were published in 1869 in *Hereditary Genius: An Inquiry Into Its Laws and Consequences.* These results, plus those of a later study of similarities and differences between members of identical and fraternal twin pairs, led Galton (1883) to conclude that "nature prevails enormously over nurture" (p. 241).

Interest in the inheritance of behavioral characters continued until the 1930s when a sharp decline occurred with the advent of behaviorism (Plomin, DeFries, & McClearn, 1990). However, a reawakening of interest in the genetics of behavior began to take place in the 1950s and received a major impetus following the publication of Fuller and Thompson's (1960) field-defining monograph *Behavior Genetics.* During the next 3 decades, research activity in behavioral genetics occurred at an accelerating rate. A focal interest of research in this new interdisciplinary field was the inheritance of general and specific cognitive abilities (e.g., Bouchard & McGue, 1981; DeFries, Vandenberg, & McClearn, 1976).

Past, present, and future research pertaining to the genetics of cognitive abilities and disabilities are described in the following five chapters. In chapter 3, "Behavioral Genetics of Cognitive Ability: A Life-Span Perspective," McGue, Bouchard, Iacono, and Lykken discuss recent evidence for an increase in the heritable nature of individual differences in general cognitive ability during adulthood, an important new finding for the nascent field of developmental behavioral genetics (see Plomin, 1986).

In chapter 4, "Continuity and Change in Cognitive Development," Fulker, Cherny, and Cardon fit a developmental path model to general cognitive ability data from the Colorado Adoption Project and the MacArthur Longitudinal Twin Study. Results of this sophisticated analysis reveal a striking diversity among the genetic and environmental processes that affect continuity and change in general cognitive ability from infancy to middle childhood. In chapter 5, "Genetics of Specific Cognitive Abilities," Cardon and Fulker subject sibling data collected in the Colorado Adoption Project to a state-of-the-art hierarchical genetic analysis. Results obtained from this analysis provide some of the first evidence for genetic influences on specific cognitive abilities, independent of general cognition.

In "Genetics of Reading Disability" (chapter 6), DeFries and Gillis review a program of research on the etiology of reading problems that was initiated at the Institute for Behavioral Genetics (University of Colorado at Boulder) in 1973. They also review some results obtained from previous family and twin studies of reading disability, as well as those from a large, ongoing twin study and a genetic linkage analysis.

Finally, Duane F. Alexander, in chapter 7 ("Cognitive Abilities and Disabilities in Perspective: Part Two Discussion"), discusses some of the methodological techniques that are currently being used to assess the nature and nurture of individual differences in cognitive abilities and disabilities.

References

Bouchard, T. J., & McGue, M. (1981). Familial studies of intelligence: A review. *Science, 212,* 1055–1059.

DeFries, J. C., Vandenberg, S. G., & McClearn, G. E. (1976). Genetics of specific cognitive abilities. *Annual Review Genetics, 10,* 179–207.

Fuller, J. L., & Thompson, W. R. (1960). *Behavior genetics.* New York: Wiley.

Galton, F. (1869). *Hereditary genius: An inquiry into its laws and consequences.* London: Macmillan.

Galton, F. (1883). *Inquiries into human faculty and its development.* London: Macmillan.

Plomin, R. (1986). *Development, genetics, and psychology.* Hillsdale, NJ: Erlbaum.

Plomin, R., DeFries, J. C., & McClearn, G. E. (1990). *Behavioral genetics: A primer* (2nd ed.). New York: Freeman.

CHAPTER 3

Behavioral Genetics of Cognitive Ability: A Life-Span Perspective

Matt McGue, Thomas J. Bouchard, Jr., William G. Iacono, and David T. Lykken

Since the time of Francis Galton, the nature–nurture controversy has been virtually synonymous with debate on whether genetic factors influence IQ. Although there still may be some who would reduce the debate to questioning the specifics of individual studies, it appears that the issue has been resolved for the vast majority of psychologists. Snyderman and Rothman (1990) recently surveyed psychologists' beliefs about intelligence and reported that over 90% of those responding agreed that IQ was, at least in part, heritable. This modern consensus, a clear reversal of opinion from the 1950s and 1960s, is a direct result of the substantial body of research compiled by behavioral geneticists over the past 50 years—a corpus that unequivocally implicates genetic factors in the development of IQ.

The preparation of this chapter was supported, in part, by U.S. Public Health Service grants AG06886 and DA05147. The chapter is based, in part, on a paper presented at the centennial meeting of the American Psychological Association in Washington, DC, August 1992.

In 1981, Bouchard and McGue (1981) published an update of the classic Erlenmeyer-Kimling and Jarvik (1963) review of familial studies of intelligence. In this update, they summarized results from over 100 separate studies reporting over 500 familial IQ correlations on over 100,000 pairs of relatives. Some of the more significant results from this review are given in Table 1. When taken in aggregate, twin, family, and adoption studies of IQ provide a demonstration of the existence of genetic influences on IQ as good as can be achieved in the behavioral sciences with nonexperimental methods. Without positing the existence of genetic influences, it simply is not possible to give a credible account for the consistently greater IQ similarity among monozygotic (MZ) twins than among like-sex dizygotic (DZ) twins, the significant IQ correlations among biological relatives even when they are reared apart, and the strong association between the magnitude of the familial IQ correlation and the degree of genetic relatedness. The correlations summarized in Table 1 also strongly implicate the existence of environmental influences: The correlation among reared-together MZ twins is less than unity; biological

TABLE 1

Average Familial IQ Correlations (*R*)

Relationship	Average *R*	Number of pairs
	Reared-together biological relatives	
MZ twins	0.86	4,672
DZ twins	0.60	5,533
Siblings	0.47	26,473
Parent–offspring	0.42	8,433
Half-siblings	0.35	200
Cousins	0.15	1,176
	Reared-apart biological relatives	
MZ twins	0.72	65
Siblings	0.24	203
Parent–offspring	0.24	720
	Reared-together nonbiological relatives	
Siblings	0.32	714
Parent–offspring	0.24	720

Note. MZ = monozygotic; DZ = dizygotic. *R* was determined using sample-size-weighted average of *z* transformations. Adapted from "Familial Studies of Intelligence: A Review" by T. J. Bouchard, Jr., and M. McGue, 1981, *Science, 250,* p. 1056; copyright 1981 by the American Association for the Advancement of Science; adapted by permission.

relatives who were reared together are more similar than biological relatives who were reared apart; and there is a significant correlation between the IQs of nonbiologically related but reared-together relatives.

In 1990, Chipeur, Rovine, and Plomin (1990) sought to estimate the extent to which IQ variability was associated with genetic, shared environmental, and nonshared environmental factors by fitting biometric models to the aggregate data summarized by Bouchard and McGue (1981). They found that a relatively simple biometric model fit the combined IQ data and yielded an estimate of 51% for the percentage of IQ variance associated with genetic factors (i.e., the heritability). Happily, their estimate of IQ heritability fell midway between the estimates advocated by some of the more strident partisans in the debate. Chipeur et al. also reported that shared environmental influences (i.e., environmental factors that are shared by individuals reared in the same home and thus contribute to their similarity) accounted for anywhere from 11% to 35%, depending on the particular relationship involved (i.e., siblings vs. DZ vs. MZ twins). The balance of IQ variance, from 14% to 38% of the total, was apportioned to nonshared environmental influences (i.e., environmental factors that are not shared by reared-together individuals and thus contribute to dissimilarities among family members). The publication of the Chipeur et al. estimates appeared to finally resolve the question of the heritability of IQ; like many psychological characteristics, it appeared to be significantly, but moderately, heritable. From a behavioral genetic perspective, IQ was distinguished from other behavioral characteristics not by its heritability but rather by the extent to which it was influenced by shared environmental factors (Plomin & Daniels, 1987).

However, a feature of the Bouchard and McGue (1981) IQ literature summary that probably has not received sufficient attention is the fact that the vast majority of the IQ kinship correlations were derived on samples of individuals who were 20 years old or younger (there are some exceptions, the most notable being the reared-apart twin studies that we discuss later). For example, in the 34 studies of reared-together MZ twins, only 1 (that of 37 MZ pairs by Shields in 1962) was based on a sample composed entirely of adults, and only 2 additional studies included in their samples any twins who were 21 years old or older. One hundred twenty-eight MZ twin pairs participated in these three studies, or 2.7% of the total number of 4,672 MZ twin pairs on which IQ correlations had been reported. Because the IQ heritability estimate of 51% is derived from studies of mostly preadolescents and adolescents, its applicability to other stages of the life span remains an open question.

There are theoretical reasons why heritability estimates derived from samples of children might not generalize to adult populations. For example, the cumulative impact of environmental effects may not be fully realized until adulthood; therefore, one might expect IQ heritability to decrease with age (cf. Baltes, 1987). Alternatively, Scarr and McCartney (1983) have argued that with age comes increasing individual control over formative experiences. Because self-selection of experience is likely to reflect, in part, underlying genetically influenced tendencies, one would expect the heritability of IQ to increase with age. There are also empirical reasons to question the generalizability of IQ heritability estimates derived from samples representing only a limited segment of the total life span. In a recent meta-analysis of twin studies (most of which were based on adolescent and preadolescent samples), McCartney, Harris, and Bernieri (1990) reported that MZ twin similarity, although decreasing with age for most personality characteristics, appeared to increase with age for intelligence. In addition, Pedersen, Plomin, Nesselroade, and McClearn (1992), in a sample of reared-together and reared-apart Swedish twins who were 50 years old and older, recently reported a heritability estimate of 81% for general cognitive ability. They speculated that the substantial difference between their estimate of IQ heritability and that derived by Chipeur et al. (1990) from the aggregate kinship correlations may be due to the relatively older nature of their sample; IQ heritability may increase with age.

The purpose of this chapter is to more fully explore the question of whether IQ heritability varies across the life span. This exploration is based on a reanalysis of published twin and adoption studies of IQ, as well as on results from a preliminary analysis of an ongoing cross-sectional twin study undertaken at the University of Minnesota (Minneapolis-St. Paul).

Analysis of Published Twin and Adoption Studies of IQ

Studies of Reared-Together Twins

We have updated the Bouchard and McGue (1981) review with studies published through 1989. During that time there were only two additional published studies of general cognitive ability in adult twins (Tambs, Sundet, & Magnus, 1984; Vernon, 1989). Because studies of older twins are relatively rare yet critical to addressing the question of life-span variability in heritability, we added to the updated compilation the single study by Pedersen et al. (1992) of twins who were in the latter half of the life span. This brings the total number of twins studied to 6,370 MZ pairs and 7,212 DZ pairs. Figure 1 plots the weighted average twin IQ correlation as a

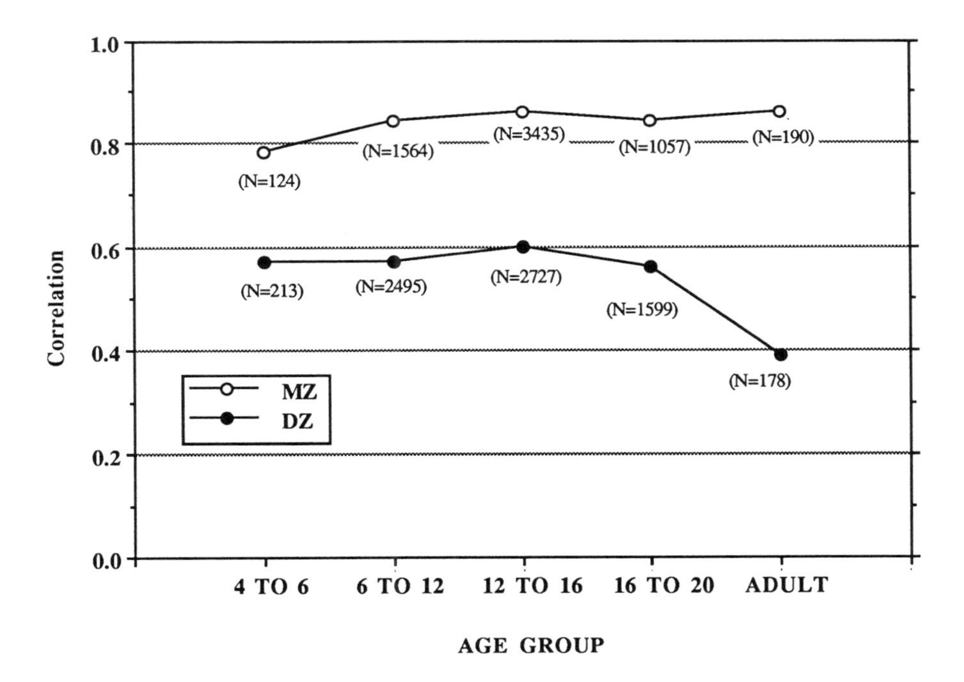

FIGURE 1. Average monozygotic (MZ) and dizygotic (DZ) reared-together twin correlations derived from published twin studies of IQ. Correlations were averaged using the Fisher *z* transformation method.

function of age. As can be seen, the MZ twin correlation increases modestly, but continuously, throughout the life span, whereas the average DZ correlation parallels the corresponding MZ average until late adolescence but then declines thereafter.

Figure 2 shows the variance component estimates derived from the average correlations given in Figure 1. Variance components were estimated with the following assumptions: All genetic variance is additive, shared environmental factors contribute equally to the similarity of MZ and DZ twins, and there is no assortative mating for IQ. Although these are the standard assumptions underlying the classical twin method (e.g., Eaves, Eysenck, & Martin, 1989), and although there is some empirical support for the validity of at least the first two assumptions as they apply to IQ (Plomin, DeFries, & McClearn, 1990), it is recognized that the variance component estimates are only approximations. Nonetheless, failure

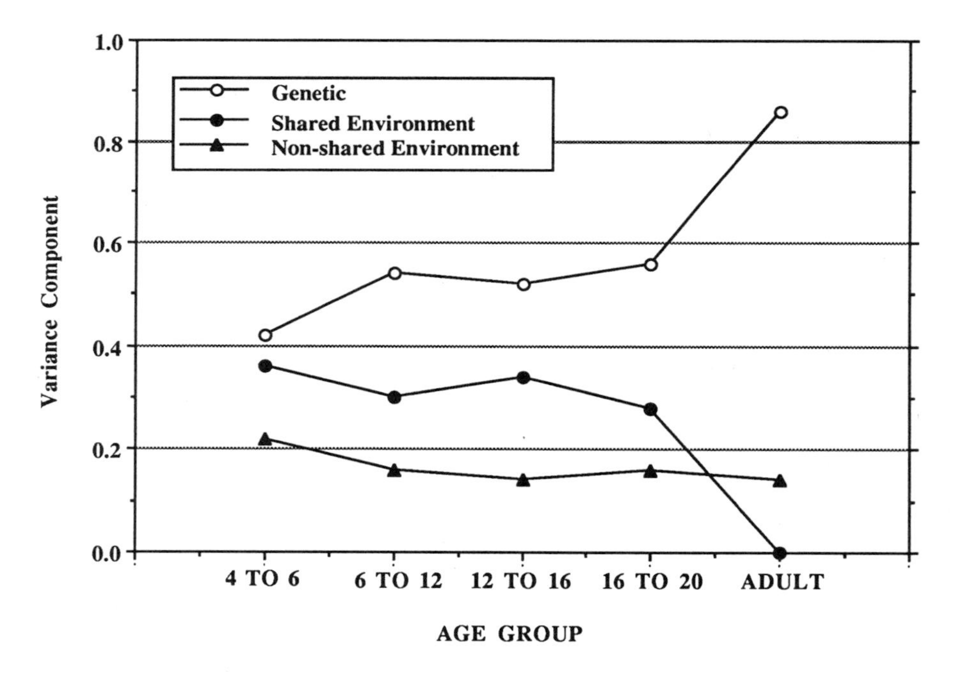

FIGURE 2. IQ variance component estimates derived from published twin IQ correlations given in Figure 1. Estimates are based on the standard assumptions used with the Falconer heritability formula (Plomin et al., 1990).

to fully meet the assumptions should not significantly bias the age comparisons that are the focus of the present investigation.

The estimated proportion of IQ variance associated with shared environmental factors is relatively constant at approximately 30% for ages up to 20 years but then drops to 0% in adulthood. The finding that shared environmental influences are greatest when the twins reside together and maintain close social contact is not unexpected. More remarkable is the observation that these shared environmental influences may not endure beyond the period of common rearing. The estimated proportion of IQ variance associated with genetic factors increases throughout development but especially after 20 years of age. From ages 4 through 20 years, the estimated heritability is clearly consistent with the summary estimate of 51% reported by Chipeur et al. (1990), whereas a heritability estimate of approximately 80% in adulthood is clearly consistent with the estimate recently reported by Pedersen et al. (1992) in their sample of older twins.

Studies of Reared-Apart Twins and Adoptees

Further support for the impressions gained from Figure 2 is provided by other aspects of published IQ–kinship correlations. Figure 3 summarizes IQ correlations from five published studies of reared-apart MZ (MZA) twins (i.e., all relevant studies except Burt, 1966). The MZA correlation provides a direct estimate of heritability (Bouchard, Lykken, McGue, Tellegen, & Segal, 1990) so that the correlations summarized in Figure 3 suggest an IQ heritability of approximately 75%, a value clearly greater than the 51% reported by Chipeur et al. (1990). But the MZA studies, except for an occasional isolated adolescent twin pair, are based on adult samples. Given the consistency of the MZA IQ correlation across studies, the relatively large MZA IQ correlation is more likely to reflect an increase in heritability in adulthood than bias in MZA sampling or assessment.

Studies of nonbiologically related, but reared-together, relatives further support the conclusion that the underlying determinants of IQ variability change during development. There have now been seven studies of IQ similarity among nonbiologically related reared-together siblings (adopted–adopted and/or adopted–biological) in childhood and three in adulthood. The correlation between reared-together nonbiologically re-

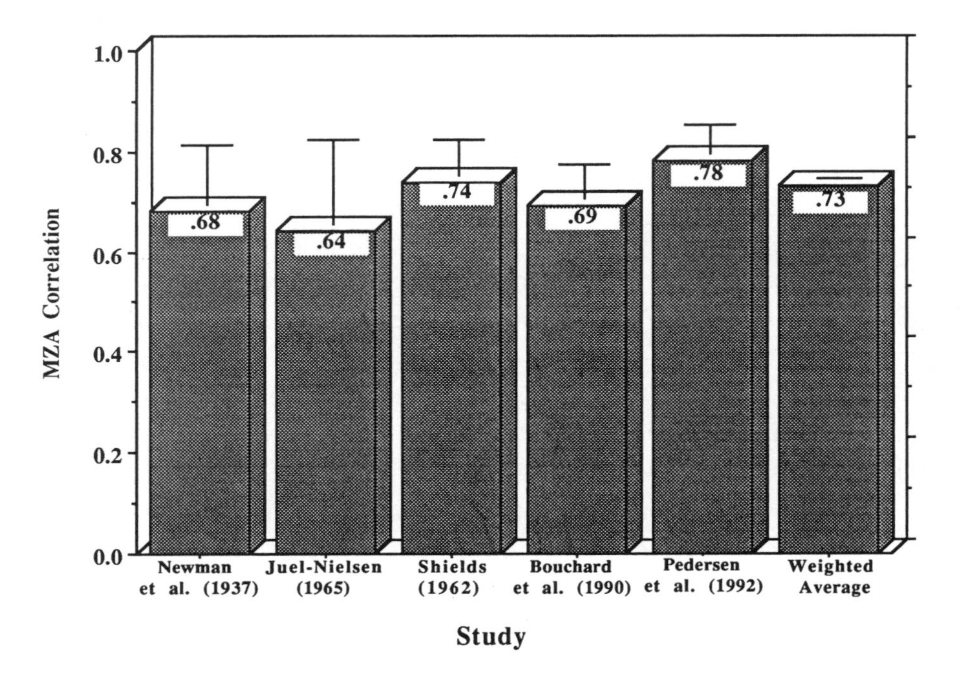

FIGURE 3. IQ correlations in studies of reared-apart monozygotic (MZA) twins. For studies that used more than one IQ measure, the reported correlation is based on the primary measure only. Although we did not include the MZA correlation of 0.771 reported by Burt (1966) because there continues to be question about the reliability of that figure (Joynson, 1990), the Burt MZA correlation is clearly consistent with those reported in the other MZA studies.

lated relatives provides a direct estimate of the proportion of IQ variance associated with shared environmental factors. Figure 4 summarizes these correlations. As can be seen, although the average correlation is moderate in childhood, which suggests that shared environmental factors account for approximately 25% of IQ variance during that stage of life, the average correlation is essentially 0% in adulthood, which suggests little enduring effect associated with common rearing. Consistent with the analysis of IQ correlations on reared-together and reared-apart twins, adoption studies suggest that once adoptive siblings leave their shared rearing home they bear no resemblance in IQ.

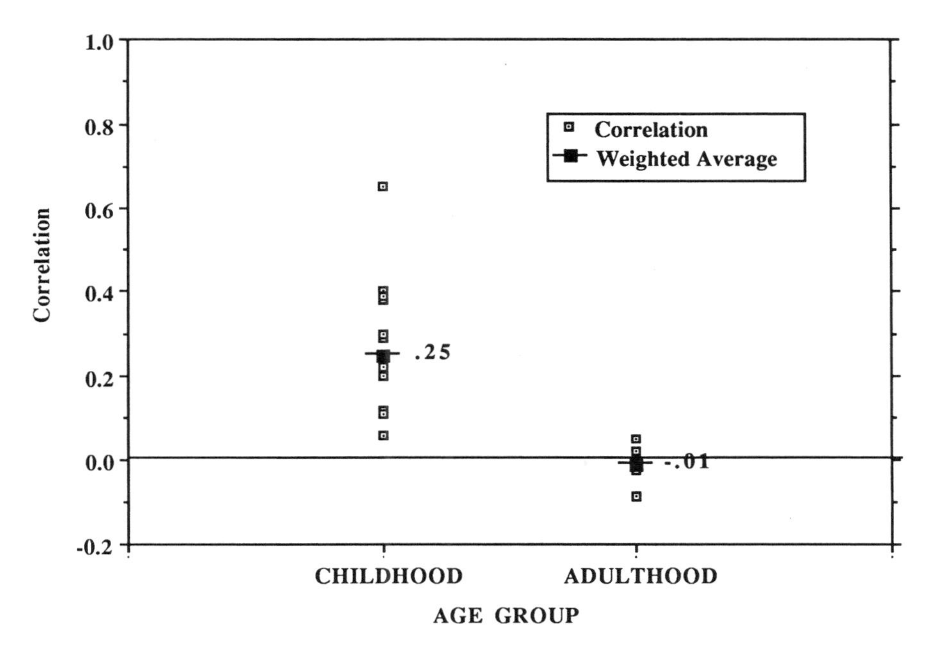

FIGURE 4. IQ correlations among nonbiologically related, but reared-together, relatives (both adopted–adopted and adopted–biological pairs). Weighted average correlations were derived using the Fisher *z* transformation method.

A Cross-Sectional Twin Study of Intellectual Resemblance Across the Life Span

Caution is, of course, needed when comparing results from different twin and adoption studies. Differences in twin and adoptee correlations reported on different-age samples from different studies may be due to actual changes in resemblance over time or to differences in study design. Studies of children compared with those of adults are likely to differ in ascertainment scheme and approach to intellectual assessment—differences that could affect familial correlations. Ideally, one would want to address the question by following a large and genetically informative sample from early childhood into adulthood; this ideal has not nearly been achieved. A pragmatic, yet satisfactory, alternative to a life-span longitudinal study is a cross-sectional twin study that uses a consistent ascertainment and assessment protocol. We report in this chapter preliminary results from such a study that is currently underway at the University of Minnesota (Minneapolis-St. Paul).

Method

Over the past 5 years, as part of a series of studies of reared-together twins undertaken at the University of Minnesota, we have administered subscales from the Weschler intelligence tests (Matarazzo, 1972) to same-sex twin pairs who were 11–88 years old. All twin pairs were ascertained, starting from birth records, using methods described in Lykken, Bouchard, McGue, and Tellegen (1990). The younger twin cohorts (i.e., the 11–12-year-old and the 17–18-year-old samples) were ascertained and assessed as part of the ongoing longitudinal Minnesota twin family study (McGue et al., 1991). The adult samples were ascertained and assessed as part of the ongoing Minnesota twin study of adult development and aging (McGue, Hirsch, & Lykken, 1993). Twins in the younger cohort were all males and participated in a study that involves a full day of psychological and psychophysiological assessment completed in our laboratories at the University of Minnesota. The adult samples include both male and female same-sex twin pairs who completed 6–7 hours of psychological, medical, and physiological assessment in their homes. Although the gender composition of the younger and older samples was different, gender has not been found to be an important moderator of twin IQ similarity in earlier research (Bouchard & McGue, 1981) and, in fact, did not moderate twin resemblance in the adult samples reported on here. Consequently, between-samples comparisons appear justified.

The adult twins were administered all subscales of the Wechsler Adult Intelligence Scale–Revised (WAIS-R) (Wechsler, 1981) except Similarities, and the younger twins were administered only four of the subtests (Information, Vocabulary, Block Design, and Picture Arrangement) from either the WAIS-R (for the 17–18 year olds) or the Wechsler Intelligence Scale for Children–Revised (Wechsler, 1974) (for the 11–12 year olds). This combination of two Verbal and two Performance subscales is one of the most widely recognized short forms of the Weschler scales and yields composite scores that correlate in excess of 0.90 with the full-scale score (Sattler, 1989). The comparisons we report here are based on the

TABLE 2
Characteristics of Cross-Sectional Twin Sample

Variable	Age Range (Years)			
	11–12	17–18	30–59	60–88
		MZ Twins		
No. of pairs	160	79	51	91
% Males	100	100	24	39
Mean IQ	104	103	104	104
SD IQ	12.9	14.5	11.9	13.5
		DZ Twins		
No. of pairs	75	39	35	68
% Males	100	100	29	56
Mean IQ	104	100	100	103
SD IQ	13.5	15.2	11.7	10.4

Note. MZ = monozygotic; DZ = dizygotic.

four subscales, either taken individually or in terms of the Verbal, Performance, and Total IQ composites. Table 2 gives a descriptive breakdown of the twin sample. In this ongoing study, intelligence test data is currently available on 381 pairs of MZ twins and 217 pairs of same-sex DZ twins. The samples are largest for the youngest and oldest twin groups and particularly small for the two intermediate-age DZ samples. The IQ means and standard deviations (*SD*s) are comparable across groups but differ slightly from the normative values of 100 and 15, respectively. The slightly elevated mean IQs and slightly depressed IQ *SD*s can be attributed largely to an underrepresentation in the sample of individuals with IQs less than 70 (cf. Bouchard et al., 1990).

Results

Twin intraclass correlations were estimated from a one-way analysis of variance and are summarized in Figure 5 (top panel, MZ twins; bottom panel, DZ twins). A clear and consistent pattern emerges with the MZ twins—IQ similarity increases with age. In contrast, for DZ twins, no consistent pattern is evident. The absence of a consistent trend with the DZ twins appears to be due largely to the results of the relatively small-size samples from the two intermediate groups. If one focuses on the youngest and oldest DZ samples only (i.e., those with the relatively large-

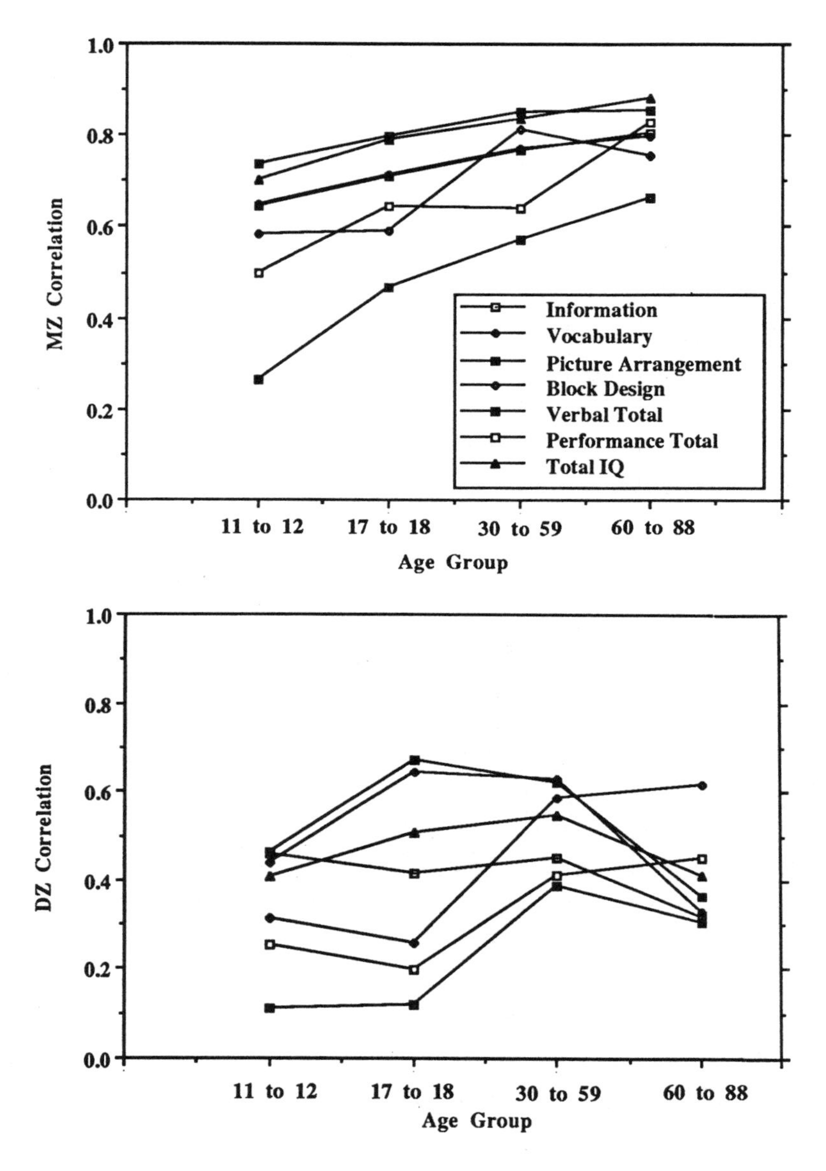

FIGURE 5. Monozygotic (MZ) (top panel) and dizygotic (DZ) (bottom panel) twin correlations for subscales and composites from the Weschler (1974, 1981) tests in the ongoing University of Minnesota cross-sectional study of reared-together twins.

size samples), then an interesting pattern does emerge. Correlations for the two verbal subscales, as well as the Verbal composite, decrease; whereas correlations for the two Performance subscales, as well as the Performance composite, increase with age. Whether this is a replicable pattern awaits future increases in the size of the adolescent and early-to-middle adulthood samples.

Variance component estimates were computed from the twin correlations using the LISREL software system (Jöreskog & Sörbom, 1986) following procedures described by Chipeur et al. (1990). Because of the relatively small size of the adolescent and early-to-middle adulthood DZ samples, we report estimates for the preadolescent and late adulthood samples only. Heritability estimates are given in the top panel of Figure 6, and the estimated proportion of IQ variance associated with shared environmental effects is given in the bottom panel of Figure 6. Both plots suggest consistent changes with age, but note again that these trends would not appear as consistent if the relatively unreliable estimates derived from the intermediate-age samples were also included. In any case, the plots suggest that IQ heritability increases from approximately 50% in preadolescence to approximately 80% in adulthood. In contrast, shared environmental influences appear to decrease from approximately 20% in preadolescence to near 0% in adulthood. Preliminary results from this ongoing cross-sectional twin study at the University of Minnesota consequently support observations from published kinship correlations on the changing heritability of IQ across the life span.

Discussion

Analysis of published kinship correlations, as well as the preliminary results from the just-mentioned ongoing cross-sectional twin study, indicate that intellectual resemblance among MZ twins increases with age. The evidence concerning DZ twin resemblance is less consistent, although there exists some support for the proposition that IQ resemblance among DZ twins, like that among adoptive siblings, declines after the age of common rearing. Taken together, these findings suggest a substantial increase in genetic influences and a declining influence of shared environmental factors during the transition from late adolescence to early

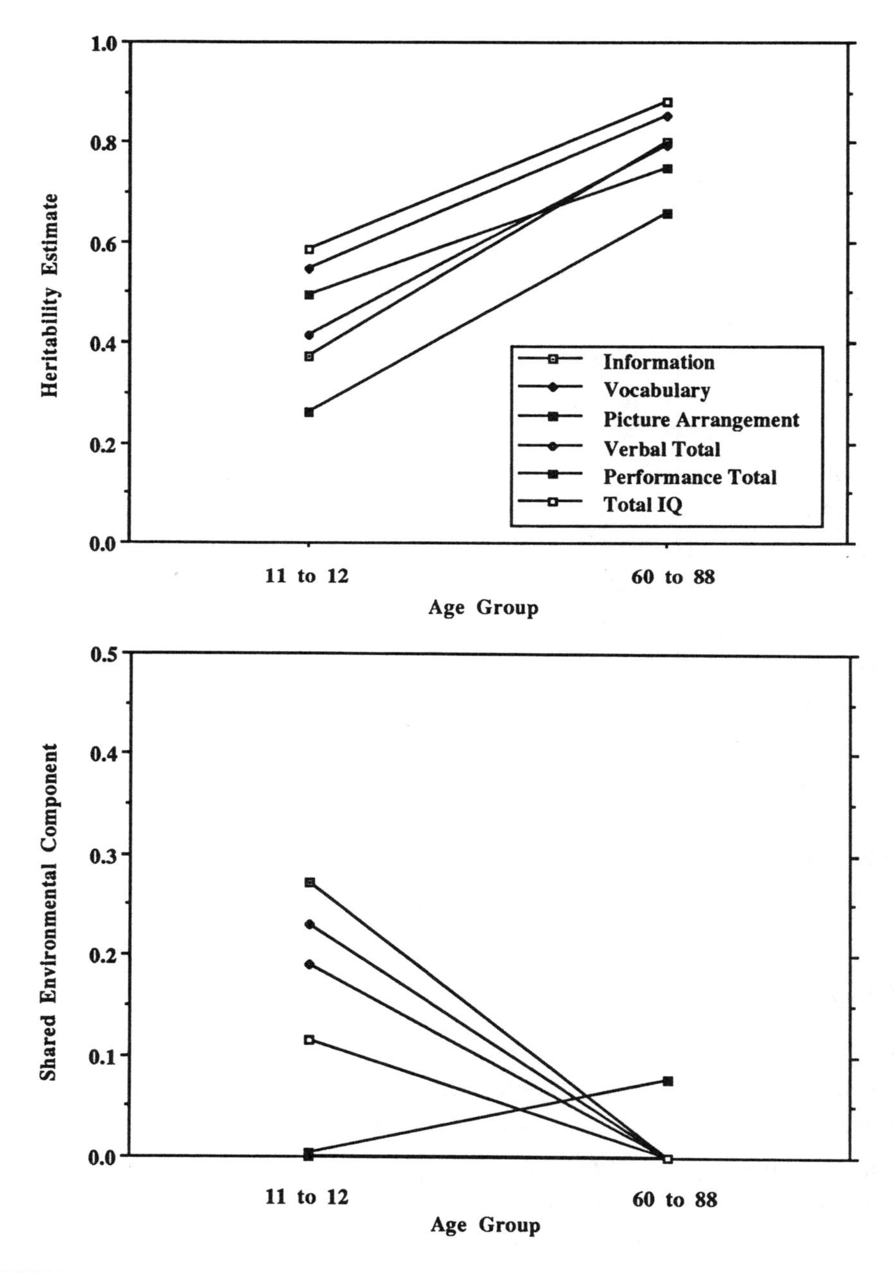

FIGURE 6. Proportion of Weschler (1974, 1981) test performance associated with heritability estimates (top panel) and shared environmental components (bottom panel) derived from the ongoing University of Minnesota cross-sectional study of reared-together twins.

adulthood. We emphasize, however, the preliminary nature of these findings. Our results are based almost exclusively on cross-sectional observations, and there is a need for additional study, especially of DZ twins in the intermediate-age years. Moreover, as with most behavioral genetic research, individuals at the extreme of the IQ distribution are underrepresented in the present studies. Our findings may not generalize to the extremes of environmental deprivation that characterize many of those whose IQs are less than 70. Despite these limitations, the present findings have implications for understanding the nature of intellectual development. In fact, they may tell us more about the effect of experience on cognitive development than about the salience of physiological factors across the life span.

Scarr and McCartney (1983) hypothesized that underlying the observed changes in familial resemblance is a changing relationship between genotype and experience. Experience exerts a substantial influence on intellectual development, but this influence is seen to be, in part, genetically directed through the mechanism of genotype–environment correlation (Plomin, DeFries, & Loehlin, 1977). For the young child, intellectual experience is largely determined by others (e.g., parents and teachers), and because the child who inherits genes conducive to high intellectual achievement is also likely to develop within a family that provides effective intellectual stimulation, genetic and experiential effects are passively correlated during the early stages of development. During adulthood, however, experience is largely self-directed and reflects, in part, inherited abilities, interests, and dispositions so that the correlation between genotype and experience is largely actively generated. Throughout development, experience is thought to not only reflect but reinforce underlying genetic differences among individuals. But, this reinforcement will be strongest when experience is self-selected rather than constructed by others. Under the Scarr and McCartney model, MZ twins become more similar in adulthood because, when given the opportunity, they select similar levels of intellectual stimulation.

A decline in DZ twin and adoptive sibling resemblance in adulthood may also reflect changes in the relationship between genotype and experience. Families, like the larger society, struggle in dealing with large achievement differences among their members. A child with modest intelligence, if reared by high-achieving parents or paired with a twin with an IQ of 150, is more likely to be perceived as having academic difficulties and singled out for special attention than if he or she were reared in a home with parents and siblings of similar ability. Exceptional parental efforts at intellectual stimulation are more likely to be aimed at minimizing rather than enhancing differences among family members. A decrease in DZ twin or adoptive sibling resemblance in adulthood may reflect the discontinuation of the leveling influence of common rearing. Because intellectual differences between two members of an MZ twin pair are not likely to be substantial, however, a similar decrease in MZ twin correlation is not expected.

Concluding Comment

Psychology appears ready to move beyond the acrimony that has marked the past century of debate on the nature–nurture issue (Lykken, Bouchard, McGue, & Tellegen, 1992). But resolution will not be achieved by proclaiming one side victorious over the other but rather by recognizing the artificial dichotomy implicit to the question. A heritability of 60%, 80%, or even 100% indicates only that IQ differences are highly predictable from genetic differences among individuals; it does not indicate that nongenetic factors are unimportant. Central to the developing rapprochement between hereditarians and environmentalists is the recognition that the circumstances of an individual's existence (i.e., his or her environment) can be distinguished from those aspects of the environment that the individual actually engages (i.e., experience). In a permissive society, individuals, especially adults, are faced with a wide array of experiential choices. How those choices are made and how the individual actively constructs his or her experiences will largely reflect the individual's abilities, interests, and temperament. Genes exert a distal influence on IQ test performance, whereas the proximal mechanism underlying that influence likely involves the individual production of experience (Bouchard, Lykken, Tellegen, & McGue, 1993; Hayes, 1962).

References

Baltes, P. B. (1987). Theoretical propositions of life-span developmental psychology: On the dynamics between growth and decline. *Developmental Psychology, 23*, 611–626.

Bouchard, T. J., Jr., Lykken, D. T., McGue, M., Segal, N. L., & Tellegen, A. (1990). Sources of human psychological differences: The Minnesota study of twins reared apart. *Science, 250*, 223–228.

Bouchard, T. J., Jr., Lykken, D. T., Tellegen, A., & McGue, M. (1993). Genes, drives, environment and experience: EPD theory-revised. In C. P. Benbow & D. Lubinski (Eds.), *From psychometrics to giftedness: Essays in honor of Julian Stanley*. Manuscript submitted for publication.

Bouchard, T. J., Jr., & McGue, M. (1981). Familial studies of intelligence: A review. *Science, 250*, 223–238.

Burt, C. (1966). The genetic determination of differences in intelligence: A study of monozygotic twins reared together and apart. *British Journal of Psychology, 57*, 137–153.

Chipeur, H. M., Rovine, M., & Plomin, R. (1990). LISREL modelling: Genetic and environmental influences on IQ revisited. *Intelligence, 14*, 11–29.

Eaves, L. J., Eysenck, H. J., & Martin, N. G. (1989). *Genes, culture and personality: An empirical approach*. New York: Academic Press.

Erlenmeyer-Kimling, L., & Jarvik, L. F. (1963). Genetics and intelligence: A review. *Science, 142*, 1477–1479.

Hayes, K. J. (1962). Genes, drives, and intellect. *Psychological Reports, 10*, 299–342.

Jöreskog, K. G., & Sörbom, D. (1986). *LISREL VI* [Computer program]. Mooresville, IN: Scientific Software.

Joynson, R. B. (1990). *The Burt affair*. New York: Academic Press.

Juel-Nielsen, N. (1965). Individual and environment: A psychiatric–psychological investigation of monozygous twins reared apart. *Acta Psychiatrica et Neurologica Scandinavica*, Suppl. 183.

Lykken, D. T., Bouchard, T. J., Jr., McGue, M., & Tellegen, A. (1990). The Minnesota twin registry: Some initial findings. *Acta Geneticae Medicae et Gemellologicae, 39*, 35–70.

Lykken, D. T., Bouchard, T. J., Jr., McGue, M., & Tellegen, A. (1992). *Nature via nurture*. Unpublished manuscript, University of Minnesota.

Matarazzo, J. D. (1972). *Weschler's measurement and appraisal of adult intelligence* (5th ed.). Baltimore: Williams & Wilkins.

McCartney, K., Harris, M. J., & Bernieri, F. (1990). Growing up and growing apart: A developmental meta-analysis of twin studies. *Psychological Bulletin, 107*, 226–237.

McGue, M., Iacono, W., Lykken, D. T., Tellegen, A., Ficken, J., & Goff, M. (1991). High-risk twin studies of alcoholism [Abstract]. *Behavior Genetics, 21*, 580.

McGue, M., Hirsch, B., & Lykken, D. T. (1993). Age and the self-perception of ability: A twin study analysis. *Psychology and Aging, 8*, 72–80.

Pedersen, N. L., Plomin, R., Nesselroade, J. R., & McClearn, G. E. (1992). A quantitative genetic analysis of cognitive abilities during the second half of the life span. *Psychological Sciences, 3*, 346–353.

Plomin, R., & Daniels, D. (1987). Why are children in the same family so different from one another? *Behavioral and Brain Sciences, 10*, 1–16.

Plomin, R., DeFries, J. C., & Loehlin, J. C. (1977). Genotype–environment interaction and correlation in the analysis of human behavior. *Psychological Bulletin, 84*, 309–322.

Plomin, R., & DeFries, J. C., & McClearn, G. E. (1990). *Behavioral genetics: A primer* (2nd ed.). New York: Freeman.

Sattler, J. M. (1989). *Assessment of children's intelligence* (2nd ed.). Philadelphia: W. B. Saunders.

Scarr, S., & McCartney, K. (1983). How people make their own environments: A theory of genotype → environment effects. *Child Development, 54*, 424–435.

Shields, J. (1962). *Monozygotic twins brought up apart and brought up together.* New York: Oxford University Press.

Snyderman, M., & Rothman, S. (1990). *The IQ controversy, the media and public policy.* New Brunswick, NJ: Transaction.

Tambs, K., Sundet, J. M., & Magnus, P. (1984). Heritability analysis of the WAIS subtests: A study of twins. *Intelligence, 8*, 283–293.

Vernon, P. A. (1989). The heritability of measures of speed of information-processing. *Personality and Individual Differences, 10*, 573–576.

Wechsler, D. (1974). *Manual for the Wechsler Intelligence Scale for Children–Revised.* New York: Psychological Corporation.

Wechsler, D. (1981). *Manual for the Wechsler Adult Intelligence Scale–Revised.* New York: Psychological Corporation.

CHAPTER 4

Continuity and Change in Cognitive Development

David W. Fulker, S. S. Cherny, and Lon R. Cardon

We know more about the etiology of individual differences in cognitive abilities than just about any other psychological domain. Bouchard and McGue (1981) noted over 140 studies of this domain, which yielded the largely consistent result that genetic differences account for approximately 50% of the observed variability in general cognitive ability. However, until fairly recently, the genetic and environmental models used by psychologists have done little more than partition variation in individual differences into two major components—the effects of nature and the effects of nurture, with each represented by a single parameter, h^2 and e^2, respectively. How could one hope, with just two parmeters, to

The preparation of this chapter was supported, in part, by National Institute of Child Health and Human Development (NICHD) grants HD-10333, HD-18426, and HD-19802; by National Institute of Mental Health grant MH-43899; by National Institutes of Health Biomedical Research Support grant RR-07013-25; and by a grant from the John D. and Catherine T. MacArthur Foundation. During the preparation of this chapter, Lon R. Cardon was supported by NICHD training grant HD-07289, which was awarded to David W. Fulker. S. S. Cherny was supported, in part, by the Natural Sciences and Engineering Research Council of Canada.

understand the complex relationships between the component parts of the cognitive phenotype, its development over time, and its relationship to variables with social relevance. In the past, behavioral geneticists could point to the importance of genetic factors but could say little about how they related to psychological complexity.

It was not until the 1960s and 1970s that the situation changed appreciably: Advances in quantitative genetics, psychometrics, and multivariate modeling brought a new level of sophistication to the field and provided an opportunity to explore complex issues in a searching and rigorous manner. The history of these developments is involved; its origins go back to the very beginnings of genetics, with the work of Francis Galton (1869) and Gregor Mendel (1866), in the last century. The subsequent development of the biometrical school by Ronald Fisher (1918), Sewell Wright (1921), Kenneth Mather (1949), and others laid the foundation for the assessment of quantitative variation. The psychometric tradition, as exemplified by Galton (1883), Charles Spearman (1908), Cyril Burt (1949), and Louis Thurstone (1938), provided the necessary techniques for measuring the phenotype. Later, the research of Jöreskog and Sörbom (1989) welded the psychometric approach to the more general problem of covariance structures, an approach that unified structural modeling in the social sciences. It is perhaps not appropriate to trace these developments in detail here, but suffice it to say that current statistical modeling in human behavior genetics owes its origin to these sources (Neal & Cardon, 1992).

Psychologists in North America and Europe subsequently built upon these bases (Cloniger, Rice, & Reich, 1979a, 1979b; Eaves, Last, Young, & Martin, 1970; Jinks & Fulker, 1970; Rice, Cloniger, & Reich, 1978). The thorough study of such central issues as genotype–environment correlations and interactions (introduced by Cattell, 1960, 1965), multivariate genetic structure (introduced by Loehlin & Vandenberg, 1968), and developmental processes (pioneered by Plomin & Defries, 1979; Wilson, 1978), is now feasible with current methodology. Advances in computer technology, particularly in the last decade, have contributed enormously to this enterprise. Multivariate analysis has now replaced the simplistic analysis of nature versus nurture (e.g., Loehlin & Vandenberg, 1968).

The purpose of this chapter and the next (Cardon & Fulker, chapter 5) is to illustrate this new approach by describing some aspects of general and specific cognitive development in young children who are participating in three ongoing developmental studies of twins and adoptees being conducted at the Institute for Behavioral Genetics at the University of Colorado, Boulder: (a) the Colorado Adoption Project (CAP) (Fulker, DeFries, & Plomin, 1988; Plomin & DeFries, 1985; Plomin, DeFries, & Fulker, 1988), (b) the MacArthur Longitudinal Twin Study (MALTS) (Emde et al., in press; Plomin et al., 1990) and (c) the Twin Infant Project (TIP) (Benson, Cherny, Haith, & Fulker, 1993; DiLalla et al., 1990).

In this chapter, we are concerned with the development of individual differences in general cognitive ability from 1 through 9 years of age. We ask two main questions: (a) How important are genetic and environmental influences at each age of assessment? and (b) What is the relationship among these influences over time? The first question is no more than a simple nature–nurture question, asked at each time point. The second question is the more important one and involves the notions of continuity and change, which are fundamental to the idea of development. The extent to which phenotypic differences are correlated over time implies continuity in development; the extent to which they are not correlated implies change. Thus, the relationship among genetic and environmental influences across time indicates the degree to which these processes of continuity and change are driven by the genotype or the environment. To conclude the chapter, we discuss the future directions in behavioral genetic research on cognitive development, including recent findings of molecular marker-linked polygenes that influence reading disability.

Method

Subjects

CAP is a longitudinal prospective adoption study of genetic and environmental determinants of behavioral development. The study began in 1975, and the initial cohort of children was evaluated at regular intervals from 1 through 16 years of age. CAP has a "full" adoption design, in that measures have been administered to both adoptive and biological parents, as well as to the adopted children. Furthermore, nonadoptive control families have been matched to the adoptive families for age, education, and occupational status of the father. However, only sibling data (i.e., pairs of children that occur in the same family) are described in this chapter.

MALTS, which began in 1986, is a longitudinal study of individual differences in the development of personality and cognition, in which identical (monozygotic [MZ]) and fraternal (dizygotic [DZ]) twins have been evaluated between 14 months and 3 years of age. TIP, which began in 1984, is a longitudinal study of continuity and change in infant intelligence that involves evaluation of MZ and DZ twin pairs not only during infancy but at 1–4 years of age. The cognitive evaluation of twins in both of these projects has been integrated since 1991 with the evaluation of the adopted children in CAP as part of a comprehensive project of behavioral development in early childhood at the Institute for Behavioral Genetics. In all studies, children were followed prospectively from birth, given individually administered tests of intelligence, and tested at regular intervals. Together, these studies constitute a unique study of the genetic and environmental determinants of cognitive development.

For the present study, general cognitive ability was measured at 1, 2, 3, and 4 years of age in 92–201 MZ twin pairs (the number depending on age) and 75–175 same-sex DZ twin pairs, who were drawn from MALTS[1] and TIP, and at 1, 2, 3, 4, 7, and 9 years of age in 32–87 adoptive (genetically unrelated) and 43–102 nonadoptive (natural) sibling pairs and in 278–300 singletons, who were drawn from CAP. Because of the ongoing nature of these studies, the number of participating twin and sibling pairs decreased over time. A total of 1,437 children from 891 families were assessed. The specific tests included the Bayley Mental Development Index (Bayley, 1969) at 1 and 2 years of age, the Stanford-Binet IQ test (Terman & Merrill, 1973) at 3 and 4 years of age, the Wechsler Intelligence Scale for Children–Revised (Wechsler, 1974) at 7 years of age, and the first principal component from a telephone-administered cognitive test battery at 9 years of age (Cardon, Corley, DeFries, Plomin, & Fulker, 1992). Table 1 shows the

[1]In MALTS, the twins were assessed at 14 months of age rather than at 12 months of age.

TABLE 1

Number of Siblings and Twins by Age and Relationship

	Age (years)					
Relationship	1	2	3	4	7	9
Nonadoptive siblings						
Probands	246	229	214	214	216	173
Natural siblings	102	93	89	88	67	45
Maximum pairs	102	91	87	88	65	43
Adoptive siblings						
Probands	243	216	205	195	194	180
Unrelated siblings	90	87	78	78	56	35
Maximum pairs	87	80	73	74	50	32
MZ twins						
Probands	203	162	122	94	—	—
Cotwins	202	164	119	94	—	—
Maximum pairs	201	162	118	92	—	—
DZ twins						
Probands	176	144	101	76	—	—
Cotwins	175	145	100	76	—	—
Maximum pairs	175	142	98	75	—	—

Note. MZ = monozygotic; DZ = dizygotic.

number of adoptive and nonadoptive siblings and the number of MZ and DZ twins at each age, along with the maximum number of pairings.

The Model

The basic behavioral genetic model that is applicable at each time point recognizes three sources of individual differences in general intelligence. In this model, shown in Figure 1, *G* represents additive genetic differences among individuals, *CE* represents common environmental influences shared by children reared together in the same home, and *SE* represents specific environmental influences unique to the individual. None of these three variables are directly observable—in the terminology of structural modeling they are latent variables—but their importance may be assessed through behavioral genetic strategies such as the twin and adoption design. The genetic correlation between genotypes for pairs of individuals varies from 0.0, in the case of adoptive or genetically unrelated siblings, to 0.5, in the case of nonadoptive or natural siblings and DZ twins, and finally to 1.0 in the case of MZ twins, as shown in Figure 2. For this reason,

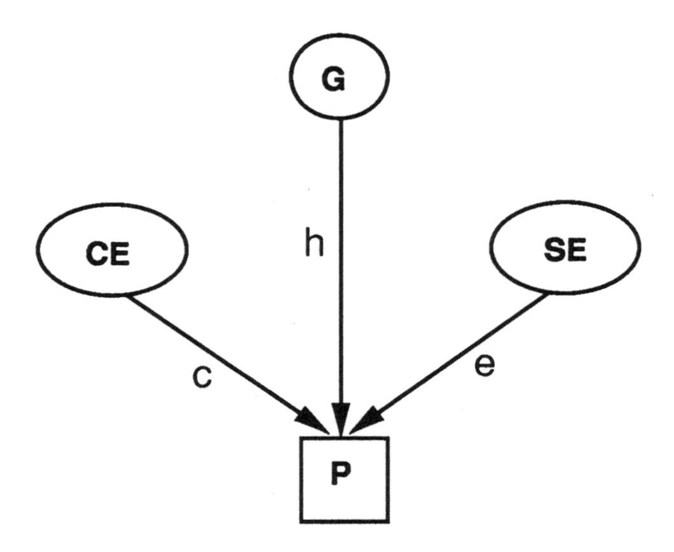

FIGURE 1. Genetic (G), common (or shared) environmental (CE), and specific (or unique) environmental (SE) influences on the phenotype (P). The impacts of the three sources of variation—genetic, shared environmental, and unique environmental—are h, c, and e, respectively.

the combined twin–adoption design, in which a key theoretical parameter varies throughout the full range from 0.0 to 1.0, is optimally powerful. Indeed, behavioral genetic designs, in general, are uniquely powerful in this respect among the designs used in the field of structural modeling in the social sciences, for which there are few instances of theoretical parameters with known values. Structural models in behavioral genetics are built on the foundations of Mendelian genetics.

The impacts of these three sources of variation—genetic, shared environmental, and unique environmental—are shown as h, c, and e, respectively, in the regression model; and the variance explained by each is the square of these quantities, h^2, c^2, and e^2. The quantity h^2 is referred to as the narrow-sense heritability. In the absence of Mendelian dominance or epistasis (gene–gene interaction), this parameter describes the total variation due to genetic differences between individuals. If the sources of nonadditive genetic variation are important, then the design of twins and siblings permits their evaluation. In addition, information

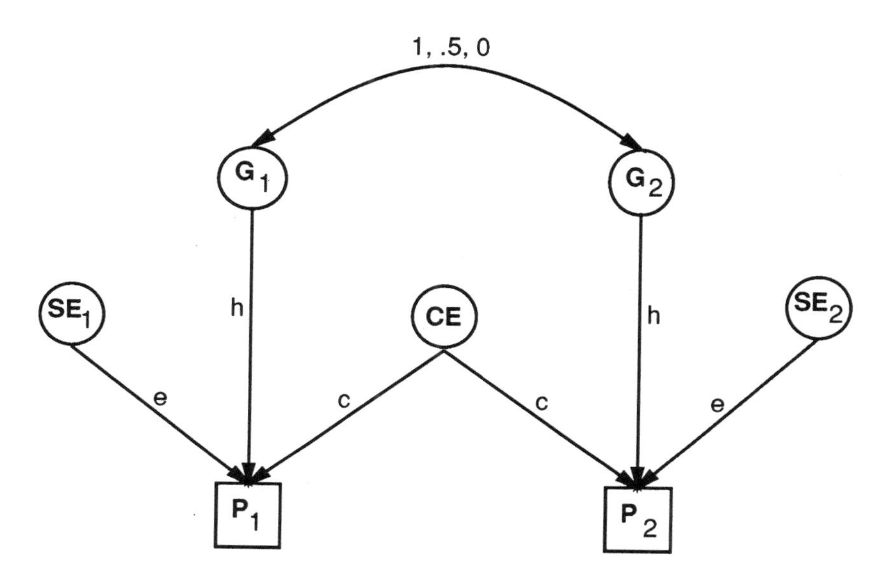

FIGURE 2. Genetic (G_1 and G_2), shared environmental (CE), and unique environmental (SE_1 and SE_2) influences on phenotypes (P_1 and P_2) of sibling and twin pairs. Genetic correlations are 1.0 for monozygotic twins, 0.5 for dizygotic twins and nonadoptive siblings, and 0.0 for adoptive siblings. The impacts of the three sources of variation—genetic, shared environmental, and unique environmental—are h, c, and e, respectively.

from the parents is capable of resolving the effects of assortative mating and genotype–environment correlation, if these prove to be important. The studies that are being carried out at the Institute for Behavioral Genetics are unique in this respect by combining a variety of informative behavioral genetic designs to better evaluate and validate the basic models used in data analysis. This methodological perspective, that of meta-analysis in behavioral genetics, originated with the research of Jinks and Fulker (1970).

Given the assessment of these basic sources of variation at each age point, what is required is a developmental model, in addition to the basic genetic and environmental model, to evaluate the relationships among the genetic and environmental variables over time. A variety of developmental models are available for exploring these relationships. However, for the present analysis, a very simple developmental model was chosen, namely, a Cholesky decomposition, which is often referred to in the factor analytic literature as triangular factorization (Gorsuch, 1983). This model is shown in Figure 3 for all six time points. The model is fully saturated; that is, it estimates as many parameters as there are independent variances and covariances. Although the model does not make any strong developmental assumptions, it can easily be interpreted in a developmental context by examining the factors that are essential for an adequate fit and the measures upon which those factors load most heavily. In general, the first factor will load most heavily on Year 1 and progressively less heavily on subsequent years. These loadings indicate the relative importance of four influences at Year 1 on later years. The second factor represents additional influences at Year 2 over and above those at Year 1 and their subsequent impact at later ages. Later factors operate in the same manner, representing influences independent of those at prior ages. Thus, the model allows the evaluation of continuity in development from one age to another and changes in development when new variation arises. Separate genetic, shared environmental, and unique environmental Cholesky decompositions, which allow partitioning of the sources of variation, were incorporated into a comprehensive model of development, shown in Figure 4. The phenotypic covariance structure is partitioned into the

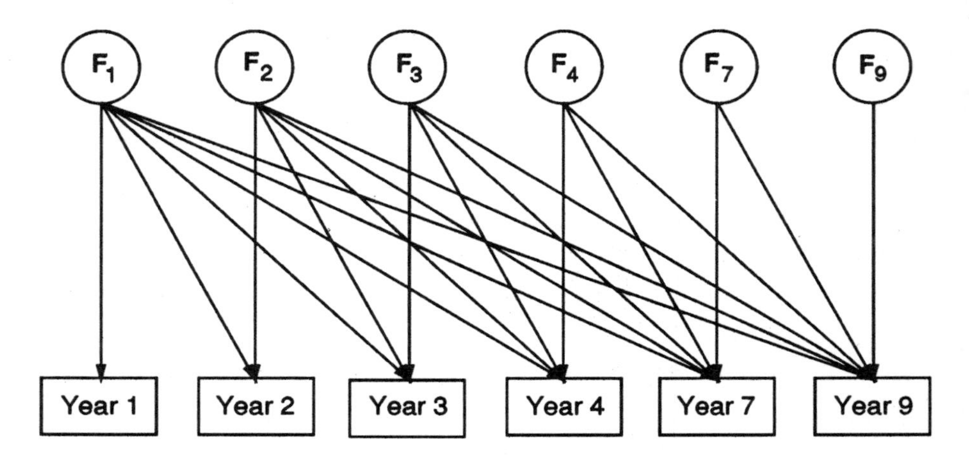

FIGURE 3. Phenotypic Cholesky decomposition model. F = factor.

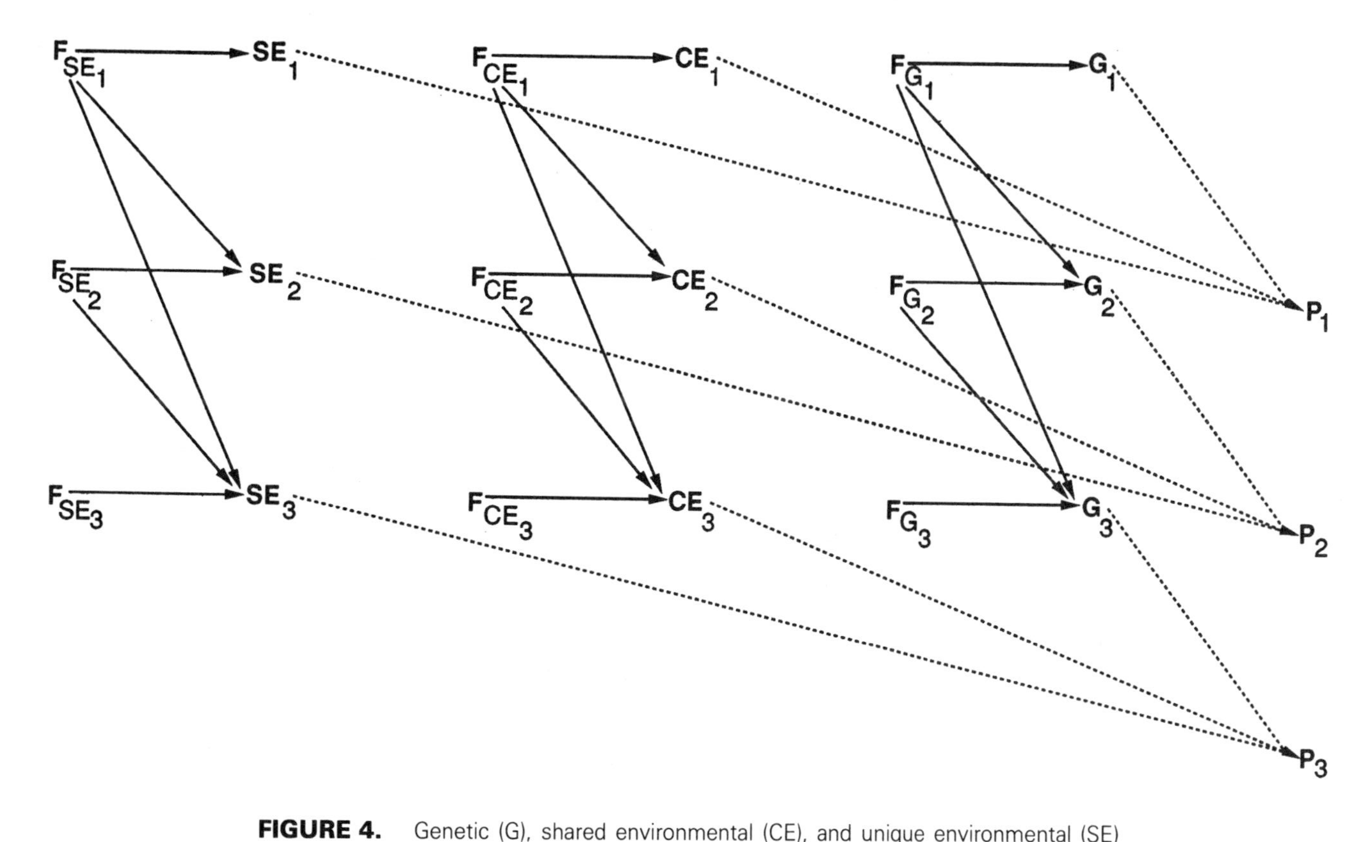

FIGURE 4. Genetic (G), shared environmental (CE), and unique environmental (SE) Cholesky decomposition model, shown only for three time points. Dotted lines imply coefficients fixed at unity. F = factor; P = phenotype.

three major components, and a separate Cholesky decomposition is modeled on each of those three components.

The technical aspects of fitting such a model to data are quite complex and fall in the realm of structural modeling familiar to psychologists through such packages as LISREL (Jöreskog & Sörbom, 1989) and EQS (Bentler, 1989). However, with twin and sibling data, the models take a special form; and with longitudinal data, which is usually incomplete, special problems arise that require fitting the model directly to the raw data rather than to covariance matrices. As with any structural modeling exercise (e.g., in confirmatory factor analysis), the factor loadings are entered into a matrix; and the form of the expected covariance matrix is specified by suitable operations, such as multiplication and addition, on the matrices of loadings. The loadings are estimated by forcing a match between the expected and the observed covariance matrix (or in the present case to the raw data), using optimization routines on the computer. The details of this procedure are given later, although an understanding of these procedures is not essential for an appreciation of the results given in the next section.

The parameters of the full Cholesky decomposition can be written into three separate Λ matrices, one for each of the genetic, shared environmental, and unique environmental levels. Each Λ matrix takes the following form:

$$\Lambda = \begin{pmatrix} \lambda_{1,1} & 0 & 0 & 0 & 0 & 0 \\ \lambda_{2,1} & \lambda_{2,2} & 0 & 0 & 0 & 0 \\ \lambda_{3,1} & \lambda_{3,2} & \lambda_{3,3} & 0 & 0 & 0 \\ \lambda_{4,1} & \lambda_{4,2} & \lambda_{4,3} & \lambda_{4,4} & 0 & 0 \\ \lambda_{5,1} & \lambda_{5,2} & \lambda_{5,3} & \lambda_{5,4} & \lambda_{5,5} & 0 \\ \lambda_{6,1} & \lambda_{6,2} & \lambda_{6,3} & \lambda_{6,4} & \lambda_{6,5} & \lambda_{6,6} \end{pmatrix}. \quad (1)$$

The first subscripts refer to the observed measures, whereas the second subscripts refer to the latent factors. For example, $\lambda_{3,1}$ is the loading of the first factor on the third measure.

The expected variance/covariance matrices (Σ) for MZ and DZ twin pairs and adoptive and nonadoptive sibling pairs, which are implied by the full genetic Cholesky decomposition, are given by:

$$\Sigma = \begin{pmatrix} \Lambda_G\Lambda_G' + \Lambda_C\Lambda_C' + \Lambda_E\Lambda_E' & \kappa \otimes \Lambda_G\Lambda_G' + \Lambda_C\Lambda_C' \\ \kappa \otimes \Lambda_G\Lambda_G' + \Lambda_C\Lambda_C' & \Lambda_G\Lambda_G' + \Lambda_C\Lambda_C' + \Lambda_E\Lambda_E' \end{pmatrix}, \tag{2}$$

where the matrices are partitioned into four equal quadrants and represents the coefficient of relationship (1.0 for MZ twins, 0.5 for DZ twins and nonadoptive siblings, and 0.0 for adoptive siblings). The top left and bottom right quadrants contain the within-pairs, or phenotypic, covariances, whereas the other two quadrants contain the between-pairs, or cross-sibling, covariances. The Λ matrices are the Cholesky loadings at the genetic (G), shared environmental (C), and unique environmental (E) levels.

Because of the incomplete nature of the ongoing developmental studies, a maximum-likelihood (ML) pedigree approach must be used to make optimal use of the data. The models were, therefore, fitted to the observed data rather than to covariance matrices, whereby the negative of the following ML pedigree log-likelihood (LL) function was minimized:

$$LL = \sum_{i=1}^{N} [-\tfrac{1}{2}\log|\Sigma_i| - \tfrac{1}{2}(x_i - \mu)'\Sigma_i^{-1}(x_i - \mu)] - \text{constant}, \tag{3}$$

where x_i is the vector of scores for sibling pair i; Σ_i is the appropriate MZ, DZ, adoptive, or nonadoptive expected covariance matrix; N is the total number of sibling pairs; μ is the vector of means; and where

$$2(LL_1 - LL_2) = \chi^2 \tag{4}$$

for testing the difference between two alternative models. The vector of means can either be modeled or, as in the present case for which we postulate no theory of mean structure, simply fixed to the observed means. The use of this fit function, as opposed to the more common ML function used by such programs as LISREL (Jöreskog & Sörbom, 1989) and EQS (Bentler, 1989), allows all of the data to be analyzed. It is assumed that the missing data are missing at random, which is a reasonable assumption in the present case. The use of the more common fit function for complete data would necessitate eliminating from the analysis those pairs that were not measured at all time points, which would mean losing information unnecessarily. Furthermore, had this been done, the assumption that the data were missing at random would still need to be made. In the case for which there are no missing data, this pedigree function yields the same results as the ML function for covariance matrices.

The data were first standardized within each age, across all individuals as a single group. This standardization procedure effectively eliminates age differences in variances, which most likely are merely a result of using different tests at different ages, while preserving MZ, DZ, adoptive, nonadoptive, Sibling 1, and Sibling 2 variance differences. Because the ML estimation procedures were performed on data that were not standardized within sibling type within each group, the resulting parameter estimates were standardized so that all variables, latent and observed, had unit variance, for ease of interpretability. Each of the genetic, shared environmental, and unique environmental Λ matrices were standardized in the following manner:

$$\Lambda_G^* = \mathrm{diag}(\Sigma_P)^{-1/2}\,\Lambda_G, \quad (5)$$

$$\Lambda_C^* = \mathrm{diag}(\Sigma_P)^{-1/2}\,\Lambda_C, \text{ and} \quad (6)$$

$$\Lambda_E^* = \mathrm{diag}(\Sigma_P)^{-1/2}\,\Lambda_E, \quad (7)$$

where Σ_P is the expected phenotypic covariance matrix, which can be obtained from either the upper left or bottom right quadrants of any of the expected covariance matrices, and the Λ^* are the standardized Λ matrices.

Results

Estimates of h^2, c^2, and e^2, obtained from fitting the full Cholesky model to the data, are presented in Table 2. Although heritability and environ-

TABLE 2

Estimates of h^2, c^2, and e^2 at Each Age

	Age (years)					
Variance component	1	2	3	4	7	9
h^2	.51	.60	.45	.50	.51	.52
c^2	.12	.18	.24	.17	.11	.24
e^2	.37	.22	.31	.33	.38	.24

mental influences appear relatively constant across all ages, the Cholesky decomposition was examined to determine to what extent the same influences persist across ages and to what extent new influences appear at each age. Fitting the full Cholesky model resulted in the following unique environmental loadings:

$$\Lambda_{SE} = \begin{pmatrix} .61 & .00 & .00 & .00 & .00 & .00 \\ .04 & .46 & .00 & .00 & .00 & .00 \\ -.01 & -.03 & .56 & .00 & .00 & .00 \\ .05 & -.03 & .04 & .57 & .00 & .00 \\ .22 & .13 & .14 & .04 & .54 & .00 \\ .05 & .18 & .30 & .09 & -.13 & .30 \end{pmatrix}. \quad (8)$$

The relatively low off-diagonal loadings imply that the unique (specific) environment (SE) is contributing very little to the observed continuity in general cognitive ability and only change.

In contrast, the shared (common) environmental (CE) loadings imply that this component of variance is contributing mostly to the observed continuity and little to change. There appear to be, at most, two common factors because the loadings of the fourth through sixth factors are zero and those of the third factor are not appreciable as a set:

$$\Lambda_{CE} = \begin{pmatrix} .34 & .00 & .00 & .00 & .00 & .00 \\ .36 & .24 & .00 & .00 & .00 & .00 \\ .21 & .35 & .27 & .00 & .00 & .00 \\ .24 & .31 & .12 & .00 & .00 & .00 \\ .13 & .27 & .16 & .00 & .00 & .00 \\ .01 & .48 & -.09 & .00 & .00 & .00 \end{pmatrix}. \quad (9)$$

The loadings observed at the genetic (G) level imply a picture that is a bit more complicated:

$$\Lambda_{G} = \begin{pmatrix} .71 & .00 & .00 & .00 & .00 & .00 \\ .37 & .68 & .00 & .00 & .00 & .00 \\ .31 & .48 & .35 & .00 & .00 & .00 \\ .20 & .43 & .46 & .24 & .00 & .00 \\ .10 & .33 & .02 & .63 & .00 & .00 \\ .12 & .15 & -.23 & .66 & .00 & .00 \end{pmatrix}. \quad (10)$$

There is a strong genetic common factor present, as seen by the substantial loadings for the first factor. There also appears to be a strong second factor common to Years 2, 3, 4, 7, and 9. A third factor mostly common to Years 3 and 4 may also be present. Finally, a fourth factor common to Years 7 and 9 may be present, as suggested by the substantial loadings on those years. There is clearly both continuity and change implied by the genetic factor loadings.

These impressions of the patterns of loadings obtained from fitting the full model can be tested by comparing alternative models using the likelihood ratio χ^2. The model comparisons are presented in Table 3. A test of the genetic component of the model, as a whole, indicated that the model fit significantly worse when all genetic loadings were omitted from the full model (Model 2). The shared environmental loadings could, however, be omitted without significant decrement in model fit (Model 3); but the change in χ^2 was not trivial (20.640), in light of the fact that the last three factors in the Cholesky all had zero loadings, which suggests that probably this omnibus test is not doing the data justice.

In an attempt to arrive at a more parsimonious model that can adequately explain these data, a series of model comparisons were per-

TABLE 3

Model Comparisons

Model	Form	$-2LL$	NPAR	χ^2	*df*	*p*
1	Full Cholesky	13631.070	63			
2	Model 1, drop all G	13778.249	42	147.179	21	$<.001$
3	Model 1, drop all CE	13651.710	42	20.640	21	$>.40$
4	Model 1, drop all SE off-diagonal loadings	13640.321	48	9.251	15	$>.80$
5	Model 4, drop all CE except 1st factor	13647.805	33	7.484	15	$>.90$
6	Model 5, equate CE loadings	13652.475	28	4.670	5	$>.40$
7	Model 6, drop common CE factor	13659.835	27	7.360	1	$<.01$
8	Model 6, drop last G factor	13652.475	27	0.000	1	$=1.00$
9	Model 8, drop 5th G factor	13658.031	25	5.556	2	$>.05$
10	Model 9, drop Year 4 loading on 4th G factor	13659.005	24	0.974	1	$>.30$
11	Model 10, drop all of 4th G factor	13714.049	22	55.044	2	$<.001$

Note. *LL* = maximum-likelihood pedigree log-likelihood function; NPAR = number of parameters.

formed, beginning with tests of the unique environmental processes. All the off-diagonal factor loadings at the unique environmental level could be omitted from the model without significant decrement in fit (Model 4). Next, all shared environmental loadings except the first common factor were dropped from the model without a significant loss of fit (Model 5). The next step was to equate the shared environmental common factor loadings, which could be done and still have an adequate fit to the data (Model 6). However, this single shared environmental parameter could not be dropped from the model without a significant reduction in model fit (Model 7).

The genetic component of the model was tested next. First, the zero loading of the sixth factor on the last age was dropped from the model (Model 8). The fifth factor in the current reduced model really is a factor common to Years 7 and 9, although it appeared as the fourth factor in the full model because that fourth factor had a relatively low loading on Year 4. This, what is now, fifth factor, was quite strong, with standardized loadings at Years 7 and 9 of .39 and .80, respectively. However, it could be dropped from the model, although the change in χ^2 approached significance (p = .06, Model 9). After dropping this fifth factor, the loadings on the fourth factor, on Years 4, 7, and 9, are .08, .52, and .69, respectively, which indicates that this factor is now really a Year 7 factor. The Year 4 loading could easily be dropped from the model (Model 10). However, the remaining two loadings on Years 7 and 9 could not be dropped without a highly significant decrement in model fit (Model 11). Further tests of the genetic Cholesky are not required because the first three factors all have high loadings, and the overall test of the genetic component of the model indicated that it was highly significant. Although we could go on to attempt to drop the few low loadings and would, no doubt, find some of those nonsignificant, it would not affect our substantive interpretations. The final reduced model of cognitive development is presented in Figure 5.[2]

[2]Loadings for the shared environmental common factor are not exactly equal due to the equality constraint being imposed before standardization of the parameter estimates.

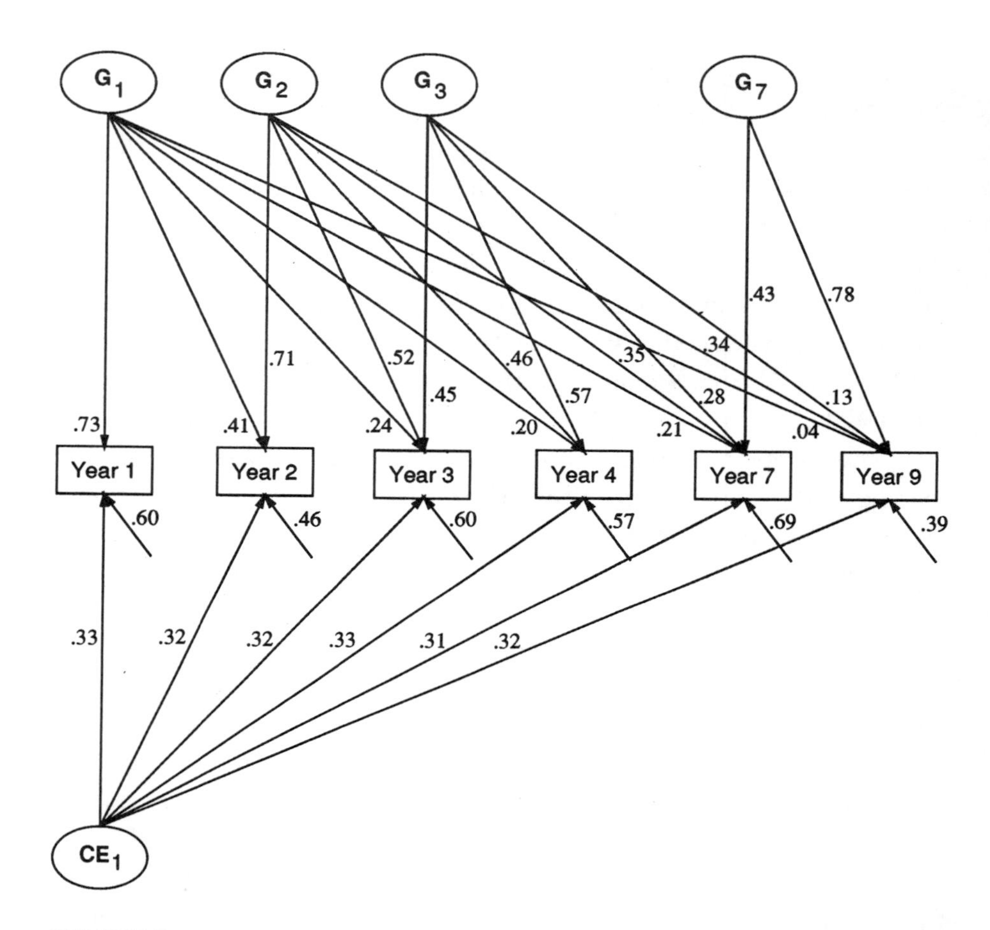

FIGURE 5. Final reduced model of cognitive development for Years 1, 2, 3, 4, 7, and 9. The unlabeled factors are unique environmental time-specific influences. G = genetic influence; CE = shared environmental influence.

Discussion

The outcome of these analyses reveals a striking diversity among the genetic and environmental processes that determine continuity and change in individual differences in general cognitive ability during the developmental period from infancy to middle childhood. The nature of these processes could not have been inferred from the phenotypic structure alone.

Shared, or home environmental, influences are completely continuous throughout the entire period: They are represented by a single factor with a single loading and no specific variances. That is, shared influences are completely correlated over time. Further analysis would be required to identify the exact nature of these influences, but one would hazard a guess that socioeconomic factors, which remain more or less constant during the period under investigation, might be largely responsible.

In complete contrast is the picture that emerges from the unique, or specific, environment. In this component, the major influences are unique to each time point, and there is no evidence of continuity in the results. That is, influences unique to the individual that either enhance or attenuate cognitive performance are entirely transitory and do not persist from one time point to the next. Here one would hazard a guess that these influences are, in large part, what psychometricians would refer to as measurement error and fluctuation in state.

By far the most interesting picture emerges at the level of the genotype. Here, there is both the continuity and change, which is characteristic of a genuine developmental process. There is a strong common factor, as in the shared environmental component, that is evident at Year 1 and operates throughout the entire developmental period, but with decreasing effect over time. New variation arises at Years 2 and 3, both with decreasing impact on subsequent years. At Year 4, however, no new genetic variation appears to arise, which suggests that there is considerable genetic stability laid down in the first 3 years of life. Interestingly, new variation arises at Year 7 and persists to Year 9, with no new variation arising at this age. Thus, genetic influences on cognitive development appear to stabilize by 4 years of age, with new variation appearing at 7 years of age, possibly in response to the novel intellectual demands imposed by schooling at this age. It is commonplace in plant and animal genetics to observe new genetic variation in response to novel environmental challenges, and it is interesting to observe the same kind of genotype–environment interaction in higher mental processes. Although one could not infer this result from analysis of the phenotypic variation only, it is interesting that cognitive developmentalists since Piaget have argued that a fundamental shift in cognition occurs during the transition from early childhood to middle childhood. The present results implicate genetic factors in this developmental change.

In conclusion, the developmental analysis that we have used clearly shows that there is no single developmental process that determines relative intellectual ability from 1 through 9 years of age. Of the three sources of variation that we have identified—shared environmental, unique environmental, and genetic influences—each appears to act in a rather different manner, with genetic influences showing the greatest complexity.

General Discussion

We have, so far, addressed genetic differences in terms of the polygenic model; that is, we have not attempted to isolate individual genetic loci, which is, of course, not possible with these kinds of data alone. We believe that even more complexity would emerge if we could determine the effects of particular loci on the phenotype. Recent developments in behavioral genetics analysis indicate how one might proceed to identify separate components of the genotype, using molecular genetic markers to locate regions of the chromosome where individual polygenes might reside. Studies of this kind, which investigate complex phenotypes, are underway in many laboratories, although only a few are concerned with cognitive development. At the Institute for Behavioral Genetics, we have developed a simple form of regression analysis applicable to highly selected samples that shows promise in this endeavor.

The method, which is based on that of Haseman and Elston's (1972) sibling-pair analysis, is modified for use with selected samples. When selecting extreme probands, one would expect their siblings to regress further back to the population mean the less they share genes identical by descent. We applied this method to sibling data on reading disability (Smith, Kimberling, Pennington, & Lubs, 1983). The average number of restriction fragment-length polymorphism (RFLP) alleles that the siblings shared identical by descent was determined from data on themselves and their parents. We then used a multiple regression to evaluate regression back to the mean to see if any polygenic variation for reading ability was associated with any of the nine RFLPs on Chromosome 15 that were chosen for analysis (Fulker et al., 1991). Significant variation appeared to be associated with three of the nine RFLP markers on the chromosome, and we are currently collecting new data on sibling pairs in an attempt to replicate and extend this finding. Methods such as this modified Haseman and Elston approach, and related methods of association, appear to offer great promise in dissecting genotypic influences on cognitive abilities.

References

Bayley, N. (1969). *Manual for the Bayley Scales of Infant Development.* New York: Psychological Corporation.

Benson, J. B., Cherny, S. S., Haith, M. M., & Fulker, D. W. (1993). Rapid assessment of infant predictors of adult IQ: The midtwin/midparent analysis. *Developmental Psychology, 29,* 434–447.

Bentler, P. M. (1989). *EQS structural equations program manual.* Los Angeles: BMDP Statistical Software.

Bouchard, T. J., Jr., & McGue, M. (1981). Familial studies of intelligence: A review. *Science, 212,* 1055–1059.

Burt, C. (1949). Alternative methods of factor analysis and their relations to Pearson's method of Principal Axes. *British Journal of Psychological Statistics, 2,* 98–121.

Cardon, L. R., Corley, R. P., DeFries, J. C., Plomin, R., & Fulker, D. W. (1992). Factorial validation of a telephone test battery of specific cognitive abilities. *Personality and Individual Differences, 13,* 1047–1050.

Cattell, R. B. (1960). The multiple abstract variance analysis equations and solutions for nature–nurture research on continuous variables. *Psychological Review, 67,* 353–372.

Cattell, R. B. (1965). Methodological and conceptual advances in evaluating heredity and environmental influences and their interaction. In S. J. Vandenberg (Ed.), *Methods and goals in human behavior genetics* (pp. 95–130). New York: Academic Press.

Cloninger, C. R., Rice, J., & Reich, T. (1979a). Multifactorial inheritance with cultural transmission and assortative mating: II. A genetical model of combined polygenic and cultural inheritance. *American Journal of Human Genetics, 31,* 176–198.

Cloninger, C. R., Rice, J., & Reich, T. (1979b). Multifactorial inheritance with cultural transmission and assortative mating: III. Family structure and the analysis of separation experiments. *American Journal of Human Genetics, 31,* 366–388.

DiLalla, L. F., Thompson, L. A., Plomin, R., Phillips, K., Fagan III, J. F., Haith, M. M., Cyphers, L. H., & Fulker, D. W. (1990). Infant predictors of preschool and adult IQ: A study of infant twins and their parents. *Developmental Psychology, 26,* 759–769.

Eaves, L., Last, K., Young, P., & Martin, N. (1978). Model fitting approaches to the analysis of human behavior. *Heredity, 41*, 249–320.

Emde, R. N., Plomin, R., Robinson, J., Reznick, J. S., Campos, J., Corley, R., DeFries, J. C., Fulker, D. W., Kagan, J., & Zahn-Waxler, C. (in press). Temperament, emotion, and cognition at 14 months: The MacArthur Longitudinal Twin Study. *Child Development.*

Fisher, R. A. (1918). The correlation between relatives on the supposition of Mendelian inheritance. *Transactions of the Royal Society of Edinburgh, 52*, 399–433.

Fulker, D. W., Cardon, L. R., DeFries, J. C., Kimberling, W. J., Pennington, B. F., & Smith, S. D. (1991). Multiple regression analysis of sib-pair data on reading to detect quantitative trait loci. *Reading and Writing: An Interdisciplinary Journal, 3*, 299–313.

Fulker, D. W., DeFries, J. C., & Plomin, R. (1988). Genetic influence on general mental ability increases between infancy and middle childhood. *Nature, 336*, 767–769.

Galton, F. (1869). *Hereditary genius: An inquiry into its laws and consequences.* London: Macmillan.

Galton, F. (1883). *Inquiries into human faculty and its development.* London: Macmillan.

Gorsuch, R. L. (1983). *Factor analysis* (2nd ed.). Hillsdale, NJ: Erlbaum.

Haseman, J. K., & Elston, R. C. (1972). The investigation of linkage between a quantitative trait and a marker locus. *Behavior Genetics, 2*, 3–19.

Jinks, J. L., & Fulker, D. W. (1970). Comparison of the biometrical genetical, mava, and classical approaches to the analysis of human behavior. *Psychological Bulletin, 73*, 311–349.

Jöreskog, K. G., & Sörbom, D. (1989). *LISREL 7: A guide to the program and applications* (2nd ed.). Chicago: SPSS.

Loehlin, J. C., & Vandenberg, S. G. (1968). Genetic and environmental components in the covariation of cognitive abilities: An additive model. In S. G. Vandenberg (Ed.), *Progress in human behavior genetics* (pp. 261–285). Baltimore: Johns Hopkins University Press.

Mather, K. (1949). *Biometrical genetics.* London: Methuen.

Mendel, G. J. (1866). Versuche uber Pflanzen-Hybriden [Experiments with plant hybrids]. *Verhandlungen des Naturforshunden in Breuen, 4*, 3–47.

Neale, M. C., & Cardon, L. R. (1992). *Methodology for genetic studies of twins and families* [NATO ASI series]. Norwell, MA: Kluwer Academic.

Plomin, R., Campos, J., Corley, R., Emde, R. N., Fulker, D. W., Kagan, J., Reznick, J. S., Robinson, J., Zahn-Waxler, C., & DeFries, J. C. (1990). Individual differences during the second year of life: The MacArthur Longitudinal Twin Study. In J. Columbo & J. Fagan (Eds.), *Individual differences in infancy: Reliability, stability, and predictability* (pp. 431–455). Hillsdale, NJ: Erlbaum.

Plomin, R., & DeFries, J. C. (1979). Multivariate behavioural genetic analysis of twin data on scholastic abilities. *Behavior Genetics, 9*, 505–517.

Plomin, R., & DeFries, J. C. (1985). *Origins of individual differences in infancy: The Colorado Adoption Project.* Orlando, FL: Academic Press.

Plomin, R., DeFries, J. C., & Fulker, D. W. (1988). *Nature and nurture in infancy and early childhood.* London: Cambridge University Press.

Rice, J., Cloninger, C. R., & Reich, T. (1978). Multifactorial inheritance with cultural transmission and assortative mating: I. Description and basic properties of the unitary models. *American Journal of Human Genetics, 30,* 618–643.

Smith, S. D., Kimberling, W. J., Pennington, B. F., & Lubs, H. A. (1983). Specific reading disability: Identification of an inherited form through linkage analysis. *Science, 219,* 1345–1347.

Spearman, C. (1904). General intelligence objectively determined and measured. *American Journal of Psychology, 15,* 201–293.

Terman, L. M., & Merill, M. A. (1973). *Stanford-Binet Intelligence Scale: 1972 norms edition.* Boston: Houghton-Mifflin.

Thurstone, L. L. (1938). *Primary mental abilities.* Chicago: University of Chicago.

Wechsler, D. (1974). *Manual for the Wechsler Intelligence Scale for Children–Revised.* New York: Psychological Corporation.

Wilson, R. S. (1978). Synchronies in mental development: An epigenetic perspective. *Science, 202,* 939–948.

Wright, S. (1921). Systems of mating. *Genetics, 6,* 111–178.

CHAPTER 5

Genetics of Specific Cognitive Abilities

Lon R. Cardon and David W. Fulker

Behavioral geneticists have been interested in the study of individual differences in general intelligence since the emergence of the discipline, which dates to Francis Galton's (1883) tracings of "human faculties" in British families. In the century since Galton's initial observations, over 140 studies of the nature and nurture of intelligence have been conducted (Bouchard & McGue, 1981), yielding the largely consistent result that heritable and environmental factors are important determinants of mental ability. Although current research in this domain is doubly indebted to Galton—for pioneering both the investigation of individual

The preparation of this chapter was supported, in part, by National Institute of Child Health and Human Development (NICHD) grants HD-10333, HD-18426, and HD-19802; by National Institute of Mental Health grant MH-43899; by National Institutes of Health Biomedical Research Support grant RR-07013-25; and by a grant from the John D. and Catherine T. MacArthur Foundation. During the preparation of this chapter, Lon R. Cardon was supported by NICHD training grant HD-07289, which was awarded to David W. Fulker.

differences in cognitive ability and the use of twins and families to do so—advancements in genetic theory, statistical methodology, and computer designs now permit evaluation of questions concerning the etiology of cognition that move beyond the simple division of phenotypic variation into the two very broad categories of nature and nurture. A brief description of some of the most notable advancements of the past century and the key researchers involved is given in the preceding chapter (Fulker, Cherny, & Cardon, chapter 4).

One area of research that has benefited considerably from the advancements of the past century, and the last decade in particular, is the study of individual differences in specific cognitive abilities. In this chapter, we present a brief historical outline of psychological and behavioral genetic research on specific abilities and describe some recent research that we have conducted at the Institute for Behavioral Genetics at the University of Colorado (Boulder) that attempts to make full use of the recent advancements and unite some psychological theories with quantitative genetic methodology and application.

Specific Cognitive Abilities in Psychology

The concept of multiple, separate mental abilities, as opposed to one single, general ability, has been traced back to the Greek philosophers (Burt, 1955), but empirical investigations of specific-versus-general abilities began about at the beginning of the 20th century and have continued unabated since. Charles Spearman provided the initial impetus for empirical research on specific abilities when he developed his theory of general intelligence of *g* (Spearman, 1904), which proposes that intelligence is a functional unity comprised of different related cognitive processes, and therefore, performance on several related intellectual tasks share something in common. It is somewhat ironic that Spearman is associated with early research on specific cognitive abilities because he endorsed general intelligence as more fruitful for study and classification than specific abilities. But a key element of Spearman's general intelligence theory was his formal distinction between intelligence as a functional unity and intelligence as a group of distinct capacities. Spearman formalized this distinction through the use of factor analysis, which he invented and which has since become an essential tool for nearly all research, psychological and genetic, on the structure of human cognition.

Although the use of *g* as a measure of intellectual ability continues to receive considerable theoretical and empirical support (e.g., Jensen, 1984), many researchers have approached intelligence from the specific-ability standpoint. This perspective originally was advanced by Thomson (1919) and especially Thurstone (1938), who held that human cognition is too rich and variegated to be expressed as a single *g* factor. In constrast, they posited that intelligence could be better defined in terms of several uncorrelated "primary" abilities, such as verbal comprehension (V), spatial visualization (S), memory (M), perceptual speed (P), and others. This perspective has also received, and continues to receive, empirical support (e.g., Cattell, 1971), although the expectation of uncorrelated primary abilities has been found untenable (Thurstone & Thurstone, 1941).

The competing theories of Spearman and Thurstone emphasize a primary question driving much cognition research of the last 100 years: Is intelligence a unified capacity comprising many elements or is it a collection of largely distinct specific attributes? This question of the structure of cognition has fueled much controversy among psychologists, which is typically centered around issues of precise definition, measurement instruments, and appropriate satistical methodology. Although at present, the debate continues between some followers of Spearman and Thurstone, a hierarchical theory of intelligence—one that combines elements of both of the competing approaches—is endorsed by many contemporary intelligence theorists (Humphreys, 1989; Vernon, 1979).

The general principle of the hierarchical model of intelligence, introduced by Burt (1949), is that the general intelligence or *g* factor is superordinate to the primary abilities, each of which is, in turn, comprised of several increasingly specific abilities. Thus, intelligence is manifest in both the higher order *g* factor of common attributes and the uncorrelated aspects of the primary factors. A hypothetical path diagram of this type of hierarchical model is presented in Figure 1. The diagram shows specific ability measures (M) as indicators of primary mental abilities (PA), which serve to define general intelligence (G).

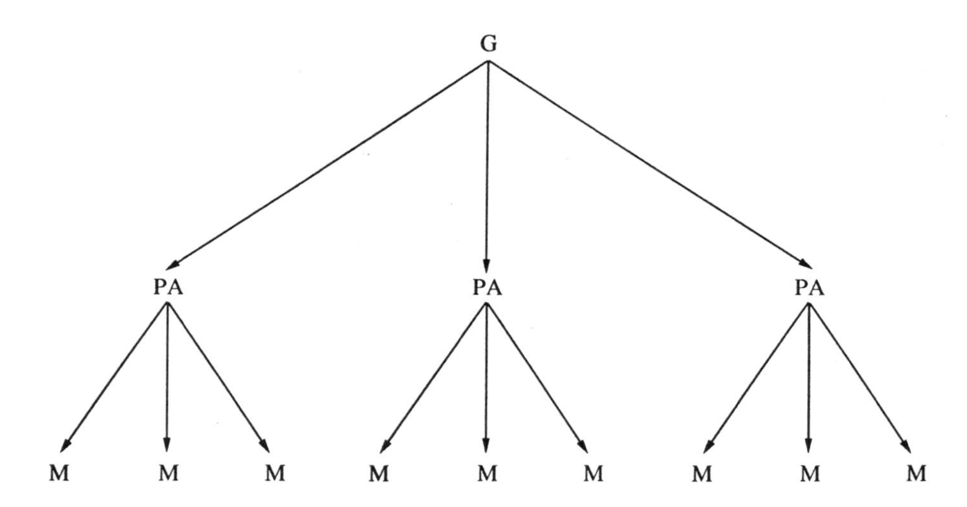

FIGURE 1. Hypothetical model of intelligence showing hierarchical influences. Shown are specific ability measures (M) as indicators of primary mental abilities (PA), which serve to define general intelligence (G).

Scarr (1989) summarized the emergent interest in hierarchical models in the following manner:

> Rather than multiple, independent intelligences . . . most psychologists and laypeople seem to have a hierarchical model in mind, a model with "g" and several levels of more specific but correlated abilities. The idea of a hierarchical model of some kind, with general intelligence at the apogee of the pyramid, has been entrenched in all theories of intelligence since Thurstone's allegedly independent, primary mental abilities failed to replicate in population samples. (p. 96)

Behavioral Genetic Approaches to Specific Cognitive Abilities

Many studies have been devoted to studying the nature and nurture of general cognitive ability; however, specific abilities have received scant attention. Of the comparatively few treatments, most studies of specific abilities have adopted a somewhat Thurstonian perspective of identifying and measuring several ability groups, to the exclusion of general ability, then using psychometric and quantitative genetic methods to assess the genetic and environmental influences on each ability group and the correlations among them. Although this approach has involved several different test batteries and genetic/psychometric procedures, nearly all studies of specific abilities have shared the common goal of determining whether specific, rather than general, mental abilities are influenced by heritable factors. In other words, for example, are there genes for verbal abilities that do not also influence spatial, memory, and perceptual speed abilities? Conversely, do the genes that contribute to individual differences in verbal ability exhibit a generalized effect on other abilities as well? These questions also are posed at the level of the environment: Are there environmental factors, such as schooling and socioeconomic status, that impact only particular abilities, or do environmental factors broadly influence several different abilities?

Empirical assessments of these questions were pioneered in large part by Steven Vandenberg, John Loehlin, and their colleagues in the 1960s using the behavioral genetic twin design. In early research, Vandenberg (1968a, 1968b) obtained evidence for genetic influence on some abilities (e.g., verbal, spatial, and language abilities) but not on others (e.g., memory, numerical, reasoning skills), which suggests that genetic factors play a larger role in determining ability scores in some domains than in others; that is, that specific abilities are differentially heritable. Vandenberg (1968a) then formalized the Spearman-versus-Thurstone, or general-versus-specific, distinction from a genetic perspective, positing a hypothesis of a *genetic g* with environmental specificity as follows: "With so many different abilities showing hereditary aspects, one may ask whether the genetic component is, perhaps, the same in all tests, with the nongenetic part determining its specific character" (p. 37). Empirical assessments of different samples and tests yielded evidence both for and against this hypothesis because some outcomes (Bock & Vandenberg, 1968; Loehlin & Vandenberg, 1968) indicated a genetic basis to the shared portion of different abilities, whereas others (Vandenberg, 1968a) indicated that genetic influences contribute to ability group specificity rather than to general commonality.

Unfortunately, the lack of congruence apparent in these first genetic studies of specific abilities has persisted in subsequent investigations, in contrast to the greater similarity among measures and samples often noted in studies of general cognition. Later studies of twins (Eaves & Gale, 1974; Loehlin & Nichols, 1976; Martin & Eaves, 1977; Martin, Jardine, & Eaves, 1984; Plomin & DeFries, 1979), nuclear families (DeFries, Johnson et al., 1979; Spuhler & Vandenberg, 1980; Williams, 1975), and adoptive and nonadoptive families (Horn, Loehlin, & Willerman, 1982; Plomin, 1988; Rice, Carey, Fulker, & DeFries, 1989) reported that some specific abilities are heritable (typically verbal and/or spatial) but revealed little agreement on the magnitude of genetic influence on the different abilities and the degree to which heritable effects on one ability are shared with those on other abilities (DeFries, Vandenberg, & McClearn, 1976).

The absence of consistent outcomes in specific ability studies is likely due to one or more of several factors, including the measures used to assess performance, the size and composition of samples, the research design, and the statistical procedure. This last factor is especially important in investigations of specific abilities because statistical inadequacies in some of the early studies hindered adequate assessment of some of the issues of interest, primarily complete disentanglement of ability-specific genetic influences from those that determine multiple abilities or *g* (Cardon, 1992). The studies have been further restricted by their uniform adherence to a Thurstonian measurement model of specific abilities; other conceptions about the structure of intelligence that have emerged from intelligence theorists have not been extensively explored.

Recent research on specific cognitive abilities in the ongoing Colorado Adoption Project (CAP) (Fulker, DeFries, & Plomin, 1988; Plomin & DeFries, 1985; Plomin, DeFries, & Fulker, 1988) has been directed toward incorporating one of the more contemporary intelligence theories, the hierarchical model, and using some of the latest advancements in genetic and psychometric methodology to develop rigorous statistical models of individual differences in mental abilities. We also have extended the hierarchical model to track the development of specific abilities over time and explore the relationships between genetic and environmental effects throughout early childhood. There is very little known about the continuity and change of genetic and environmental influences on specific abilities in childhood, and the longitudinal hierarchical model, in conjunction with the prospective longitudinal adoption data of CAP, provides a unique opportunity to explore this area. In this chapter, we describe the hierarchical genetic model and some preliminary outcomes from applications to adopted and nonadopted sibling data in CAP.

The Colorado Adoption Project

CAP is a longitudinal prospective study of individual differences in behavioral development that has been ongoing for over 15 years. The sample comprises parents and offspring from 245 adoptive and 245 nonadoptive (control) families and the biological parents of the adoptive children. In addition, approximately 100 genetically unrelated and 100 natural siblings of the adoptive and control children participate in the study. The present analyses use data from CAP siblings exclusively, although the methods described in this chapter also have been developed for application to measures from parents and their offspring simultaneously (Cardon, 1992). The siblings in the present analysis are all younger siblings of the proband children, many of whom have reached 9 years of age. Our analyses were conducted on all measurements presently available at 3, 4, 7, and 9 years of age. The full CAP sample has been described in detail by Plomin et al. (1988) (see also Fulker et al., chapter 4).

The nature and nurture of cognitive development is one of the primary research foci of CAP, and standardized measures of general and specific abilities are, and have been, administered at regular intervals to the siblings in the study. At the inception of the project, specific ability measures were selected to assess four broad cognitive domains at each measurement age, including V, S, P, and M. These measures were based on a set of adult tests used earlier in the large Hawaii Family Study of Cognition (DeFries, Vandenberg, McClearn, Kuse, & Wilson, 1979). The child measures in CAP were chosen with the primary aim of being as isomorphic as possible across time. Therefore, comparisons of verbal, spatial, perceptual speed, and memory abilities over time should reflect, in large part, real developmental growth rather than age-based differences in measurement instruments. At each age there are typically two tests for each primary ability factor, although at Years 3 and 4 some factors are defined by one or three items. Further descriptions of specific properties of these tests have been presented by Cardon (1992); Cardon, Corley, DeFries, Plomin, and Fulker (1992); Rice, Corley, Fulker, and Plomin (1986); and Singer, Corley, Guiffrida, and Plomin (1984).

For the present analysis of specific abilities, we combined the longitudinal adoption data with data obtained from identical (monozygotic [MZ]) and fraternal (dizygotic [DZ]) twins who are participating in the MacArthur Longitudinal Twin Study (MALTS; Plomin et al., 1990) and the Twin Infant Project (TIP; DiLalla et al., 1990) at the Institute for Behavioral Genetics. Descriptions of these two projects are given in chapter 4. At the time of analysis, specific ability data were available from approximately 50 MZ and 50 DZ pairs at 3 and 4 years of age.

The Model

The basic genetic model that we used for this analysis is the same as that described by Fulker et al. (chapter 4). In this model, we assume that observed scores on specific ability tests, or phenotypes (*P*), are fully determined by additive genetic effects (*G*), environmental effects that are shared by siblings reared together (*CE*), and individually specific environmental effects (*SE*). This phenotypic expectation is depicted in diagrammatic form in Figure 2a, where the unidirectional arrows characterize the causal influence of the latent variables (*G*, *CE*, and *SE*) on the observed variable (*P*). The absence of any two-headed arrows between the latent variables makes explicit our assumption of no genotype–environment correlation in this analysis. Analysis of genetically informative family designs such as adopted/nonadopted siblings or MZ/DZ twins permits assessment of the relative impact of *G*, *CE*, and *SE* on the observed measure. Sibling and twin relationships are shown in Figure 2b, which illustrates the genetic correlations between siblings—known from Mendelian genetics to be 1.0 for MZ twins, 0.5 on average for DZ twins and natural siblings, and 0.0 for adopted siblings—and the shared environmental sibling correlations, which we assume to be 1.0 for all siblings/twins reared together. This diagram makes explicit the so-called "equal-

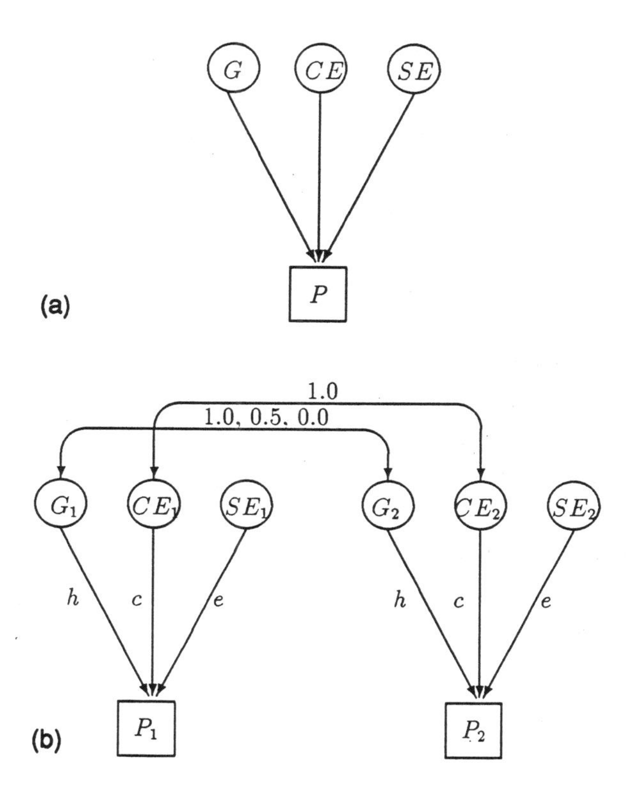

FIGURE 2. Path diagrams showing (a) latent genetic (*G*), shared or common environment (*CE*), and unique or specific environment (*SE*) causes of an observed measure (*P*); and (b) expected relationships between siblings and twins for the latent variables. Genotypic correlations are 1.0 for identical twins, 0.5 for fraternal twins and natural siblings, and 0.0 for adopted siblings.

environments" assumption of the design; we assume that rearing environments do not differ between siblings and twins in their effect on specific abilities. Previous studies have shown this to be a reasonable assumption for cognitive data (Loehlin & Nichols, 1976).

With multiple observations (e.g., multiple measures of specific abilities in the present study), the basic behavioral genetic design of Figure 2 is easily extended. Instead of inferring the genetic and environmental influences on a single measure, P, we infer the genetic and environmental effects on groups of measures, $\mathbf{P}$ (variables representing matrices are shown in boldface in this notation). Similarly, in multivariate form, $\mathbf{G}$,

CE, and **SE** represent groups of genetic and environmental factors. One primary aim of multivariate genetic analysis is to determine the number of unique genetic and environmental factors and the relationships among them: Are the different measures influenced by one or many sets of genes? and if the latter, Are the different genes related to one another or do they operate independently? Of course, these questions apply at the level of the environment as well. These particular questions have motivated genetic studies of specific cognitive abilities since the 1960s.

As noted earlier, our recent attempts to answer these questions have been directed toward bridging current intelligence theory in psychology and sophisticated multivariate modeling techniques in behavioral genetics. Accordingly, we examined our specific ability **P** variables in a hierarchical framework and estimated the role of genes and the environment in determining the hierarchical structure. Our hierarchical model of CAP data is shown in Figure 3, where it may be seen that two ability tests define each of the four primary abilities, verbal (V), spatial (S), perceptual

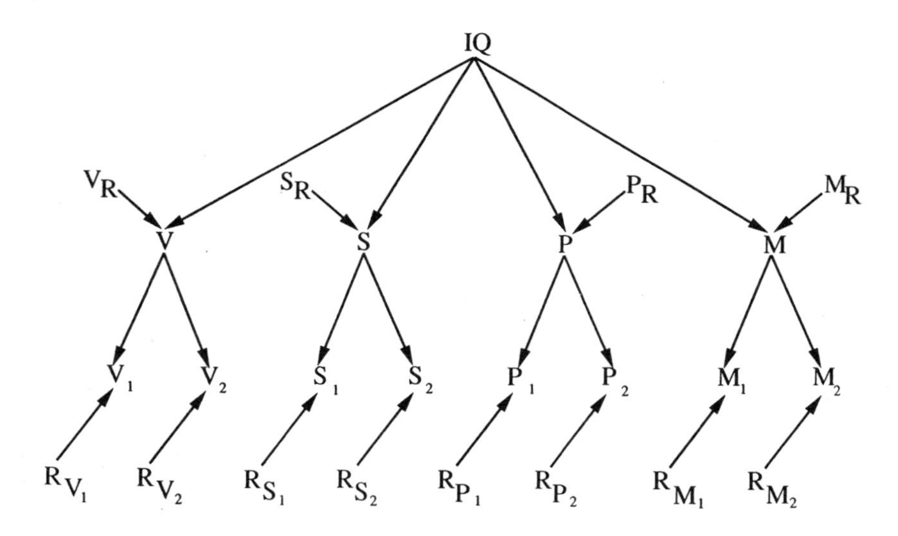

FIGURE 3. Hierarchical model of specific cognitive abilities in CAP. Symbols V, S, P, and M denote verbal, spatial, perceptual speed, and memory abilities, respectively. Residual effects are symbolized as R. General intelligence is represented by the higher order factor IQ.

speed (P), and memory (M), which are correlated by a higher order general *g* factor (shown as IQ in Figure 3 for distinction from genetic G). The specific tests and the primary abilities also have residual effects (e.g., R_{V1} and R_{S1} on the tests and V_R and S_R on the primary factors) that are uncorrelated with other tests or ability groups. By using the sibling/twin design, we can impose this hierarchical model on **G**, **CE**, and **SE** and, therefore, estimate genetic and environmental effects that correspond exactly to a hierarchical theory of intelligence. That is, we can estimate the extent to which a single set of genetic or environmental factors determine the hierarchical relationships and the extent to which independent, residual, genetic and environmental effects impact each ability group and measure.

The longitudinal structure of CAP data affords even further investigation of the nature and nurture of specific abilities. With longitudinal observations, we can examine the persistence of ability-specific or shared genetic and environmental effects over time and track the pattern of influences during childhood. Phillips and Fulker (1989) and Cardon, Fulker, DeFries, and Plomin (1992a) have used a "simplex model" (Guttman, 1954) of continuity and change in childhood development for investigation of cognitive ability data in CAP. The simplex model was originally developed for genetic application by Eaves, Long, and Heath (1986), where it was used to estimate age-to-age stability (shown as the direct effects of P_i on P_{i+1} in Figure 4) and growth or change (shown as residual effects at each age i) of genetic and environmental effects on cognition. Although the genetic simplex formulation is statistically different from the Cholesky approach used in chapter 4 (see Neale & Cardon, 1992), it shares the global aim of describing continuity and change in cognitive development.

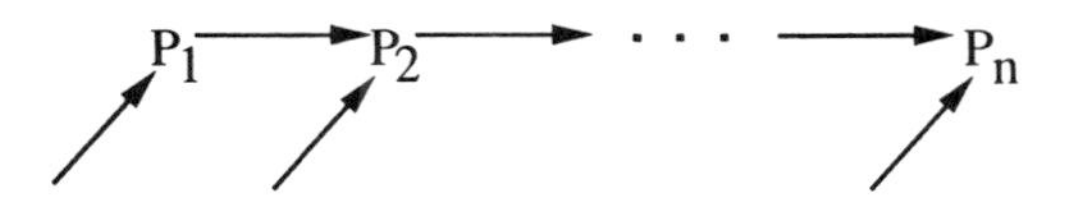

FIGURE 4. Path diagram of autoregressive simplex model. (P = phenotype.)

Combining the developmental simplex and the hierarchical models provides a general model of cognitive abilities whereby the processes underlying individual differences in abilities may be elucidated in some detail. Genetic and environmental effects that are ability specific may be distinguished from those that are common to multiple abilities, and the continuous impact of the distinct or shared effects may be assessed over time. In this manner, we can begin to address questions concerning the long-term effects of early childhood experiences on specific attributes or general intelligence later in life, the time at which genes "turn-on" and the subsequent impact of those genes on general and specific abilities, or perhaps the differentiation of genetic or environmental factors from amalgamated determinants having general effects in infancy into specialized components having focused effects in later childhood (Garrett, 1946). Our longitudinal hierarchical model is presented in Figure 5. For simplicity, this model shows only two primary factors at each of three ages; application to the present sibling and twin measures actually involves four ability factors (V, S, P, and M) at each of four ages (Years 3, 4, 7, and 9; see Cardon, 1992, for a detailed description of this model).

Results

At a single measurement occasion, the specific paths in the hierarchical model of Figure 3 relate directly to the important question of whether specific abilities are influenced by the same or different genes and environments. The impact of the residual factors (V_R, S_R, etc.) on the primary abilities represent underlying variation that is ability specific, whereas the paths between the latent g factors and primary abilities reflect shared genetic or environmental effects. Application of this hierarchical genetic model to CAP, MALTS, and TIP data at 3, 4, 7, and 9 years of age suggests that specific cognitive abilities are related by general intelligence genes, as one might expect from the results of many genetic studies of IQ, but certain abilities are additionally influenced by ability-specific genes.

Genetic and environmental factor loadings from model-fitting applications are presented in Table 1. Parameter estimates are tabulated in terms of ability-specific and general components for each observed primary ability factor at each age. The columns representing proportions of

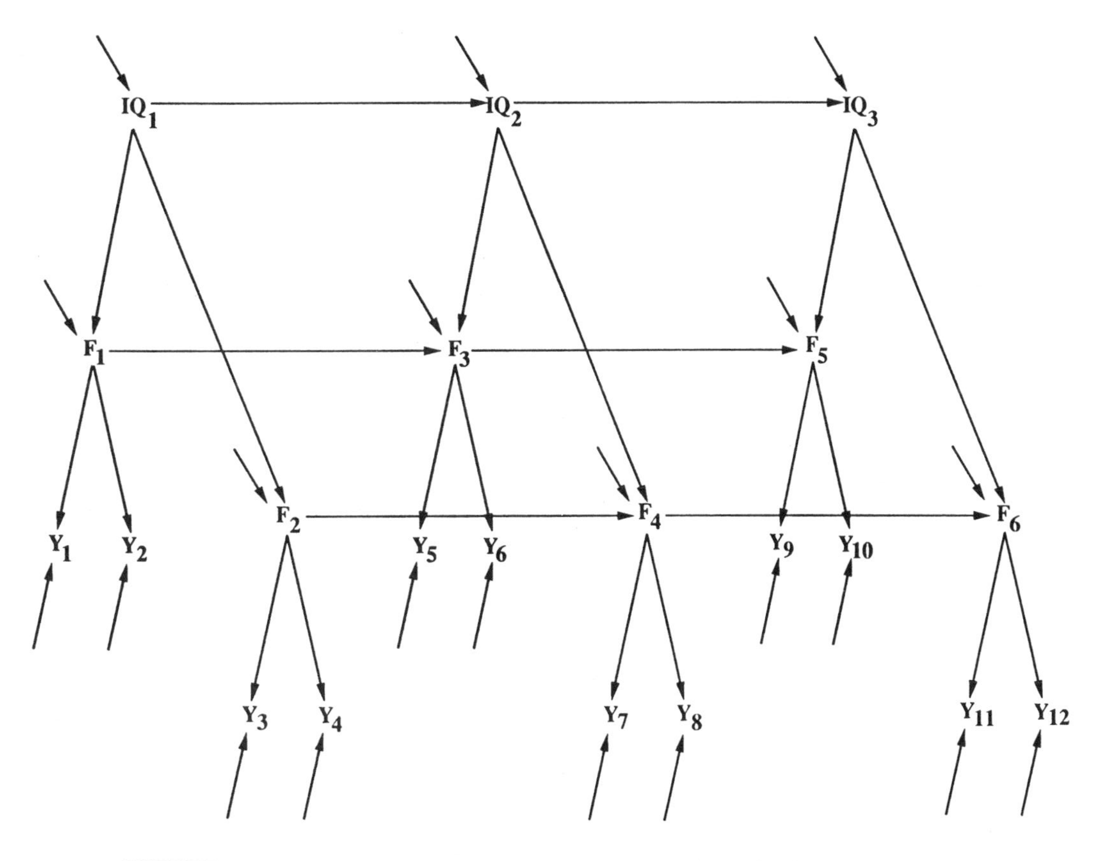

FIGURE 5. Simplified longitudinal hierarchical model of specific abilities showing two primary factors (F) at each of three occasions (Y).

TABLE 1
Parameter Estimates From Hierarchical Genetic Model

Primary ability	SE		CE		G				
	Gen.	Spc.	Gen.	Spc.	Gen.	Spc.	e^2	c^2	h^2
				3 years of age					
V[a]	.06	.64	.18	.04	.73	.42	.41	.04	.56
S[a]	.78	.41	.23	.03	.40	.20	.78	.05	.17
P	.03	.17	.40	.39	.73	.35	.03	.31	.66
M	.06	.11	.12	.00	.88	.44	.02	.01	.97
				4 years of age					
V	.30	.65	.01	.16	.57	.38	.51	.03	.46
S	.265	.85	.14	.02	.36	.23	.80	.02	.19
P	.80	.04	.16	.01	.41	.40	.65	.03	.33
M	.09	.49	.09	.00	.54	.68	.24	.01	.75
				7 years of age					
V	.04	.01	.59	.00	.29	.75	.00	.35	.65
S	.06	.00	.07	.15	.88	.43	.00	.03	.97
P	.46	.57	.02	.00	.58	.02	.66	.00	.34
M	.03	.41	.07	.00	.33	.84	.17	.00	.82
				9 years of age					
V	.24	.38	.00	.73	.73	.47	.20	.06	.74
S	.78	.01	.25	.35	.35	.05	.61	.26	.13
P	.53	.20	.80	.19	.19	.00	.32	.64	.04
M	.28	.80	.06	.01	.22	.01	.95	.00	.05

Note. For primary abilities: V = verbal; S = spatial; P = perceptual speed; M = memory; SE = unique (i.e., specific) environmental; CE = shared (i.e., common) environmental; G = genetic; Gen. = general; Spc. = specific.
[a]Residual parameters fixed for model identification.

overall variance (e^2, c^2, and h^2) are noteworthy in this table because, in contrast to traditional calculations of heritabilities and environmentalities, the measurement portion of the hierarchical model effectively removes the test measurement error that usually confounds e^2 estimates. Thus, estimates of h^2 and/or c^2 may be higher than those typically reported because of the reduction in overall variance of the primary factors.

At the early ages, Years 3 and 4, specific abilities reveal substantial impact from both genetic and environmental sources (with the possible exception of the spatial factor, which seems largely determined by environmental effects at these ages). The general genetic factor, or genetic *g*, (shown in the fifth column of Table 1) accounts for most of the genetic effects on the four ability groups, with all abilities showing additional specific influence. Environmental effects, though substantial, arise predominantly from features that are unique to individuals (e^2) and not shared by siblings (c^2). These unique environment effects are primarily ability specific at 3 years of age, with an emergent general factor at 4 years of age.

At 7 years of age, the general genetic factor column shows moderate to large loadings for all ability factors, which indicates that genetic influences on all primary abilities are to some extent shared with genetic influences on *g*. This is similar to the pattern observed at the earlier ages but is accompanied by greater proportions of ability-specific genetic variance than at 3 or 4 years of age. Unique environment effects show the Year 3 pattern of mainly specific variation and a lack of generalized influence. Shared environmental effects do not contribute appreciably to either variances or covariances at 7 years of age, as is apparent at the early ages, although these effects do appear substantial for V at this later occasion (see Cardon et al., 1992a, for a detailed analysis of the Year 7 data). At 9 years of age, shared environment effects emerge for S and P, and the genetic components appear to generalize into a common genetic factor, with only V showing independent influence.

The hierarchical results strongly indicate that specific abilities are more than simple subsets of general intelligence. They are complex attributes that are related to one another to some degree yet also have elements of specificity. CAP data provide evidence for genetic and environmental origins of both specificity and commonality, thus yielding partial support for the theories of Spearman and Thurstone at the latent determinant level. With respect to genetic theory, the present data yield an affirmative answer to the question of different genes for different abilities posed (albeit in the opposite direction) by Vandenberg in the 1960s.

It is exciting that specific and general latent components are discernable at such early ages in children, given the unavoidable difficulties with assessment of cognitive abilities in young children, particularly as young as 3 years of age. However, it is important to recognize that the results in Table 1 do not imply that specific abilities, or their underlying causes, are static and unchanging during childhood development. Some of the estimates differ considerably at different occasions, which suggests that genetic and environmental influences may have fleeting or lasting effects on different abilities across time. Of course, the differences in estimates may also reflect incongruities in the phenotypes defined by the different measurement batteries, imprecise resolution of the models with the present sample sizes, or developmental differences between twins and nontwin siblings, although a concerted effort has been made in CAP to minimize and/or statistically test such factors. Our longitudinal extension of the hierarchical model is designed for exploration of the systematic change and continuity in the etiology of specific abilities.

To examine the long-term effects of genes and the environment on specific cognitive abilities, we fitted the longitudinal hierarchical model to CAP, MALTS, and TIP data at all ages simultaneously. Because of the complexity of this model, we also fitted several submodels in search of a parsimonious account of the data. Our model-fitting comparisons, involving global tests of parameters (e.g., testing all longitudinal correlations among the genetic g factors as a group), indicated that shared environmental effects do not exert sufficient impact to be detectable in the sibling/twin sample and that unique environmental effects do show continuity over time but only through the general factor. Ability-specific environmental persistence was not apparent. These environmental findings suggest that unique experiences in childhood have long-term consequences for cognitive abilities but in a general capacity rather than in specific outcomes. Both general and specific genetic factors appear to have lasting effects because neither could be omitted from the model without significant loss of model fit.

The genetic parameter estimates from the final reduced longitudinal model are presented in Figure 6 (showing only the primary and general factors for simplicity). There are several striking features of this diagram that warrant mention. First, there appears to be a great deal of age-to-age stability of the genetic influences that are specific to certain abilities. For example, path coefficients for V range from .61 to .74 between occasions, which indicates that ability-specific genes expressed at early ages

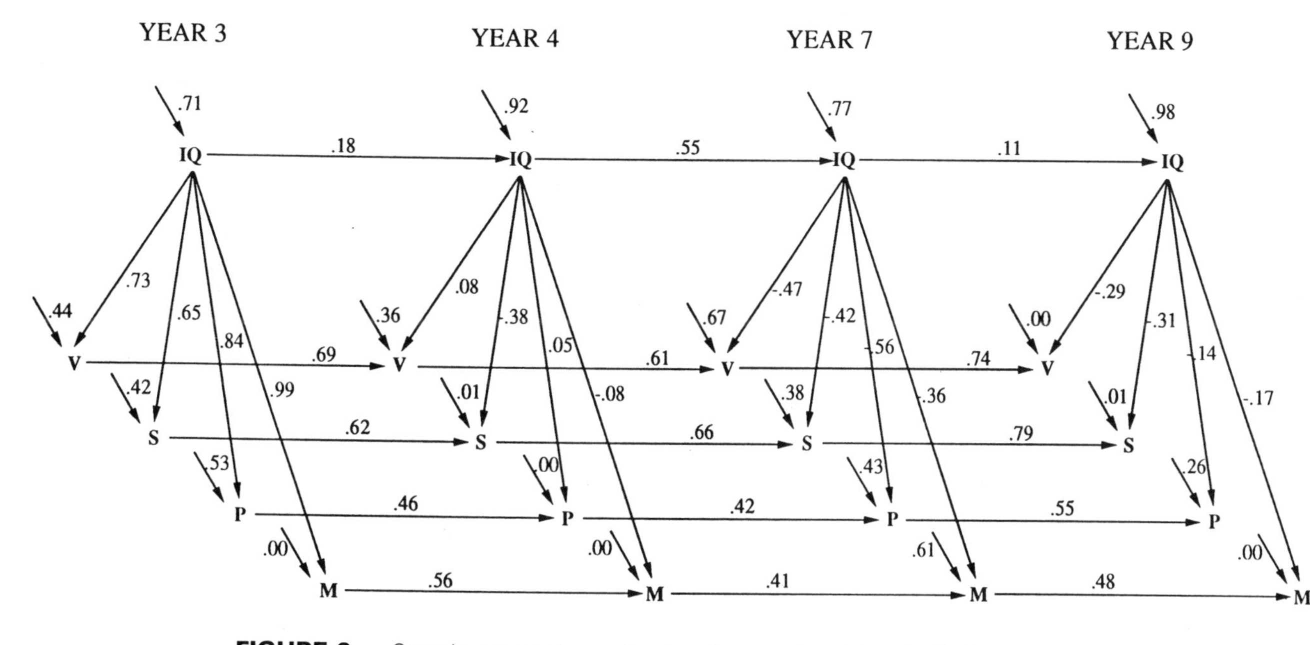

FIGURE 6. Genetic parameter estimates from reduced longitudinal hierarchical model. Measurement loadings and residuals have been omitted to simplify illustration. Symbols V, S, P, and M denote verbal, spatial, perceptual speed, and memory abilities, respectively.

tend to have considerable persistence and impact at later ages. Second, the age-to-age coefficients for the genetic g factors are moderate, which suggests that the genetic continuity of general intelligence overlaps considerably with that of specific abilities. And third, the residual estimates on the primary factors are generally moderate to large at 3 and 7 years of age and small at 4 and 9 years of age. Similarly, the general factor loadings are larger at Years 3 and 7 than at the other ages (note that the negative factor loadings shown at Years 4, 7, and 9 yield correlations in the expected positive direction, although they are reflective of some instabilities in the model and/or data). This pattern suggests that genes may be turned on at two occasions in childhood: the first in infancy (3 years of age or earlier) and the second between 4 and 7 years of age. The genes expressed at the first occasion continue to influence abilities at subsequent occasions, and these effects are augmented and broadened by new heritable variation at about 7 years of age, which in turn persists through 9 years of age. The apparent change between 4 and 7 years of age corresponds closely to the developmental shift at these ages that has long been observed by psychologists (e.g., Piaget, 1962), which indicates that the observed shift may be genetically driven.

Conclusions

Although behavioral geneticists have extensively explored the nature and nurture of general cognition, the etiology of specific abilities has received comparatively little attention. Conclusions about the relative impact of genes and the environment on different abilities are elusive because the few genetic studies of specific abilities have not yielded consistent findings as in the case of general intelligence. The lack of congruence seems attributable at least in part to differences and inadequacies of the statistical procedures used, as well as to differences in samples and measures. A major research focus of the ongoing adoptive and twin studies at the Institute for Behavioral Genetics has been to develop and apply mental ability models that have more rigorous statistical attributes and correspond more closely to alternative intelligence theories emerging from psychology.

The present results, obtained from applications of hierarchical genetic models of intelligence, provide some of the first evidence for genetic influences on specific abilities in childhood that are unrelated to those determining general cognition. Our assessments of V, S, P, and M indicate that, in part, these different abilities are influenced by the same genes, and in part, the abilities are determined by genes operating independently of one another and independently of *g*. The longitudinal outcomes extend the findings of genetic commonality and specificity, which suggests that the ability-specific genes are pervasive throughout young childhood. In addition, the genetic persistence underlying observed continuity is accompanied by transitions in the genotype that lead to observed change. Our results indicate that the persistent infant genes are augmented by a novel genetic component that emerges at Year 7 and continues to influence ability variation at Year 9. This same pattern of genetic continuity and change was noted for general cognition in the previous chapter (Fulker et al., chapter 4), which points strongly to a genetic basis for the developmental shift at these ages.

The environmental outcomes indicate that although shared sibling environments do not exert a measurable effect in our sample, nonshared environmental factors play a large role in specific cognitive abilities. The unique environmental factors are important at each age and exhibit some lasting effects over childhood. The finding of environmental persistence through *g* and not specific factors implies that childhood experiences may have important consequences for a particular ability at the time of occurrence but may have generalized effects on mental skills at later occasions. For example, educational or rearing changes that might facilitate verbal learning at one age also may facilitate verbal *and* performance abilities in later childhood. Continuity of unique environmental effects is rarely seen in family studies of cognition, but the present environmental findings are not confounded with test measurement error as in most behavioral genetic models. Therefore, these effects should reflect real environmental influences, and their persistence should reflect a genuine developmental trend.

Continuing research at the Institute for Behavioral Genetics will be directed toward further exploration and characterization of the factors underlying individual differences in mental abilities—specificity as well as commonality, change as well as continuity. The adoptive and twin samples are well placed for these types of investigations because they are longitudinal, prospective, and ongoing. Our aim is to examine the processes of development throughout childhood right up to young adulthood, with concurrent development of alternative genetic models to account for some of the subtle complexities manifest even at the young ages of the present samples.

References

Bock, R. D., & Vandenberg, S. G. (1968). Components of heritable variation in mental test scores. In S. G. Vandenberg (Ed.), *Progress in human behavior genetics* (pp. 233–260). Baltimore: Johns Hopkins University Press.

Bouchard, T. J., & McGue, M. (1981). Familial studies of intelligence: A review. *Science, 212*, 1055–1059.

Burt, C. (1949). Alternative methods of factor analysis and their relations to Pearson's method of principal axes. *British Journal of Psychological Statistics, 2*, 98–121.

Burt, C. (1955). The evidence for the concept of intelligence. *British Journal of Educational Psychology, 25*, 158–178.

Cardon, L. R. (1992). *Multivariate path analysis of specific cognitive abilities in the Colorado Adoption Project.* Unpublished doctoral dissertation, University of Colorado, Boulder.

Cardon, L. R., Corley, R. P., DeFries, J. C., Plomin, R., & Fulker, D. W. (1992). Factorial validation of a telephone test battery of specific cognitive abilities. *Personality and Individual Differences, 13*, 1047–1050.

Cardon, L. R., Fulker, D. W., DeFries, J. C., & Plomin, R. (1992a). Continuity and change in general cognitive ability from 1 to 7 years. *Developmental Psychology, 28*, 64–73.

Cardon, L. R., Fulker, D. W., DeFries, J. C., & Plomin, R. (1992b). Multivariate genetic analysis of specific cognitive abilities in the Colorado Adoption Project at age 7. *Intelligence, 16*, 383–400.

Cattell, R. B. (1971). *Abilities: Their structure, growth, and action.* Boston: Houghton-Mifflin.

DeFries, J. C., Johnson, R. C., Kuse, A. R., McClearn, G. E., Polovina, J., Vandenberg, S. G., & Wilson, J. R. (1979). Familial resemblance for specific cognitive abilities. *Behavior Genetics, 9*, 23–43.

DeFries, J. C., Vandenberg, S. G., & McClearn, G. E. (1976). Genetics of specific cognitive abilities. *Annual Review of Genetics, 10*, 179–207.

DeFries, J. C., Vandenberg, S. G., McClearn, G. E., Kuse, A. R., & Wilson, J. R. (1979). Near identity of cognitive structure in two ethnic groups. *Science, 183*, 338–339.

DiLalla, L. F., Thompson, L. A., Plomin, R., Phillips, K., Fagan, J. F., Haith, M. M., Cyphers, L. H., & Fulker, D. W. (1990). Infant predictors of preschool and adult IQ: A study of infant twins and their parents. *Developmental Psychology, 26*, 759–769.

Eaves, L. J., & Gale, J. S. (1974). A method for analyzing the genetic basis of covariation. *Behavior Genetics, 4*, 253–267.

Eaves, L. J., Long, J., & Heath, A. C. (1986). A theory of developmental change in quantitative phenotypes applied to cognitive development. *Behavior Genetics, 16*, 143–162.

Fulker, D. W., DeFries, J. C., & Plomin, R. (1988). Genetic influence on general mental ability increases between infancy and middle childhood. *Nature, 336*, 767–769.

Galton, F. (1883). *Inquiries into human faculty and its development*. London: Macmillan.

Garrett, H. E. (1946). A developmental theory of intelligence. *American Psychologist, 1*, 372–378.

Guttman, L. (1954). A new approach to factor analysis: The radex. In P. F. Lazarsfeld (Ed.), *Mathematical thinking in the social sciences* (pp. 258–349). Glencoe, IL: Free Press.

Horn, J. M., Loehlin, J. C., & Willerman, L. (1982). Aspects of the inheritance of intellectual abilities. *Behavior Genetics, 12*, 479–516.

Humphreys, L. G. (1989). Intelligence: Three kinds of instability and their consequences for policy. In R. L. Linn (Ed.), *Intelligence: Measurement, theory, and public policy* (pp. 193–216). Chicago: University of Illinois Press.

Jensen, A. R. (1984). Test validity: g versus the specificity doctrine. *Journal of Social and Biological Sciences, 7*, 93–118.

Loehlin, J. C., & Nichols, R. C. (1976). *Heredity, environment, and personality*. Austin: University of Texas Press.

Loehlin, J. C., & Vandenberg, S. G. (1968). Genetic and environmental components in the covariation of cognitive abilities: An additive model. In S. G. Vandenberg (Ed.), *Progress in human behavior genetics* (pp. 261–285). Baltimore: Johns Hopkins University Press.

Martin, N. G., & Eaves, L. J. (1977). The genetical analysis of covariance structure. *Heredity, 38*, 79–95.

Martin, N. G., Jardine, R., & Eaves, L. J. (1984). Is there only one set of genes for different abilities? A reanalysis of the National Merit Scholarship Qualifying Test (NMSQT) data. *Behavior Genetics, 14*, 355–370.

Neale, M. C., & Cardon, L. R. (1992). *Methodology for genetic studies of twins and families*. Norwell, MA: Kluwer Academic.

Phillips, K., & Fulker, D. W. (1989). Quantitative genetic analysis of longitudinal trends in adoption designs with application to IQ in the Colorado Adoption Project. *Behavior Genetics, 19*, 621–658.

Piaget, J. (1962). *Play, dreams, and imitation in childhood*. New York: Norton.

Plomin, R. (1988). The nature and nurture of cognitive abilities. In R. Sternberg (Ed.), *Advances in the psychology of human intelligence* (Vol. 4). Hillsdale, NJ: Erlbaum.

Plomin, R., Campos, J., Corley, R., Emde, R., Fulker, D. W., Kagan, J., Reznick, J. S., Robinson, J. L., Zahn-Waxler, C., & DeFries, J. C. (1990). Individual differences during the second year of life: The MacArthur Longitudinal Twin Study. In J. Colombo & J. Fagen (Eds.), *Individual differences in infancy* (pp. 431–455). Hillsdale, NJ: Erlbaum.

Plomin, R., & DeFries, J. C. (1979). Multivariate behavioural genetic analysis of twin data on scholastic abilities. *Behavior Genetics, 9*, 505–517.

Plomin, R., & DeFries, J. C. (1985). *Origins of individual differences in infancy: The Colorado Adoption Project.* Orlando, FL: Academic Press.

Plomin, R., DeFries, J. C., & Fulker, D. W. (1988). *Nature and nurture in infancy and early childhood.* London: Cambridge University Press.

Rice, T., Carey, G., Fulker, D. W., & DeFries, J. C. (1989). Multivariate path analysis of specific cognitive abilities in the Colorado Adoption Project: Conditional path model of assortative mating. *Behavior Genetics, 19*, 195–207.

Rice, T., Corley, R., Fulker, D. W., & Plomin, R. (1986). The development and validation of a test battery measuring specific cognitive abilities in four-year-old children. *Educational and Psychological Measurement, 46*, 699–708.

Scarr, S. (1989). Protecting general intelligence. In R. L. Linn (Ed.), *Intelligence: Measurement, theory, and public policy* (pp. 74–118). Chicago: University of Illinois Press.

Singer, S., Corley, R., Guiffrida, C., & Plomin, R. (1984). The development and validation of a test battery to measure differentiated cognitive abilities in three-year-old children. *Educational and Psychological Measurement, 49*, 703–713.

Spearman, C. (1904). General intelligence objectively determined and measured. *American Journal of Psychology, 15*, 201–293.

Spuhler, K. P., & Vandenberg, S. G. (1980). Comparison of parent–offspring resemblance for specific cognitive abilities. *Behavior Genetics, 10*, 413–418.

Thomson, G. A. (1919). On the cause of hierarchical order among correlation coefficients. *Royal Society of London. Proceedings. Series A. Mathematical and Physical Sciences, 95*, 400–408.

Thurstone, L. L. (1938). *Primary mental abilities.* Chicago: University of Chicago.

Thurstone, L. L., & Thurstone, T. G. (1941). Factorial studies of intelligence. *Psychometric Monographs*, No. 2

Vandenberg, S. G. (1968a). The nature and nurture of intelligence. In D. C. Glass (Ed.), *Genetics* (pp. 3–58). New York: Rockefeller University Press.

Vandenberg, S. G. (1968b). Primary mental abilities or general intelligence? Evidence from twin studies. In J. M. Thoday & A. S. Parke (Eds.), *Genetic and environmental influences on behavior* (pp. 146–160). New York: Plenum.

Vernon, P. E. (1979). *Intelligence: Heredity and environment.* San Francisco: Freeman.

Williams, T. H. (1975). Family resemblance in abilities: The Wechsler scales. *Behavior Genetics, 5*, 405–409.

CHAPTER 6

Genetics of Reading Disability

John C. DeFries and Jacquelyn J. Gillis

Four years after the inception of the American Psychological Association, W. Pringle Morgan (1896) used the term *congenital word-blindness* to describe the case of an intelligent 14-year-old boy who was incapable of learning to read. Within a decade, the familial nature of this condition was described by Thomas (1905) as follows:

> In this connection it is to be noted that it frequently assumes a family type; there are a number of instances of more than one member of the family being affected, and the mother often volunteers the statement that she herself was unable to learn to read, although she had every opportunity. (p. 381)

This research was supported, in part, by National Institute of Child Health and Human Development program project and center grants HD-11681 and HD-27802 to John C. DeFries. This chapter was prepared while Jacquelyn J. Gillis was supported by National Institute of Mental Health training grant MH-16880. The invaluable contributions of staff members of the many Colorado school districts and of the families who participated in this study are gratefully acknowledged.

The brief description of "Case 8" that Thomas (1905) provided is especially noteworthy:

> J. H., aged 14, has been five years in a Special School. In March, 1901, it was noted: "Improving in everything but reading; cannot interpret any word."
>
> In November, 1904, no progress has been made in reading, although his attainments in other respects normal. He did difficult problems in mental arithmetic with ease; drawing is good; and manual subjects excellent.
>
> He cannot read the word "cat," although when spelt aloud, he recognised it at once.
>
> A sister, S. H., passed through this school, and her final note states that she could do everything but read on leaving.
>
> *The mother states that she herself could never learn to read, although she had every opportunity.* Five other children in the same family have been unable to learn to read. (pp. 383–384)

Various other terms were subsequently used for this condition, including *specific reading disability* (Eustis, 1947), *specific developmental dyslexia* (Critchley, 1970), and *unexpected reading failure* (Symmes & Rapoport, 1972). In deference to Samuel A. Kirk (1962), a clinical psychologist and pioneer in the diagnosis and remediation of learning disabilities (Kirk & Bateman, 1962), the term *reading disability* is used in this chapter. Although Kirk conceptualized reading disability as being only one of several learning disabilities, reading is the primary academic problem in about 80% of the children with a diagnosed learning disability (Lerner, 1989).

Results obtained from several family studies (e.g., DeFries, Vogler, & LaBuda, 1986; Finucci & Childs, 1983; Gilger, Pennington, & DeFries, 1991) have confirmed the familial nature of reading disability, and our ongoing twin study has provided compelling evidence that reading disability is due at least in part to heritable influences (e.g., DeFries, Fulker, & LaBuda, 1987). The primary objectives of this chapter are threefold: (a) to review previous family and twin studies of reading disability; (b) to summarize recent results obtained from our twin study; and (c) to discuss future research directions, including attempts to determine the chromosomal location of the individual genes that cause reading disability.

Previous Family Studies

Familial transmission for reading disability has been frequently reported, and a number of different modes of inheritance have been proposed to account for this observed familiality. For example, in the 112 families included in Hallgren's (1950) classic study, 88% of the probands (the index cases through which other family members were ascertained) had one or more relatives who were also affected. Based on the results of this study, Hallgren concluded that the familial transmission of reading disability is due to an autosomal dominant gene; that is, a single copy of a specific allele (an alternative form of a gene) is sufficient to cause the disorder (Plomin, DeFries, & McClearn, 1990). However, there are several problems with this interpretation. First, if reading disability was inherited in an autosomal dominant manner, then at least one parent of each affected child should also have been affected. Contrary to this expectation, both parents were unaffected in 17% of the families. Second, diagnosis was problematic. Although test data were available from probands and some family members, adults were diagnosed primarily on the basis of interview data. Third, and perhaps most important, a casual review of Hallgren's case studies suggests that family history was often considered for making diagnoses. This practice could have resulted in the ascertainment of families in which parent–offspring transmission was overrepresented (for an excellent critical review of the early literature pertaining to the genetics of reading disability see Finucci, 1978).

In an attempt to account for the unequal gender ratio typically observed in referred samples of reading-disabled children (3 or 4 boys to each girl), Symmes and Rapoport (1972) proposed that the condition is caused by an X-linked recessive allele. Whereas females have two X chromosomes (in addition to 22 pairs of autosomes), males have only one X and a smaller Y chromosome. Therefore, for an X-linked recessive condition to be expressed in females, the recessive allele must be present on both X chromosomes. In males, however, only one X-linked recessive allele is sufficient to cause the condition. Consequently, X-linked recessive disorders (e.g., color blindness) are more prevalent in males than in females. For example, if the frequency of an X-linked recessive allele for some condition in a random mating population were 0.1, then the expected prevalences in males and females would be 0.1 and $(0.1)^2 = 0.01$, respectively. Moreover, parent–offspring resemblance for X-linked characters differs as a function of gender (Plomin, DeFries, & McClearn, 1990). Fathers transmit their only X chromosome to their daughters, and sons inherit their only X chromosome from their mothers. In contrast, mothers and daughters each share one of two X chromosomes, and fathers and sons have no X chromosomes in common. Thus, for characters influenced by an X-linked allele, father–daughter and mother–son resemblances should be approximately equal in magnitude, mother–daughter resemblance should be lower, and father–son resemblance should be negligible. Results obtained from analyses of data that have assessed these various relationships in families of children with reading disability have failed to provide evidence for the X-linked recessive hypothesis (e.g., DeFries & Decker, 1982).

The first family study of reading disability in which the probands and their relatives were both administered an extensive battery of psychometric tests was reported by Finucci, Guthrie, Childs, Abbey, and Childs (1976). In a sample of 20 probands (15 males and 5 females), 34 of 75 first-degree relatives were also objectively diagnosed as being reading disabled. In the 16 families of probands in which both parents were tested, 13 had one or both parents affected, a proportion (81%) very similar to that previously reported by Hallgren (1950). However, various patterns of inheritance were apparent in the pedigrees in this study; thus, Finucci et al. concluded that reading disability is genetically heterogeneous.

Colorado Family Reading Study

Our first attempt to assess the etiology of reading disability was the Colorado Family Reading Study, which was initiated in 1973. The primary objectives of that study were threefold: (a) to construct a short battery of tests that differentiates children with a diagnosed reading disability from matched controls, (b) to assess possible deficits in the parents and siblings of children with reading problems, and (c) to study the transmission of reading disability in families of affected children. Children with reading problems were ascertained by referral from local school districts in Colorado using the following selection criteria: 7.5–12 years of age; reading performance equal to or less than half of that predicted on the basis of age and grade level (e.g., a child in the 4th grade who is reading at or below the 2nd grade level); IQ of 90 or above; no uncorrected auditory or visual acuity deficits; no serious emotional or behavioral problems; and living at home with both biological parents. Control children were matched to the reading-disabled probands on the basis of age (within 6 months), gender, school, and home neighborhood. Parents and siblings (7.5–18 years of age) of probands and control children were also tested. All families were middle class, and the language spoken in the homes was standard American English.

During a 3-year period, 125 probands, their parents and siblings, and members of 125 control families were administered an extensive psychometric test battery (DeFries, Singer, Foch, & Lewitter, 1978). It is of special interest to note that this referred sample of reading-disabled children included 96 boys and 29 girls, a gender ratio of 3.3:1.

Scores obtained on eight tests administered to all subjects in the study (Reading Recognition, Reading Comprehension, Spelling, and Mathematics subtests of the Peabody Individual Achievement Test [Dunn & Markwardt, 1970]; Coding Subtest Form B of the Wechsler Intelligence Scale for Children–Revised [WISC-R] [Wechsler, 1974]; the Colorado Perceptual Speed Test; Primary Mental Abilities–Spatial Relations [Thurstone, 1962]; and the Nonverbal Culture Fair Intelligence Test [Institute for Personality and Ability Testing, 1973]) were age adjusted and subjected to principal-component analysis with Varimax rotation (Nie, Hull, Jenkins, Steinbrenner, & Bent, 1975). Three readily interpretable dimensions accounted for 77% of the common variance (Decker & DeFries, 1980): a general reading performance component that correlated highly with Reading Recognition, Reading Comprehension, and Spelling; a symbol-processing speed factor that loaded on Colorado Perceptual Speed and WISC-R Coding; and a spatial/reasoning component that correlated with

Nonverbal Intelligence and Spatial Relations. From the principal-component loadings and the correlations among the tests, factor–score coefficients were estimated and then used to compute three composite scores for each subject, with one representing each of the ability dimensions.

Although the difference between the average component scores of the probands and controls was significant ($p < .05$) for each of the three measures, the difference for reading (about 1.8 standard deviations) was substantially larger than that for symbol-processing speed (0.28) and spatial reasoning (0.33). Significant gender differences were in the expected direction, with boys obtaining higher average spatial scores (0.58 standard deviations) and girls having higher average scores for symbol-processing speed (0.64).

The difference between siblings of probands and siblings of controls was significant for both reading (0.50 standard deviations) and symbol-processing speed (0.29). Moreover, there was a significant interaction between family type and gender for reading, with the difference between brothers of probands and brothers of controls (0.79 standard deviations) exceeding that for sisters (0.21).

The pattern of main effects for the parental data was highly similar to that of the siblings. Again, the largest difference between parents of probands and parents of controls was for the reading measure (0.53 standard deviations). Fathers of probands also tended to be somewhat more affected (0.64) than mothers (0.41); however, this difference was not large enough to yield a significant interaction between family type and gender for the parental data.

The reading performance deficits of siblings and parents of probands in the Colorado Family Reading Study conclusively demonstrated the familial nature of reading disability. Thus, these family data were subsequently subjected to various genetic analyses. For example, as predicted by the gender-influenced polygenic threshold model, relatives of female probands were found to be at greater risk for reading problems than were relatives of male probands (DeFries & Decker, 1982). Several different tests for major-gene influence were also undertaken but with mixed results. For example, as predicted by a major-gene hypothesis, variances of reading performance scores obtained by relatives of probands were significantly greater than those for relatives of controls. Because genetic variance is a function of gene frequency, a rare major gene may not contribute importantly to the observed variance. However, relatives of affected individuals will have the gene in higher frequency and, thus, should manifest greater variance. In accordance with this expectation, the variances of both siblings and parents of probands in the Colorado Family Reading Study were larger than those of controls for a composite measure of reading performance (DeFries & Decker, 1982). However, little or no evidence was obtained for X linkage. Similarly, segregation analysis of reading performance data from families of male probands provided no evidence for autosomal major-gene influence (Lewitter, DeFries, & Elston, 1980). In contrast, when data from families of female probands were subjected to the same analyses, a hypothesis of single-gene, recessive inheritance could not be rejected. Based on a comparison of various test statistics, Lewitter et al. (1980) concluded that the failure to reject the recessive-allele hypothesis was not due to the smaller number of female probands in the study. Thus, results of analyses of data from the Colorado Family Reading Study also suggested that reading disability is genetically heterogeneous.

Subsequently, Pennington et al. (1991) reanalyzed data from the Colorado Family Reading Study, as well as family data from three independently ascertained samples (23 families included in a previous linkage study, 9 families from Washington State, and 39 families who were tested in Iowa). Unlike the Lewitter et al. (1980) study, Pennington et al. (1991) analyzed dichotomous data. Family members from the Colorado Family Reading Study and the linkage study were classified as reading disabled if they had either a history of reading problems or a significant discrepancy between ability and achievement. This definition was used to diagnose compensated adults with a positive history but normal reading performance. These data were subjected to complex segregation analysis, using a computer program (POINTER) that can test for a major-gene effect in the presence of a multifactorial background (the "mixed model"). In general, results of this analysis suggested the presence of major-gene (additive or dominant) transmission with gender-specific penetrances in three of the samples but polygenic transmission in the Iowa sample. Based on these results, as well as those of linkage analyses summarized later in this chapter, Pennington et al. concluded that major-gene transmission occurs in a significant proportion of families of reading-disabled children and that the condition is almost certainly etiologically heterogeneous.

Previous Twin Studies

Results of previous twin studies also suggest that reading disability is due at least in part to heritable influences. Zerbin-Rüdin (1967) reviewed six previously published case studies, a Danish twin study, and six pairs of twins who were included in Hallgren's (1950) family study. The probandwise concordance rates for the 17 monozygotic (MZ) and 34 dizygotic (DZ) twin pairs included in this combined sample were 100% and 52%, respectively. Because case studies of concordant pairs are more likely to be reported than are those of discordant pairs (Harris, 1986), these concordance rates are probably inflated at least to some extent.

Bakwin (1973) ascertained a sample of 338 same-sex twin pairs through mothers-of-twins clubs and obtained reading history information via parental interviews, telephone calls, and mail questionnaires. Defining reading disability as "a reading level below the expectation derived from the child's performance in other school subjects" (p. 184), Bakwin (1973) identified 31 pairs of MZ twins and 31 pairs of DZ twins in which at least one member was reading disabled. The probandwise concordance rates for these MZ and DZ twin pairs were 91% and 45%, respectively.

More recently, Stevenson, Graham, Fredman, and McLoughlin (1984, 1987) reported results from the first study of reading disability in which twins were administered standardized tests of intelligence, reading, and spelling. The Schonell Graded Word Reading and Spelling Tests and the Neale Analysis of Reading Ability were used to diagnose reading or spelling "backwardness" or "retardation" in a sample of 285 pairs of 13-year-old twins. *Reading backwardness* was defined as reading age 18 months below chronological age, whereas *reading retardation* was identified by marked underachievement in relation to that predicted from IQ and chronological age. Using their various diagnostic criteria, the probandwise concordance rates for MZ twin pairs (33–59%) were only slightly higher than those for DZ twin pairs (29–54%). Thus, substantial variation exists among the results of previous twin studies of reading disability.

Colorado Twin Study of Reading Disability

Because of the paucity of previous twin studies of reading disability, a twin study was initiated at the University of Colorado (Boulder) in 1982 (Decker & Vandenberg, 1985; DeFries, 1985). In this ongoing study, an extensive psychometric test battery that includes the WISC-R (Wechsler, 1974) or the Wechsler Adult Intelligence Scale–Revised (Wechsler, 1981) and the Peabody Individual Achievement Test (PIAT) (Dunn & Markwardt, 1970) is being administered to MZ and DZ twin pairs in which at least one member of each pair manifested a positive school history of reading problems and to a comparison group of twins with a negative school history.

To minimize the possibility of ascertainment bias, the sample of twins is being systematically obtained via cooperating school districts in Colorado. Without regard to reading status, all twin pairs in a school are identified. Parental permission is then sought to review the school records of each twin for evidence of reading problems (e.g., low reading achievement test scores, referral to a reading therapist because of poor reading performance). Twin pairs in which at least one member has a positive school history of reading problems are invited to be tested in laboratories at the University of Colorado and Denver University. Data from the PIAT Reading Recognition, Reading Comprehension, and Spelling subtests are then used to compute a discriminant function score for each member of the pair. (Discriminant weights were estimated from an analysis of data from an independent sample of 140 reading-disabled and 140 control nontwin children.) Twin pairs are included in the proband sample if at least one member of the pair with a positive school history of reading problems is also classified as affected by the discriminant score and has a Verbal or Performance IQ of at least 90; no diagnosed neurological, emotional, or behavioral problems; and no uncorrected visual or auditory acuity deficits. Control twin pairs with a negative school history for reading problems are matched to probands on the basis of age (within 6 months), gender, and school district.

Selected items from the Nichols and Bilbro (1966) questionnaire are used to determine twin zygosity. Zygosity is confirmed in doubtful cases by analyzing blood samples. All twin pairs were reared in English-speaking, middle-class homes, and ranged in age from 8 to 20 years at the time of testing.

As of June 30, 1992, 133 pairs of MZ twins and 98 pairs of same-sex DZ twins met our criteria for inclusion in the proband sample. In contrast to previous studies that used referred samples of reading-disabled children, the gender ratio in our 324 twin probands was 159 males to 165 females (i.e., 0.96:1). Female MZ twin pairs are often overrepresented in twin studies (Lykken, Tellegen, & DeRubeis, 1978); thus, this unexpectedly low gender ratio may be due at least in part to a differential volunteer rate of male and female MZ twin pairs. In accordance with this expectation, the gender ratio in the sample of MZ probands (0.78:1) is lower than that for the same-sex DZ probands (1.36:1). Therefore, males may be at a slightly higher risk than females for reading disability. However, both ratios are substantially lower than the ratio of 3 or 4 males to each female that is typically found in referred samples. Shaywitz, Shaywitz, Fletcher, and Escobar (1990) also recently observed that the gender ratio in a research-identified sample of nontwin, reading-disabled children was substantially lower than that in a referred sample.

The probandwise concordance rates for the MZ and same-sex DZ twin pairs are 66% and 43%, respectively. This difference is significant ($p < .001$) and, thus, confirms the evidence obtained from previous twin studies that reading disability is caused in part by heritable influences.

Multiple Regression Analysis of Twin Data

For dichotomous variables (e.g., presence or absence of a psychiatric illness), a comparison of concordance rates in MZ and DZ twin pairs provides a test for a genetic etiology; however, reading-disabled probands are ascertained because of deviant scores on continuous measures. For such characters, a multiple regression analysis of twin data facilitates an alternative test of genetic etiology, as well as an analysis of individual differences within the proband sample (DeFries & Fulker, 1985).

When MZ and DZ probands have been ascertained because of deviant scores, their cotwins are expected to regress toward the mean of the unselected population. However, to the extent that the condition is due to heritable influences, this regression to the mean should differ for MZ and DZ cotwins (see Figure 1). MZ twins are genetically identical, whereas DZ twins share only about one-half of their segregating genes on average; thus, if the deviant scores of the probands are due at least in part to genetic factors, then scores of DZ cotwins should regress more than those of MZ cotwins. Consequently, if the MZ and DZ proband means are approximately equal, then a simple t test of the difference between the means of the MZ and DZ cotwins could be used as a test of genetic etiology. However, a multiple regression analysis of such twin data provides a more general, statistically powerful, and flexible test (DeFries & Fulker, 1985, 1988).

DeFries and Fulker (1985) formulated two regression models for the analysis of selected twin data: a basic model in which the partial regression of a cotwin's score on the coefficient of relationship provides a test for genetic etiology and an augmented model that yields direct measures of the extent to which individual differences within the selected group are due to heritable and shared environmental influences. These two regression equations are as follows:

$$C = B_1P + B_2R + A \text{ and} \quad (1)$$

$$C = B_3P + B_4R + B_5PR + A, \quad (2)$$

where C symbolizes the cotwin's score, P is the proband's score, R is the coefficient of relationship ($R = 1.0$ for MZ twins and 0.5 for DZ twins), PR is the product of proband's score and relationship, and A is the regression constant.

When the basic model is fitted to selected twin data, the partial regression of cotwin's score on proband's score (i.e., B_1) provides a measure of average MZ and DZ twin resemblance. Of greater interest, B_2 estimates twice the difference between the means of the MZ and DZ

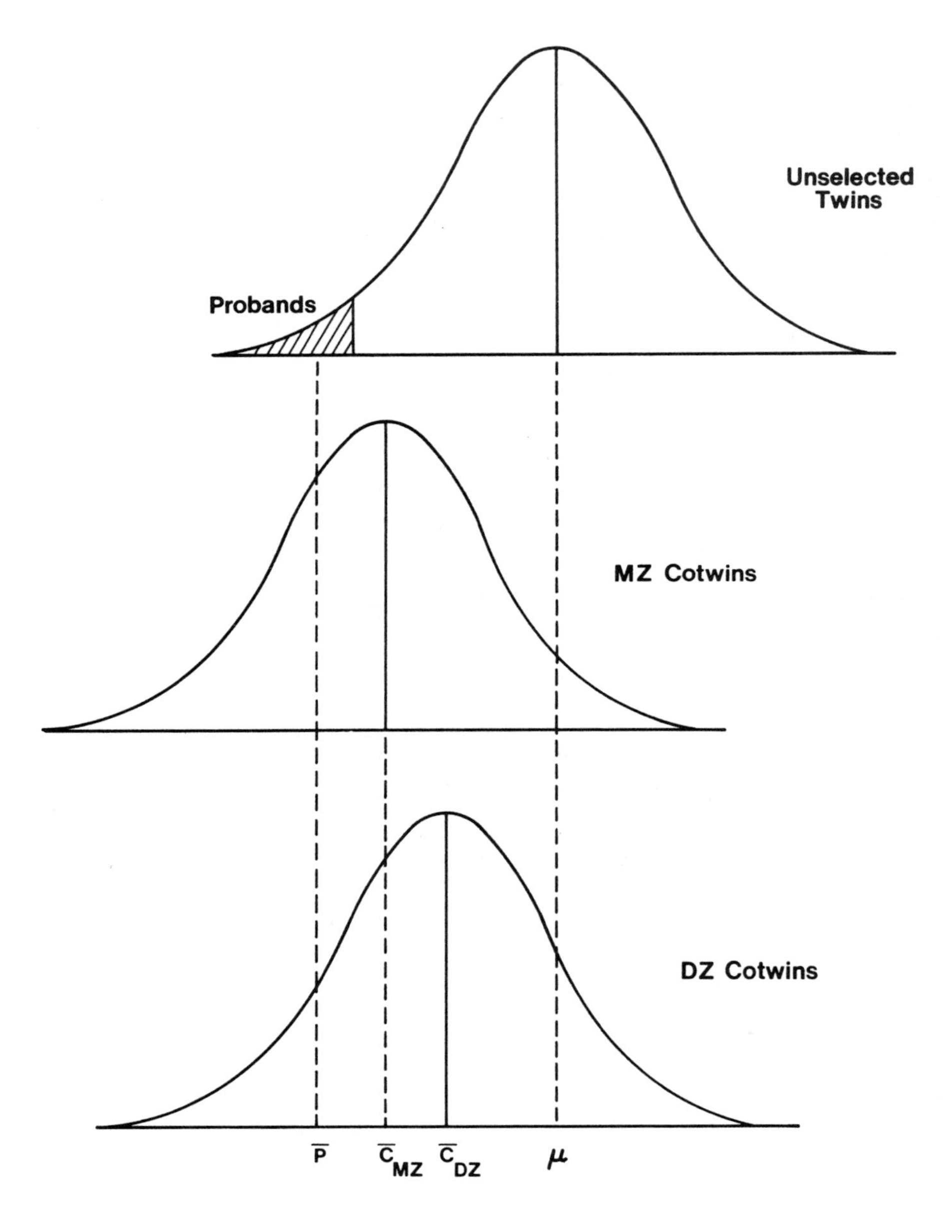

FIGURE 1. Hypothetical distributions for reading performance of an unselected sample of twins and of the identical (MZ) and fraternal (DZ) cotwins of probands with a reading disability. The differential regression of the MZ and DZ cotwin means toward the mean of the unselected population (μ) provides a test of genetic etiology. From "Evidence for a Genetic Aetiology in Reading Disability of Twins" by J. C. DeFries, D. W. Fulker, and M. C. LaBuda, 1987, *Nature, 329*, pp. 537–539; copyright 1987 by Macmillan Journals Ltd.; reprinted by permission.

cotwins after covariance adjustment for any difference between the scores of the MZ and DZ probands. Thus, B_2 provides a more general and statistically powerful test for genetic etiology than does a comparison of MZ and DZ twin concordance rates (DeFries & Fulker, 1988). When Equation 2 is fitted to twin data, B_3 estimates the proportion of variance due to environmental influences that are shared by members of twin pairs (c^2), and B_5 estimates heritability (h^2), a measure of the extent to which individual differences within the proband sample are due to heritable influences.

B_2 is a function of h_g^2, an index of the extent to which the deficit of probands is due to genetic factors (DeFries & Fulker, 1985, 1988). Therefore, a comparison of h_g^2 and h^2 can be used to test the hypothesis that the etiology of deviant scores differs from that of individual differences within the normal range of variation. Reading disability, for example, could be due to a major gene effect or environmental insult, whereas individual differences in reading performance may be due to multifactorial influences. Therefore, if the etiology of reading disability differs from that of individual differences within the normal range, then h_g^2 and h^2 should differ in magnitude. On the other hand, if individuals with reading disability merely represent the lower tail of a normal distribution of individual differences in reading performance (Shaywitz, Escobar, Shaywitz, Fletcher, & Makuch, 1992), then h_g^2 and h^2 should be similar in magnitude.

DeFries and Fulker (1988) showed that a simple transformation of twin data (each score is expressed as a deviation from the mean of the unselected population and then divided by the difference between the proband and control means) before regression analysis facilitates a direct test of the hypothesis that the etiology of extreme scores differs from that of individual differences within the normal range. When selected twin data have been transformed in this manner, $B_2 = h_g^2$ and $B_4 = h_g^2 - h^2$.

To illustrate this multiple regression analysis of selected twin data, Equations 1 and 2 were fitted to discriminant function score data from the probands and cotwins who were tested in the Colorado Twin Study of Reading Disability. Because truncate selection (Thompson & Thompson, 1986) was used to ascertain the probands, concordant pairs were double entered in a manner analogous to that used to calculate probandwise concordance rates (DeFries & Gillis, 1991). Therefore, the resultant computer-generated estimates of standard errors and tests of significance for the various regression coefficients were adjusted accordingly.

The average scores of the MZ and DZ probands and their cotwins, expressed as standardized deviation units from the mean of 442 individuals included in the control sample, are presented in Table 1. From this table it may be seen that the MZ and DZ proband means are highly similar, over 2.5 standard deviations below the mean of the control twins. In addition, it may be seen that the MZ cotwins have regressed only 0.24 standard deviation units toward the control mean, whereas the DZ cotwins have regressed 0.87 standard deviations. When Equation 1 was fitted to these data, $B_2 = -1.34 \pm 0.28$ ($p < .001$, one tailed). When transformed data were fitted to the same model, $B_2 = h_g^2 = 0.50 \pm 0.11$ ($p < .001$); thus, about half of the reading performance deficit of probands, on average, is due to heritable influences.

When Equation 2 was fitted to the transformed discriminant function data, $B_5 = h^2 = 0.73 \pm 0.35$ ($p < .05$) and $B_3 = c^2 = 0.11 \pm 0.29$ ($p = .71$). Thus, these results suggest that individual differences among probands are highly heritable and that shared environmental influences do not contribute importantly to twin resemblance. It is also important to note that the estimate of h_g^2 (0.50) is somewhat lower than that of h^2 (0.73), which suggests that the etiology of reading disability may differ from that of individual differences within the normal range of reading performance. However, $B_4 = h_g^2 - h^2 = -0.23 \pm 0.37$, which is not significant ($p > .50$). Thus, a larger sample of twin pairs will be required to test this hypothesis more rigorously.

TABLE 1

Mean Discriminant Scores of 133 Pairs of Identical Twins and 98 Pairs of Fraternal Twins in Which at Least One Member of the Pair Is Reading-Disabled[a]

Twin	Proband	Cotwin
Identical	-2.71 ± 0.78	-2.47 ± 1.03
Fraternal	-2.59 ± 0.80	-1.72 ± 1.35

[a]Expressed as standardized deviations from control mean.

Differential Genetic Etiology in Males and Females

The multiple regression analysis of twin data is a highly flexible methodology. In addition to testing hypotheses about the genetic etiology of group deficits and individual differences, it can also be used to assess differential genetic etiology as a function of various covariates such as age (Wadsworth, Gillis, DeFries, & Fulker, 1989), gender (DeFries, Gillis, & Wadsworth, in press), IQ (DeFries & Gillis, in press), and subtype (Olson, Rack, Conners, DeFries, & Fulker, 1991). This methodology is illustrated by fitting an extended model that tests for differential genetic etiology as a function of gender to the current data set.

In a commentary for the proceedings of a conference on "Sex Differences in Dyslexia," Geschwind (1981) suggested that girls may be less affected than boys "by certain environmental influences, such as quality of teaching, social class differences, or outside pressures within society" (p. xiv). Thus, the etiology of reading disability in females may differ from that in males.

Obviously, the basic regression model (Equation 1) can be fit separately to transformed discriminant function data from male and female twin pairs. However, the hypothesis of a differential genetic etiology as a function of gender can be tested directly by fitting the following extended model to the reading performance data from male and female twin pairs simultaneously:

$$C = B_6P + B_7R + B_8S + B_9PS + B_{10}RS + A, \tag{3}$$

where S symbolizes the proband's gender (a dummy variable coded 0.5 and -0.5 for males and females, respectively). The coefficients of the PS and RS interaction terms (B_9 and B_{10}) test for differential twin resemblance and differential h_g^2 as a function of gender.

Estimates of h_g^2 obtained by fitting Equation 1 separately to transformed discriminant function data from the male and female twin pairs included in the present sample are 0.45 ± 0.15 and 0.62 ± 0.14, respectively. Although the estimate of h_g^2 for males is somewhat smaller than that for females, the difference between these two estimates is not significant. When Equation 3 was fitted to the transformed discriminant function score data of male and female twin pairs simultaneously, $B_{10} = 0.17 \pm 0.21$ ($p = .40$).

Genetic Linkage Analyses

Finding evidence that a complex character such as reading disability is due at least in part to heritable influences is only the first step toward a comprehensive genetic analysis. If a character is influenced by one or more genes with discernible effects, it may be possible to determine their location on specific chromosomes. Molecular biological techniques could then be used to clone the genes, determine what proteins they produce, and thereby discover the fundamental basis of the character's etiology and development (Kidd, 1991).

Genes that are located near each other on the same chromosome tend to be transmitted together, whereas those that are far apart on the same chromosome, or on entirely different chromosomes, are inherited independently. Therefore, a gene can be localized to a specific chromosomal region by comparing its transmission pattern to those of marker genes whose locations are already known. Observed cotransmission between a putative gene for a character and a marker gene provides evidence that the gene is located near the marker.

Evidence for linkage between a gene of interest and a marker is quantified by computing a *LOD* score. LOD is an acronym for the logarithm to the base 10 of the odds, where the "odds" is the conditional probability of observing cotransmission between the character and a marker, given linkage, divided by the probability of cotransmission, given independent assortment. For example, if the probability of observing cotransmission given linkage is equal to that given independent assortment, then the odds ratio is one, and the logarithm of this ratio is zero. Thus, a LOD score of zero indicates that the two events are equally likely; that is, it provides no evidence either for or against linkage. In contrast, a LOD score of three indicates that the probability of observing cotransmission given linkage is 1,000 times greater than that given independent inheritance. However, a LOD score of three is not equivalent to a $p = .001$ significance test for linkage. Because the prior probability of linkage to a given chromosomal marker is low, the posterior odds in favor of linkage given a LOD score of three is only about 20:1 (Risch, 1992). Of course, a negative LOD score provides evidence against linkage. LOD scores are computed for individual families, assuming various possible linkage relationships, and then summed across families.

Smith, Kimberling, Pennington, and Lubs (1983) reported the first evidence that a hypothesized gene that causes reading disability may be linked to a specific chromosome. Psychometric test data were obtained from 84 individuals who were members of nine extended families in which reading disability had apparently been transmitted in an autosomal dominant manner. For children, diagnostic criteria for reading disability included a reading level at least 2 years below expected grade level and a Full-Scale IQ greater than 90. Because of the possibility of compensation in adults, self-reports were used for diagnosis if there was a discrepancy between test results and a history of reading problems. LOD scores were computed to assess the possibility of linkage between a putative gene for reading disability and various genotype and chromosomal markers. When the LOD scores representing each of the various linkage relationships were summed across the nine families, a total LOD score of 3.2 was obtained for a marker on Chromosome 15. However, about 70% of this total LOD score was due to cotransmission in only one family, and data from another family yielded a large negative score. Thus, results of this first genetic linkage analysis of reading disability also suggested that the condition is genetically heterogeneous.

Smith, Pennington, Kimberling, and Ing (1990) subsequently increased their sample to include 250 individuals in 21 families. In addition to the genotype and chromosomal markers used in their previous analysis, DNA markers were also used in this study. With the inclusion of the additional families, the total LOD score for linkage to Chromosome 15 was reduced to 1.33. However, as shown in Figure 2, the LOD scores for individual families varied greatly.

Results of recent analyses by Smith and her colleagues (see DeFries, Olson, Pennington, & Smith, 1991) suggest that only about 20% of the families of children with reading disabilities manifest apparent linkage to

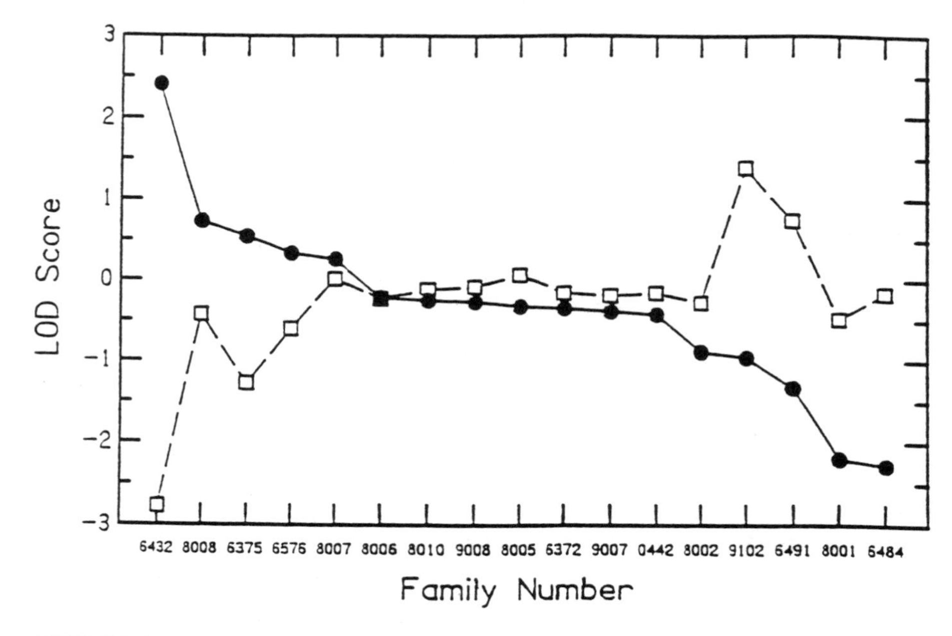

FIGURE 2. LOD scores for markers on Chromosomes 15 (filled circles) and 6 (open squares) in families of children with reading disabilities. Families are listed in descending order of their LOD scores for Chromosome 15. From "Colorado Reading Project: An Update" by J. C. DeFries, R. K. Olson, B. F. Pennington, and S. D. Smith, 1991, p. 81, in *The Reading Brain: The Biological Basis of Dyslexia,* edited by D. D. Duane and D. B. Gray, Parkton, MD: York Press; copyright 1991 by York Press; reprinted by permission.

Chromosome 15. Therefore, most cases of heritable reading disability must be caused by genes at other chromosomal locations. As also shown in Figure 2, results of a preliminary analysis indicate that a few families manifest some apparent linkage to Chromosome 6 but not to Chromosome 15. The markers for Chromosome 6 are within the human leucocyte antigen region; thus, this very tentative evidence for linkage to Chromosome 6 suggests a possible relationship between reading disability and genes that affect the immune system in a subgroup of reading-disabled children.

Linkage analyses of complex characters involving extended family pedigrees have serious limitations. For example, it is assumed that the condition is caused by a single gene with a specified mode of inheritance (e.g., autosomal dominance). However, it is likely that characters such as reading disability are influenced by genes at several chromosomal locations. Also, because each family member must be diagnosed, accurate diagnosis is required across a wide age range, as well as across generations.

Evidence for linkage may also be obtained by analyzing data from sibling pairs (Haseman & Elston, 1972). This alternative method does not assume a specific mode of inheritance, and data from only one generation are required. However, because the sibling-pair analysis is less powerful than the family study method, data may be required from more individuals to achieve statistical significance.

If a gene that influences a character is closely linked to a chromosomal marker, then pairs of siblings that are concordant for the marker should be more similar for the character than siblings that are discordant. To determine whether the chromosomal markers carried by two siblings are "identical by descent" (IBD), the parents of the siblings must be genotyped. The resulting data are then used to estimate the proportion (π) of the alleles that are IBD at the marker locus for each sibling pair. The sibling-pair linkage analysis of Haseman and Elston (1972) relates the square of the difference between the scores of siblings to their π value at each marker locus. Smith, Kimberling, and Pennington (1991) recently used this method to reanalyze data from siblings included in their extended family study, and again, they detected possible linkages to markers on both Chromosomes 6 and 15.

To detect the chromosomal location of quantitative trait loci (Gelderman, 1975; Lander & Botstein, 1989), Fulker et al. (1991) recently advocated an alternative analysis of sibling-pair data that uses a simple extension of the DeFries and Fulker (1985) multiple regression analysis of twin data. However, rather than using the coefficient of relationship (R) in the multiple regression models, π is used to index IBD status at the various marker loci. Fulker et al. (1991) asserted that this method has several advantages over the Haseman and Elston (1972) approach: (a) It is conceptually simple; (b) it can be applied using readily available computer programs; (c) it is applicable for the analysis of data from either selected or unselected samples; (d) it is highly flexible, thereby facilitating the analysis of interactions with variables such as age and gender; and (e) it is statistically powerful, especially when applied to data from selected samples.

Fulker et al. (1991) used this multiple regression analysis to reanalyze discriminant reading score data from sibling pairs included in the Smith et al. (1991) extended-family study. Several possible linkages to DNA markers on Chromosome 15 were again indicated. The results of this study also clearly demonstrated the power of the multiple regression method to detect linkage when probands have been selected because of deviant scores. Because of the potential of this new methodology to detect the chromosomal location of quantitative trait loci, we have recently initiated a linkage analysis of data from fraternal twin pairs tested in the Colorado Reading Project.

Concluding Remarks

The increasing use of the methods of molecular genetics will almost certainly revolutionize the field of behavioral genetics during the next few decades (Aldhous, 1992; Johnson, 1990). In 1980, Botstein, White, Skolnick, and Davis proposed that a map of the human genome could be constructed using restriction fragment-length polymorphisms (RFLPs). A few years later, the gene that causes Huntington's disease was localized to Chromosome 4 using RFLP markers (Gusella et al., 1983). Using the methods of molecular genetics, the gene for cystic fibrosis was recently cloned (Rommens et al., 1989). One of the goals of the human genome project (Watson, 1990) is to facilitate the localization of disease-causing genes by constructing a human genetic map with very closely spaced molecular markers.

The development of a high-density map of the human genome will also facilitate the detection of quantitative trait loci (Johnson, DeFries, & Markel, 1992). In addition to RFLPs, a new class of molecular markers using polymorphisms in microsatellite sequences has recently been developed for genetic mapping. Because such microsatellites are very polymorphic and widely dispersed throughout the genome, they will serve as excellent DNA markers for linkage analyses. Use of the polymerase chain reaction, which requires only very small amounts of DNA, will also greatly facilitate future gene mapping efforts.

Although the methodology of molecular genetics will almost certainly contribute substantially to advances in the genetic analysis of complex behavioral characters, quantitative genetic analyses of such characters will continue to be important. When results from both molecular and quantitative genetic approaches are obtained, the relative importance of quantitative trait loci can be measured by assessing their contribution to the genetic variance and covariance due to all genetic influences. Thus, single-gene and quantitative genetic analyses

> are complementary, not mutually exclusive; both should be exploited in order to attain a more complete understanding of the genetic causes of individual differences in behavior. (DeFries & Hegmann, 1970, p. 53)

References

Aldhous, P. (1992). The promise and pitfalls of molecular genetics. *Science, 257*, 164–165.

Bakwin, H. (1973). Reading disability in twins. *Developmental Medicine and Child Neurology, 15*, 184–187.

Botstein, D., White, R. L., Skolnick, M., & Davis, R. W. (1980). Construction of a genetic map in man using restriction fragment length polymorphisms. *American Journal of Human Genetics, 32*, 314–331.

Critchley, M. (1970). *The dyslexic child.* New York: Heinemann Medical Books.

Decker, S. N., & DeFries, J. C. (1980). Cognitive abilities in families with reading disabled children. *Journal of Learning Disabilities, 13*, 517–522.

Decker, S. N., & Vandenberg, S. G. (1985). Colorado Twin Study of Reading Disability. In D. B. Gray & J. F. Kavanagh (Eds.), *Biobehavioral measures of dyslexia* (pp. 123–135). Parkton, MD: York Press.

DeFries, J. C. (1985). Colorado Reading Project. In D. B. Gray & J. F. Kavanagh (Eds.), *Biobehavioral measures of dyslexia* (pp. 107–122). Parkton, MD: York Press.

DeFries, J. C., & Decker, S. N. (1982). Genetic aspects of reading disability: A family study. In R. N. Malatesha & P. G. Aaron (Eds.), *Reading disorders: Varieties and treatments* (pp. 255–279). New York: Academic Press.

DeFries, J. C., & Fulker, D. W. (1985). Multiple regression analysis of twin data. *Behavior Genetics, 15*, 467–473.

DeFries, J. C., & Fulker, D. W. (1988). Multiple regression analysis of twin data: Etiology of deviant scores versus individual differences. *Acta Geneticae Medicae et Gemellolgiae, 37*, 205–216.

DeFries, J. C., Fulker, D. W., & LaBuda, M. C. (1987). Evidence for a genetic aetiology in reading disability of twins. *Nature, 329*, 537–539.

DeFries, J. C., & Gillis, J. J. (1991). Etiology of reading deficits in learning disabilities: Quantitative genetic analysis. In J. E. Obrzut & G. W. Hynd (Eds.), *Neuropsychological foundations of learning disabilities: A handbook of issues, methods and practice* (pp. 29–47). Orlando, FL: Academic Press.

DeFries, J. C., & Gillis, J. J. (in press). Reading disability in twins. In J. H. Beitchman (Ed.), *Language, learning and behavior*. Baltimore: Williams & Wilkins.

DeFries, J. C., Gillis, J. J., & Wadsworth, S. J. (in press). Genes and genders: A twin study of reading disability. In A. M. Galaburda (Ed.), *The extraordinary brain: Neurobiologic issues in developmental dyslexia*. Cambridge, MA: Harvard University Press.

DeFries, J. C., & Hegmann, J. P. (1970). Genetic analysis of open-field behavior. In G. Lindzey & D. C. Thiessen (Eds.), *Contributions to behavior–genetic analysis: The mouse as a prototype* (pp. 23–56). New York: Appleton-Century-Crofts.

DeFries, J. C., Olson, R. K., Pennington, B. F., & Smith, S. D. (1991). Colorado Reading Project: An update. In D. D. Duane & D. B. Gray (Eds.), *The reading brain: The biological basis of dyslexia* (pp. 53–87). Parkton, MD: York Press.

DeFries, J. C., Singer, S. M., Foch, T. T., & Lewitter, F. I. (1978). Familial nature of reading disability. *British Journal of Psychiatry, 132*, 361–367.

DeFries, J. C., Vogler, G. P., & LaBuda, M. C. (1986). Colorado Family Reading Study: An overview. In J. L. Fuller & E. C. Simmel (Eds.), *Perspectives in behavior genetics* (pp. 29–56). Hillsdale, NJ: Earlbaum.

Dunn, L. M., & Markwardt, F. C. (1970). *Examiner's manual: Peabody Individual Achievement Test*. Circle Pines, MN: American Guidance Service.

Eustis, R. (1947). Specific reading disability. *New England Journal of Medicine, 237*, 243–249.

Finucci, J. M. (1978). Genetic considerations in dyslexia. In H. R. Myklebust (Ed.), *Progress in learning disabilities* (Vol. 4, pp. 41–63). New York: Grune & Stratton.

Finucci, J. M., & Childs, B. (1983). Dyslexia: Family studies. In C. A. Ludlow & J. A. Cooper (Eds.), *Genetic aspects of speech and language disorders* (pp. 157–167). New York: Academic Press.

Finucci, J. M., Guthrie, J. T., Childs, A. L., Abbey, H., & Childs, B. (1976). The genetics of specific reading disability. *Annals of Human Genetics, 40*, 1–23.

Fulker, D. W., Cardon, L. R., DeFries, J. C., Kimberling, W. J., Pennington, B. F., & Smith, S. D. (1991). Multiple regression analysis of sib-pair data on reading to detect quantitative trait loci. *Reading and Writing: An Interdisciplinary Journal, 3*, 299–313.

Gelderman, H. (1975). Investigations on inheritance of quantitative characters in animals by gene markers: I. Methods. *Theoretical and Applied Genetics, 46*, 319–330.

Geschwind, N. (1981). A reaction to the conference on sex differences and dyslexia. In A. Ansara, N. Geschwind, A. Galaburda, M. Albert, & N. Gartrell (Eds.), *Sex differences in dyslexia* (pp. xiii–xviii). Baltimore: Orton Dyslexia Society.

Gilger, J. W., Pennington, B. F., & DeFries, J. C. (1991). Risk for reading disability as a function of parental history in three family studies. *Reading and Writing: An Interdisciplinary Journal, 3*, 205–217.

Gusella, J. F., Wexler, N. S., Conneally, P. M., Naylor, S. L., Anderson, M. A., Tanzi, R. E., Watkins, P. C., Ottina, K., Wallace, M. R., Sakaguchi, A. Y., Young, A. B., Shoulson, I., Bonilla, E., & Martin, J. B. (1983). A polymorphic DNA marker genetically linked in Huntington's disease. *Nature, 306*, 234–238.

Hallgren, B. (1950). Specific dyslexia: A clinical and genetic study. *Acta Psychiatrica et Neurologica Scandinavica, 65*(Suppl.), 1–287.

Harris, E. L. (1986). The contribution of twin research to the study of the etiology of reading disability. In S. D. Smith (Ed.), *Genetics and learning disabilities* (pp. 3–19). San Diego, CA: College-Hill Press.

Haseman, J. K., & Elston, R. C. (1972). The investigation of linkage between a quantitative trait and a marker locus. *Behavior Genetics, 2*, 3–19.

Institute for Personality and Ability Testing. (1973). *Measuring intelligence with culture fair tests: Manual for Scales 2 and 3.* Champaign, IL: Author.

Johnson, D. (1990). Can psychology ever be the same after the human genome is mapped? *Psychological Science, 1*, 331–332.

Johnson, T. E., DeFries, J. C., & Markel, P. D. (1992). Mapping quantitative trait loci for behavioral traits in the mouse. *Behavior Genetics, 22*, 635–653.

Kidd, K. K. (1991). Trials and tribulations in the search for genes causing neuropsychiatric disorders. *Social Biology, 38*, 163–178.

Kirk, S. A. (1962). *Educating exceptional children.* Boston: Houghton Mifflin.

Kirk, S. A., & Bateman, B. (1962). Diagnosis and remediation of learning disabilities. *Exceptional Children, 29*, 73–78.

Lander, E. S., & Botstein, D. (1989). Mapping Mendelian factors underlying quantitative traits using RFLP linkage maps. *Genetics, 121*, 185–199.

Lerner, J. W. (1989). Educational interventions in learning disabilities. *Journal of the American Academy of Child and Adolescent Psychiatry, 28*, 326–331.

Lewitter, F. I., DeFries, J. C., & Elston, R. C. (1980). Genetic models of reading disability. *Behavior Genetics, 10*, 9–30.

Lykken, D. T., Tellegen, A., & DeRubeis, R. (1978). Volunteer bias in twin research: The rule of two-thirds. *Social Biology, 25*, 1–9.

Morgan, W. P. (1896). A case of congenital word-blindness. *British Medical Journal, 11*, 1378.

Nichols, R. C., & Bilbro, W. C. (1966). The diagnosis of twin zygosity. *Acta Genetica et Statistica Medica, 16*, 265–275.

Nie, N. H., Hull, C. H., Jenkins, J. G., Steinbrenner, K., & Bent, D. H. (1975). *Statistical package for the social sciences* (2nd ed.). New York: McGraw-Hill.

Olson, R. K., Rack, J. P., Conners, F. A., DeFries, J. C., & Fulker, D. W. (1991). Genetic etiology of individual differences in reading disability. In L. V. Feagans, E. J. Shart, & L. J. Meltzer (Eds.), *Subtypes of learning disability*. Hillsdale, NJ: Erlbaum.

Pennington, B. F., Gilger, J. W., Pauls, D., Smith, S. A., Smith, S. D., & DeFries, J. C. (1991). Evidence for major gene transmission of developmental dyslexia. *Journal of the American Medicial Association, 266*, 1527–1534.

Plomin, R., DeFries, J. C., & McClearn, G. E. (1990). *Behavioral genetics: A primer* (2nd ed.). New York: Freeman.

Risch, N. (1992). Genetic linkage: Interpreting LOD scores. *Science, 255*, 803–804.

Rommens, J. M., Ianuzzi, M. C., Kerem, B.-S., Drumm, M. L., Melmer, G., Dean, M., Rozmahel, R., Cole, J. L., Kennedy, D., Hidaka, N., Zsiga, M., Buchwald, M., Riordan, J. R., Tsui, L.-C., & Collins, F. S. (1989). Identification of the cystic fibrosis gene: Chromosome walking and jumping. *Science, 245*, 1059–1065.

Shaywitz, S. E., Escobar, M. D., Shaywitz, B. A., Fletcher, J. M., & Makuch, R. (1992). Evidence that dyslexia may represent the lower tail of a normal distribution of reading ability. *New England Journal of Medicine, 326*, 145–150.

Shaywitz, S. E., Shaywitz, B. A., Fletcher, J. M., & Escobar, M. D. (1990). Prevalence of reading disability in boys and girls. *Journal of the American Medical Association, 264*, 998–1002.

Smith, S. D., Kimberling, W. J., & Pennington, B. F. (1991). Screening for multiple genes influencing dyslexia. *Reading and Writing: An Interdisciplinary Journal, 3*, 285–298.

Smith, S. D., Kimberling, W. J., Pennington, B. F., & Lubs, H. A. (1983). Specific reading disability: Identification of an inherited form through linkage analysis. *Science, 219*, 1345–1347.

Smith, S. D., Pennington, B. F., Kimberling, W. J., & Ing, P. S. (1990). Familial dyslexia: Use of genetic linkage data to define subtypes. *Journal of the American Academy of Child and Adolescent Psychiatry, 29*, 204–213.

Stevenson, J., Graham, P., Fredman, G., & McLoughlin, V. (1984). The genetics of reading disability. In C. J. Turner & H. B. Miles (Eds.), *The biology of human intelligence* (pp. 85–97). Nafferton, UK: Nafferton Books.

Stevenson, J., Graham, P., Fredman, G., & McLoughlin, V. (1987). A twin study of genetic influences on reading and spelling ability and disability. *Journal of Child Psychology and Psychiatry, 28*, 229–247.

Symmes, J. S., & Rapoport, J. L. (1972). Unexpected reading failure. *American Journal of Orthopsychiatry, 42*, 82–91.

Thomas, C. J. (1905). Congenital "word-blindness" and its treatment. *Ophthalmoscope, 3*, 380–385.

Thompson, J. S., & Thompson, M. W. (1986). *Genetics in medicine*. Philadelphia: W. B. Saunders.

Thurstone, T. G. (1962). *Examiner's manual: Primary Mental Abilities*. Chicago: Science Research Associates.

Wadsworth, S. J., Gillis, J. J., DeFries, J. C., & Fulker, D. W. (1989). Differential genetic aetiology of reading disability as a function of age. *Irish Journal of Psychology, 10*, 509–520.

Watson, J. D. (1990). The Human Genome Project: Past, present, and future. *Science, 248*, 44–49.

Wechsler, D. (1974). *Manual for the Wechsler Intelligence Scale for Children–Revised*. New York: Psychological Corporation.

Wechsler, D. (1981). *Manual for the Wechsler Adult Intelligence Scale–Revised*. New York: Psychological Corporation.

Zerbin-Rüdin, E. (1967). Kongenitale Wortblindheit oder spezifische Dyslixic [Congenital word-blindness]. *Bulletin of the Orton Society, 17*, 47–56.

CHAPTER 7

Cognitive Abilities and Disabilities in Perspective

Duane Alexander

The recent discovery of a knock-out gene for spatial learning in mice (Silva, Paylor, Wehner, & Tonegawa, 1992) and the precise prediction of this event (Plomin & Neiderhiser, 1992) have provided validation for the study of the genetics of cognitive abilities and disabilities.

The question of the predominant influence of nature versus nurture in governing human behavior and intelligence far predates the discipline of psychology. This topic has been one of the primary foci, as well as one of the most controversial issues, of this developing discipline. The development of the necessary tools to study the nature–nurture issue has been a preoccupation for a number of leading psychologists, as the chapters in this section have documented. The authors of these chapters have demonstrated some of the most useful techniques for addressing this issue, and by applying modern statistical methodology, these researchers have advanced the discipline to a new plateau of sophistication.

The first of these techniques is family studies, which examines through interviews and testing the similarities of relatives to probands in general intelligence or specific cognitive skills. Thus, in their analysis of the Colorado Reading Project in chapter 6, DeFries and Gillis concluded that there is familial influence but, based on segregation analyses, that reading disability is genetically heterogeneous.

A second method uses twins to examine differences in intelligence or specific cognitive ability in monozygotic (MZ) versus dizygotic (DZ) twin pairs. Again in chapter 6, DeFries and Gillis reported that in the Colorado Twin Study the concordance rate for reading disability was 66% for MZ twin pairs versus 43% for DZ twin pairs, which is a statistically significant but not completely clear-cut difference.

A third technique is adoption studies, in which intelligence or specific cognitive function of adoptees is compared with that of their natural versus adoptive parents, their natural siblings in the same household versus their adoptive siblings, or their natural siblings raised in a different household. This technique is most informative when the rare opportunity exists to compare MZ and DZ twins reared together or apart. One of the largest adoption studies is the Colorado Adoption Project, which was discussed by Carden and Fulker in chapter 5. These authors concluded that there is no specific developmental process that determines relative intellectual ability in children 1–9 years of age, that certain abilities appear to be influenced by ability-specific genes, and that these abilities seem to emerge at certain ages, which thereby suggests that these genes might turn on at certain times.

Another technique for studying the nature–nurture issue is to see how intelligence correlations change with age. In chapter 3, McGue, Bouchard, Iacono, and Lykken compared correlations in IQ of MZ twins versus DZ twins over their life span. Although one might expect that genetic correlations might diverge with age as life experiences have greater influence, McGue et al. found just the opposite: Genetic influences on IQ appear to increase with age. These authors postulate as an explanation for this finding the possibility that genetically driven self-selection of similar experiences over the lifetime provides environmental reinforcement for existing genetic similarities.

Another technique for studying the nature–nurture topic, which has been made available through modern genetics, is gene linkage analysis.

This technique allows researchers to search for genetic or familial abnormalities that tend to occur together and thereby place them on the same chromosome. With this technique, researchers can even determine the distance between two linked genes. Using gene linkage analysis, Herb Lubs at the University of Miami (Smith, Kimberling, Pennington, & Lubs, 1983) has been able to demonstrate a familial type of reading disability that localizes to Chromosome 15 and accounts for about 20% of his familial cases, and in chapter 6, DeFries and Gillis also cite cases of reading disability that may be linked to Chromosome 15. Furthermore, other familial cases of reading disability may be linked to Chromosome 6.

Yet another technique is to examine subjects with specific genetic abnormalities or chromosome defects for their intelligence or specific cognitive difficulties. The appearance of chromosome analysis in the 1960s and the discovery of trisomy 21 as the cause of Down's syndrome and other abnormalities as causes of wide-spectrum mental retardation caused some excitement. In general, however, even small abnormalities seem to cause extensive intellectual dysfunction.

One exception was an area in which I had my first research experience as a medical student while working with John Money in medical psychology at Johns Hopkins University. He had discovered that girls with Turner's syndrome, who have only one X chromosome instead of two, were generally not mentally retarded but did appear to have a specific cognitive dysfunction that he termed "space-form blindness." Over the course of several years, we did extensive testing of a number of these girls and demonstrated striking abnormalities just in their ability to draw geometric shapes or mentally rotate objects in space (Alexander, Ehrhardt, & Money, 1966; Alexander, Walker, & Money, 1964). This disability was most clearly documented in a road-map test of direction sense that we developed and standardized (Money, 1965). This test showed that most of the Turner's syndrome subjects could not perform the basic tasks of map reading. The disability was relatively specific and did not hinder their ability to make the letter and word orientation necessary to learn to read. At the time we published these studies, they were the first demonstration of a specific cognitive disability that was linked to a chromosomal abnormality, which presumably results from having only a single dose of a gene on the X chromosome. Scientists are now exploring other syndromes with specific cognitive or behavioral phenotypes.

With the extensive activities of the human genome mapping project, we can anticipate the ultimate in our ability to identify genes that are associated with intelligence or specific cognitive abilities, but even then the picture will not be simple. Not only is intelligence certain to be polygenic, but the genes themselves will have various polymorphisms that alter their function and complex controls that turn them on and off. Life will get more complicated, not less.

But what about nurture? Although virtually all of the authors in this section have focused on genetics and made a strong case, something should be said about nurture. At least two points should be made. First, the striking and clear-cut findings from the North Carolina Abecedarian Project, the Infant Health and Development Project, and similar studies of the positive effect of early environmental stimulation on intellectual function of children have provided a powerful argument for nurture. Second, behavioral genetic analyses of children's IQ scores converge on the conclusion that about half of the variance is due to genetic differences among the children. But this means that half of the variance is not genetic, and thus environmental factors, as well as genetic influences, need to be studied.

But even nature and nurture may not be sufficient to explain all human behavior. My wife recently pointed out that I failed to include the "devil theory" of behavior. When I asked her for an explanation, she said, "Well, I know your parents, and I know you got good genes from them, and I know they raised you right, and I know I create a good environment for you, but sometimes you do things that are so different from that that the only explanation could be that the devil made you do it." I don't know if the devil works through nature or nurture or independently, but on reflection there are certainly things that we probably never will explain by nature and nurture alone.

References

Alexander, D., Ehrhardt, A., & Money, J. (1966). Defective figure drawing, geometric and human, in Turner's syndrome. *Journal of Nervous and Mental Disease, 142*, 161–167.

Alexander, D., Walker, H. T., & Money, J. (1964). Studies in direction sense: I. Turner's syndrome. *Archives of General Psychiatry, 10*, 337–339.

Money, J. (1965). *A standardized road-map test of direction sense.* Baltimore: Johns Hopkins University Press.

Plomin, R., & Neiderhiser, J. M. (1991). Quantitative genetics, molecular genetics, and intelligence. *Intelligence, 15*, 369–387.

Silva, A. J., Paylor, R., Wehner, J. M., & Tonegawa, S. (1992). Impaired spatial learning in alpha-calcium-calmodulin kinase II mutant mice. *Science, 257*, 206–210.

Smith, S. D., Kimberling, W. J., Pennington, B. F., & Lubs, H. A. (1983). Specific reading disability: Identification of an inherited form through linkage analysis. *Science, 219*, 1345–1347.

PART THREE

Nature–Nurture and the Development of Personality

PART THREE

Introduction

H. H. Goldsmith

The past 2 decades have witnessed major transformations in research on the genetics of personality. This introduction outlines the conclusions that have been apparent for many years, discusses current trends in the field, and ends with some suggestions for integration with other personality research realms. Over 2 decades ago, a collection of small-sample twin and family studies suggested that (a) genetic influences on individual differences in many personality dimensions were moderate (heritabilities varied widely around 40–50%) and (b) environmental factors acted chiefly to make family members different rather than similar (Jinks & Fulker, 1970). Until much more recently (Plomin & Daniels, 1987), researchers discussed the conclusion about moderate genetic effects much more extensively than the conclusion about how environmental factors operate. In the earlier period, the conclusions were circumscribed in that they applied primarily to assessment by paper-and-pencil self-reports of personality and to samples of students and young adults.

The database for drawing conclusions about the nature and nurture of personality has been almost completely refurbished during the last 2 decades. The dominant design, twin studies, now includes sample sizes ranging into the thousands. Almost all age groups are represented, including infancy and old age. Moreover, the traditional identical twin–versus–fraternal twin comparisons have been supplemented by major adoption studies and studies of twins reared apart. Assessment has often been extended to entire families. This expansion of personality research has been international in flavor, with major studies not only in the U.S. but also in Australia, Finland, Norway, Sweden, The Netherlands, the United Kingdom, and elsewhere. Whereas earlier studies typically used single-scale, single-occasion analyses, many of the current studies are longitudinal and multivariate; and they often entail sophisticated structural modeling techniques. In the course of their research, behavioral geneticists have made some advances in the modeling techniques themselves (Neale & Cardon, 1992). Another important advance has been the extension from self-report to multimodal assessment, sometimes including objective observations conducted in laboratory and home contexts.

What conclusions have emerged from these new and improved studies? The findings of moderate genetic variance and nonshared rather than shared environmental effects have been strongly reinforced. The flavor of current conclusions is exemplified by Loehlin's (1992) summary statements concerning biometric model-fitting for Surgency, Agreeableness, Conscientiousness, Emotional Stability, and Culture—the "Big Five" factors of normal personality variation:

> Additive genetic effects . . . fell generally in the moderate range, 22% to 46%; shared family environmental effects were small, 0% to 11%; and the ambiguous third factor—nonadditive genetic or special MZ [identical] twin environments—was intermediate at 11% to 19%, except for Culture, where it was a lesser 2% to 5%. The remaining variation, 44% to 55%, presumably represents some combination of environmental effects unique to the individual, genotype–environment interaction, and measurement error. (p. 68)

Some other empirical findings have important developmental implications. For instance, two studies suggest strongly that stable genetic effects account for the stable aspects of childhood temperament and early adult personality (Goldsmith & McArdle, 1992; McGue, Bacon, & Lykken, 1993). A fascinating finding from the neonatal period is that temperamentlike behavioral patterns are not at all influenced by genetic variation (Riese, 1990). Whether there is systematic change in the influence of genes during development is still under active investigation.

Besides Loehlin's (1992) just-mentioned book, the reader who is interested in sampling the current empirical literature on normal personality might consult Carey and Rice (1983); Eaves, Eysenck, and Martin (1989); Goldsmith (1989); Matheny (1989); Plomin, DeFries, and Fulker (1988); Plomin, Pedersen, McClearn, Nesselroade, and Bergeman (1988); Rose, Koskenvuo, Kaprio, Sarna, and Langinvainio (1988); Scarr, Webber, Weinberg, and Wittig (1981); and Tellegen et al. (1988). Review chapters include Goldsmith (1983); McCartney et al. (in press); and Plomin, Chipuer, and Loehlin (1990).

The chapters in this section attempt to bridge gaps between genetically oriented investigators and other personality researchers and to highlight important current and needed future research.

In chapter 9, David Rowe provides a brief orientation to behavioral genetics concepts as applied to the personality realm. He explains that basic assumptions of behavioral genetics methods have undergone testing, with largely supportive results. The treatment of the environment in behavioral genetics designs and the relation (and lack thereof) of behavioral genetics and evolutionary psychology come under scrutiny. Rowe's chapter clearly indicates that behavioral genetics techniques are continually evolving. More and more, both practitioners and critics of behavioral genetics will need to develop expertise ranging from evolution to structural modeling and from molecular genetics to environmental processes to do their jobs well.

The study of early personality—temperament—has been reinvigorated, in part, by the studies of Kagan and colleagues on inhibition (chapter 10). This temperament research has highlighted some very old questions in the personality field (e.g., Are some personality characteristics better conceived of as types rather than dimensional traits?). Studying temperament during infancy also facilitates investigation of some very contemporary issues (e.g., How do an individual's personality dispositions influence family experiences, and vice versa?). In contrast to some other aspects of personality, recent research on temperament has been notable for its consistent grounding in psychobiology, including behavioral genetics (Goldsmith, 1989). Research from Kagan's laboratory has elucidated the psychobiological underpinnings of temperamental inhibition, but in chapter 10, the emphasis is on experiential factors that modulate the development of inhibition from "high reactive" profiles in earlier infancy. The chapter also adds new documentation of ethnic differences in temperament reactivity, which suggests that these differences are partially genetically based, and speculates that mean temperamental differences contribute to the attractiveness of differing basic philosophical approaches in the East versus the West.

The overarching goal of better integration of behavioral genetics results on personality into the broader field is well served by Nathan Brody's contribution (chapter 8), which systematically compares behavioral genetics research on intelligence with that on personality. Brody points out that—among other issues—researchers need to explore the genetic and environmental bases of covariation between specific personality traits and (a) their biological correlates, (b) their measurement at different developmental stages, (c) associated socially important outcomes, and (d) associated performance in laboratory tasks. Quite appropriately, just these issues are the subject of recent publications and much research in progress, especially in the developmental area. However, more extensive integration of the fields will depend on the growing company of mainline personality researchers who are integrating the concepts and findings of behavioral genetics into their research.

The three chapters in this section capture only some of the excitement and promise of the behavioral genetics study of personality. One issue that should occupy researchers' attention is the interface of normal personality variation and psychopathology. Important facets of this issue are the relation of normal personality variation to the third, revised edition of the *Diagnostic and Statistical Manual of Mental Disorders (DSM-III-R)* (American Psychiatric Association, 1987) Axis II personality disorders; the usefulness of spectrum concepts that include personality dimensions for schizophrenia, Tourette syndrome, and possibly other disorders; the status of temperamental traits as risk-increasing or risk-decreasing factors in liability to psychopathology; and similar questions. Another set of issues that is currently under investigation involves genetic analyses of phenomena once considered to be "obviously" environmental in origin; these phenomena include life events, family emotional environment, and divorce. The recently documented heritability of such features of social interchange is probably mediated by the genetic influences on personality; however, crucial research lies ahead. A third source of excitement is the prospect of studying genetically related persons using some of the "new" methods of personality assessment—narrative techniques, time-sampling methods (e.g., beeper studies), examination of autobiographical material, "live-in" naturalistic observation, and laboratory-based assessment of various sorts.

In broader perspective, this section reflects the minor renaissance that personality research has recently undergone; it is clear that behavioral genetics research is playing a continuing role in that welcome renaissance. That role is likely to increase as behavioral genetics methodology increasingly integrates both molecular genetics and measurements of specific environmental processes.

References

American Psychiatric Association. (1987). *Diagnostic and statistical manual of mental disorders* (3rd ed. revised, *DSM-III-R*). Washington, DC: Author.

Carey, G., & Rice, J. (1983). Genetics and personality temperament: Simplicity or complexity? *Behavior Genetics, 13*, 43–63.

Eaves, L. J., Eysenck, H. J., & Martin, N. G. (1989). *Genes, culture, and personality.* New York: Academic Press.

Goldsmith, H. H. (1983). Genetic influences on personality from infancy to adulthood. *Child Development, 54*, 331–355.

Goldsmith, H. H. (1989). Behavior–genetic approaches to temperament. In G. A. Kohnstamm, J. E. Bates, & M. K. Rothbart (Eds.), *Temperament in childhood* (pp. 111–132). New York: Wiley.

Goldsmith, H. H., & McArdle, J. J. (1992). *Longitudinal biometric twin analyses of temperamental change.* Manuscript submitted for publication.

Jinks, J. L., & Fulker, D. W. (1970). Comparison of the biometrical, genetical, MAVA, and classical approaches to the analysis of human behavior. *Psychological Bulletin, 73*, 311–349.

Loehlin, J. C. (1992). *Genes and environment in personality development.* Newbury Park, CA: Sage.

Matheny, A. P. (1989). Children's behavioral inhibition over age and across situations: Genetic similarity for a trait during change. *Journal of Personality, 57*, 215–235.

McCartney, K., Biryukov, S., Borkenau, P., Buss, D., Goldsmith, H. H., Kline, P., Loehlin, J. C., Maier, W., Nothen, M., Pedersen, N., Schepank, H., & Waller, N. (in press). New directions in twin research on personality. In T. J. Bouchard & P. Propping (Eds.), *Twins as a tool of behavioral genetics* (Dahlem Workshop Report LS 53). New York: Wiley.

McGue, M., Bacon, S., & Lykken, D. T. (1993). Personality stability and change in early adulthood: A behavioral genetic analysis. *Developmental Psychology, 29*, 96–109.

Neale, M. C., & Cardon, L. C. (1992). *Methodology for genetic studies of twins and families.* Norwell, MA: Kluwer Academic.

Plomin, R., Chipuer, H. M., & Loehlin, J. C. (1990). Behavioral genetics and personality. In L. A. Pervin (Ed.), *Handbook of personality theory and research* (pp. 225–243). New York: Guilford.

Plomin, R., & Daniels, D. (1987). Why are children in the same family so different from each other? *Behavioral and Brain Sciences, 10*, 1–16.

Plomin, R., DeFries, J. C., & Fulker, D. W. (1988). *Nature and nurture in infancy and early childhood.* New York: Cambridge University Press.

Plomin, R., Pedersen, N. L., McClearn, G. E., Nesselroade, J. R., & Bergeman, C. S. (1988). EAS temperaments during the last half of the life span: Twins reared apart and twins reared together. *Psychology and Aging, 3*, 43–50.

Riese, M. L. (1990). Neonatal temperament in monozygotic and dizygotic twin pairs. *Child Development, 61*, 1230–1237.

Rose, R. J., Koskenvuo, M., Kaprio, J., Sarna, S., & Langinvainio, H. (1988). Shared genes, shared experiences, and the similarity of personality: Data from 14, 288 adult Finnish twins. *Journal of Personality and Social Psychology, 54*, 161–171.

Scarr, S., Webber, P. L., Weinberg, R. A., & Wittig, M. A. (1981). Personality resemblance among adolescents and their parents in biologically related and adoptive families. *Journal of Personality and Social Psychology, 40*, 885–898.

Tellegen, A., Lykken, D. T., Bouchard, T. J., Jr., Wilcox, K. J., Segal, E. L., & Rich, S. (1988). Personality similarity in twins reared apart and together. *Journal of Personality and Social Psychology, 54*, 1031–1039.

CHAPTER 8

Intelligence and the Behavioral Genetics of Personality

Nathan Brody

In this chapter, I derive suggestions for the study of genetic and environmental influences on personality traits from a consideration of research on intelligence. I argue that more is known about the behavioral genetics of intelligence than is known about the behavioral genetics of personality. And, what is known about the behavioral genetics of intelligence provides a basis for addressing unresolved issues in personality.

The Measurement of Traits

There is a fundamental difference between research on intelligence and research on personality traits. Intelligence is measured behaviorally, whereas personality traits are usually measured by self-reports or ratings of behavior. If intelligence were measured in the same way as personality traits are measured, then a person would be asked to rate his or her vocabulary. Scores on an omnibus measure of intelligence are aggregates of a person's responses to items that sample intellectual behavior in a variety of relevant situations. It is difficult to obtain indices of behavior in a variety of relevant situations that would provide a comparable index for personality traits such as extraversion, impulsivity, or neuroticism. There is no inventory of appropriate situations in which behavior should be sampled for any of our major personality traits. Ratings and self-reports are used as surrogates for the aggregated samplings of behavior in relevant situations that would provide an index for a personality trait that is comparable to a score on an omnibus measure of intelligence. From this perspective, there are no behavioral genetic studies of adult personality. There are a small number of behavioral genetic studies of children and a small number of studies of ratings of children's behavior (e.g., Goldsmith & Gottesman, 1981). I cannot recall a single twin study or adoption study of adult personality that includes behavioral measures. The first version of this chapter was written in August 1992. In that version I wrote, "Nor, for that matter, can I recall a single behavioral genetic study of adult personality that relies on ratings of personality rather than on self-reports." Then I received the July 1992 issue of the *Journal of Personality and Social Psychology*, which included an article by Heath, Neale, Kessler, Eaves, and Kendler (1992) that reported the results of a twin study of adult personality using both self-reports and cotwin ratings of personality questionnaire data. With the singular exception of this study, there are virtually no adequate behavioral genetic analyses of ratings of adult personality. It is also the case that behavioral genetic studies of aggregate indices of adult personality are virtually nonexistent. The behavioral genetics of adult personality have not been studied, but the behavioral genetics of self-reports about personality have been studied. This is a serious omission. It is not necessarily the case that self-reports are invalid. It is simply that they are subject to methodologically specific sources of variance. Ever since Campbell and Fiske (1959) developed the logic of the heteromethod–heterotrait matrix, researchers have understood that the true-score variance of a hypothetical latent trait is defined by heteromethod agreement for different indices of that trait.

One can only speculate about the possible results of behavioral genetic analyses of aggregate indices for personality traits. My hunch is that such analyses would yield evidence of higher heritability than do current estimates of self-reports about traits. This hunch is supported by the following speculations: Aggregate indices would contain more true-score variance for the trait and would remove methodologically specific sources of variance that are associated with self-report indices. Current analyses conflate nonshared environmental sources of variance with error variance. If there are dependent variables with more true-score variance, then estimates of nonshared environmental variance would decrease, which would lead to possible increases in the magnitude of genetic sources of variance. This hunch is supported by the results of Heath et al. (1992). The design of this study, which included both self-reports and ratings by cotwins, permitted control for measurement error. Furthermore, the heritability estimates that were found in this study were somewhat higher than usual. They obtained heritabilities for neuroticism and extraversion of 0.63 and 0.73, respectively. Heath et al. recognized that their study was only a limited examination of the utility of behavioral genetic analyses of personality ratings. Ratings were made solely by cotwins and were made at the same time as self-reports were obtained. Additional research using aggregated personality ratings as well as additional heteromethod analyses should provide a better understanding of the behavioral genetics of personality.

The Structure of Traits

Intelligence

Behavioral genetic analyses provide information about individual differences in the structure of traits. Behavioral genetic analyses also provide evidence for three generalizations about the structure of intellect. First, ordered relationships among components of aggregate indices of intelligence are partially derivable from the heritability of components. Omnibus measures of intelligence are based on aggregates of responses to diverse intellectual tasks. A common basis of constructing an ordered relationship among the components of the aggregate is to calculate the loading of each of the components on a general factor—the *g* loading. Pedersen, Plomin, Nesselroade, and McClearn (1992) administered a battery of tests of intelligence to subjects in the Swedish adoption/twin study of aging.

Their sample included monozygotic (MZ) and dizygotic (DZ) twins who were reared together and apart. They used these data to obtain estimates of the heritabilities of each of the component tests in the composite battery of tests of intelligence. They obtained a correlation of 0.77 between the heritability values for the tests and the *g* loadings of the tests. These results indicate that *g* loadings are not arbitrary or meaningless characteristics of tests; they are strongly related to the heritability of test components.

Second, individual differences in the patterning of intellectual abilities are heritable. *g* does not exhaust all of the variance in a correlation matrix of diverse ability measures. Contemporary analyses suggest that individual differences in the structure of intellect include specific ability factors in addition to a common general ability factor. Cardon, Fulker, DeFries, and Plomin (1992) used data from the Colorado Adoption Project to demonstrate that verbal, spatial, and memory factors that are independent of *g* are heritable. Thus, individual differences in the profile of group factors as well as individual differences in general intelligence may be heritable (see also Segal, 1985; Wilson, 1986).

Third, the degree to which general intelligence is heritable may vary among different groups of individuals. There is tentative evidence that the heritability of intelligence may differ for different groups of individuals. For example, Detterman, Thompson, and Plomin (1990) reported that the heritability of intelligence was inversely related to IQ.

Personality

What is the status of evidence for personality traits of each of the three just-presented generalizations about intelligence? First, there are no clearly defined variations among components of personality traits that are comparable to variations in *g* loadings. As a result, it is not known whether variations in an ordering principle for personality traits are related to the heritability of the components. In principle, it is possible to define variations in the loadings of behavioral referents of a personality trait either by using ratings of the extent to which different behaviors are "prototypical" for the trait (Buss & Craik, 1980) or by factor analyzing various behavioral measures that are assumed to be related to the personality trait (see Jackson & Paunonen, 1985, for an example of a factor analysis of behavioral measures of conscientiousness). In practice, however, there have been few attempts to order the alleged behavioral manifestations of personality traits. Therefore, the relationship between the ordering of aggregate components of personality traits and their heritability remains indeterminate.

Heath, Eaves, and Martin (1989) reported the results of a twin study of responses to items on the Eysenck Personality Questionnaire (EPQ) (Eysenck & Eysenck, 1968). They obtained correlations separately for MZ and DZ twin pairs for twin response to one item and cotwin response to all other items. Using these data, they were able to determine common genetic influences on items that separately defined the extraversion and neuroticism factors on the EPQ. The latent genotypic structure of the Psychoticism scale was not congruent with the conventional factor structure of the scale. Although these analyses demonstrate agreements between behavioral genetic analyses and factor structures for personality test items, in two respects these results are not the same as those reported by Pedersen et al. (1992) for intelligence. The data are not behavioral, and there is no indication that variations in factor loadings for item scores are correlated with the heritability of different items.

Second, if personality traits, like intelligence, are aggregate measures, then it is possible that profile characteristics among components to the aggregates may be heritable. Relatively little is known about the heritability of profiles of trait components. Dworkin (1979) reported the results of a twin study using a situation–response inventory of anxiety that asked individuals to rate their characteristic anxiety responses in different situations. Dworkin obtained MZ correlations of 0.36 and DZ correlations of 0.19. These data imply that some aspects of the patterning of trait scores may be heritable.

Hershberger, Plomin, Pedersen, and McClearn (1993) used another approach to study the heritability of profile characteristics of responses to personality questionnaires. They obtained interitem standard deviation scores (metatrait measures) for responses to several personality traits that were assessed by self-reports using data from the Swedish adoption/twin study of aging. Their analyses indicated that metatrait scores are heritable and that some of the genetic influences on the metatrait are independent of the genetic influences on the trait. Metatrait scores may be interpreted as an index of the consistency of response to trait-relevant indices. These data suggest that the extent to which individuals are consistent in their expression of trait-relevant behaviors may be partially attributable to their genotypes.

Third, does the heritability of personality traits vary for different groups of individuals? Perhaps extreme scores on some personality traits may have higher heritabilities than average scores. Thus, introversion and extraversion might be more heritable than ambiversion. Such results might indicate something about the expressivity of different genotypes for personality and the extent to which the influence of the genotype on the phenotype is responsive to different environmental events. Evidence for individual differences in the heritability of traits for different groups of individuals would provide evidence against a strictly nomothetic theory of traits.

Studies of the heritability of orderings of aggregate components of traits, patterning of trait components, and differential heritability for individuals with different scores on a trait may contribute to a deeper understanding of the nature of the genotypes that influence personality trait phenotypes. Individuals may inherit dispositions to respond in specific ways to specific situations (e.g., phobic responses triggered by specific situations) or they may inherit rather broad dispositions to respond in similar ways to a large class of situations (e.g., a disposition to respond fearfully). In either case, traits might be heritable, but the reasons for the heritability of traits might be different. In the former case, specific narrow response dispositions generalize to create a broader disposition that is heritable because of the heritability of one or more of its core constituents; in the latter case, the heritability of the trait derives from the heritability of a relatively broad nonsituationally specific disposition.

Longitudinal Behavioral Genetics

Intelligence

Twin studies and adoption studies both provide evidence for an increase in the heritability of intelligence with age. Wilson (1983) found that the correlations in intelligence for MZ and DZ twins exhibited increasing divergence from infancy to 15 years of age, together with evidence of declining DZ correlations that did not appear to reach a plateau at 15 years of age. McGue and colleagues (McGue, Bouchard, Iacono, & Lykken, chapter 3; see also McCartney, Harris, & Bernieri, 1990) reported comparable results for analyses of twin correlations for intelligence across the life span. The pattern of declining DZ correlations and MZ correlations that either drift upward or remain constant over the life span implies that the heritability of intelligence increases with age.

Personality

Although there is an emerging understanding of the behavioral genetics of changes in intelligence, there is relatively little comparable information about changes in personality. Twin studies of self-report measures of personality do not provide evidence for increasing heritability of personality over the life span (Loehlin, 1992). Data from the Swedish adoption/twin study of aging (Pedersen et al., 1992) suggest marginally lower heritabilities in its sample of older twins (average age = 59 years) than were obtained in studies of younger adult twins.

Twin studies of adult personality do not provide clear evidence that phenotypes change by an increasing resemblance to genotypes as they do for intelligence. Eaves and Eysenck (1976) obtained EPQ Neuroticism Scale scores for twins in a longitudinal study. For the 2-year period of their investigation, there was no evidence that change scores were heritable. That is, correlations between change scores for MZ twin pairs were not conspicuously higher than correlations for DZ twin pairs. By contrast, analyses of personality ratings obtained in the Texas adoption study (Loehlin, Willerman, & Horn, 1987) provide evidence for changes in personality that increased the resemblance of the mean scores of adoptees to the mean scores of their biological mothers. The biological mothers of the adoptees had lower EPQ emotional stability ratings than did the adopted mothers. In relation to natural children of adopted mothers, adopted children exhibited a decrease in emotional stability ratings from initial testing to retesting 10 years later. These data suggest that changes in personality lead to an increase in phenotypic and genotypic resemblance (Loehlin et al., 1987). Most of the data on changes in personality do not provide clear evidence for genetic influences on personality changes after childhood. Most of the available data are based on self-report measures. Whether measures of personality traits that are based on heteromethod aggregates would provide evidence for increasing heritability over the life span is unknown.

It is possible to argue on a priori grounds that the heritability of personality would increase or decrease with age. The following is the argument for the increase in heritability. Caspi and Moffitt (1991, 1993) argued that dispositional continuities in personality are enhanced by responses to novel, stressful, and unexpected events. They studied the response of adolescent girls to early menarche. They found that dispositional differences were exaggerated following the onset of early menarche. Adolescents who were inclined to be troubled before the onset of menarche became more troubled after menarche. Traitlike differences between relatively well-adjusted girls and poorly adjusted girls were enhanced following exposure to a stressful experience. One can speculate that genetic differences that predispose individuals to respond differentially to stressful and novel events increase their influence on phenotypes as a result of cumulative exposures. As individuals age, they have more opportunity to encounter novel and stressful events. These cumulative experiences may enhance the relationship between genotype and phenotype. This model of development would imply that changes in personality lead to increases in the heritability of personality traits. If personality traits are construed as having a genotypic core, then this model of development construes personality changes as increasing the extent to which phenotypes evolve in the direction of their preexistent genotypes. Genotypic influences not expressed at one age may be expressed at a later age. Thus, changes in personal characteristics may be continually influenced by genotypes. Such a model seems to be required for intelligence. It may or may not be appropriate for personality.

I now present an argument for decreasing heritability of personality traits as a function of age. All current behavioral genetic analyses of personality traits indicate that personality is influenced by environmental events not shared by individuals reared together (Plomin, Chipuer, &

Loehlin, 1990). These events may be idiographic, occurring infrequently to more than one individual, yet critical in determining the life course of a particular individual. If personality is shaped by such events and individuals change in response to idiographic unrepeatable events, then individuals with common genotypes will cumulatively drift apart as a result of idiosyncratic experiences. The relatively low correlations for older MZ twins that were obtained in the Swedish adoption/twin study of aging are compatible with the assumption of declining heritability for personality traits as individuals age (Plomin et al., 1990).

It is possible that both of the processes just outlined may exist, and their relative influence in determining continuity and change may differ for different individuals. Research combining behavioral genetic analyses and longitudinal studies should help in understanding genetic and environmental influences on change in personality over the life span.

Nomological Networks

The nomological network of relations that collectively serve to specify the meaning of a trait typically extends beyond the demonstration that traits exemplify consistencies in behavior in diverse settings that are derived from the ordinary language meaning of the trait name. There are at least three broad classes of relationships that demonstrate the broader meaning of traits: (a) Traits are related to biological measures, (b) traits are related to socially significant outcomes, and (c) traits are related to behavior in laboratory situations that are designed to obtain measures of behavior under highly controlled conditions.

Relations Between Biological Measures and Traits

There are relationships between intelligence and biological indices. Some of these relationships are based on very extensive databases and are very well established (e.g., those relating IQ to head circumference and myopia) (see Jensen & Sinha, 1992). The causal relationship between such biological indices and intelligence is ambiguous. The relationships may be attributable to the influence of environmental events that independently influence both intelligence and a biological index. For example, both head circumference and intelligence may be influenced by nutrition (see Lynn,

1990). There is very preliminary evidence that the covariance between IQ and these biological indices may be attributable to genetic influences. Jensen (as cited in Jensen & Sinha, 1992) obtained a between-families MZ correlation for the relationship between head circumference and IQ of 0.39. The DZ correlation was 0.15. These data imply that the covariance between head circumference and IQ is primarily mediated by genetic covariance. Cohn, Cohn, and Jensen (1988) found that relationships between myopia and IQ were found within families. They studied a sample of high-IQ adolescents who had IQs that were 14 points higher than their siblings on the Ravens test (Raven, 1938). The high-IQ adolescents had myopia scores that were 0.39 standard deviations higher than those of their siblings. These data indicate that the variables that cause adolescents to develop IQs that are higher than those of their siblings are correlated with the variables that cause siblings to develop myopia. Strictly speaking, these data do not imply that genetic covariance is responsible for the relationship between myopia and IQ. Genetic covariance, however, is a possible explanation for the relationship between correlated sibling differences in IQ and myopia.

Personality traits are also related to biological indices (see Zuckerman, 1991, for a comprehensive review of the biological basis of personality). Although there is ample evidence relating personality traits to various biological indices, I am not aware of any studies of the genetic covariance between measures of personality and biological indices. It is often assumed that personality traits are derived from heritable characteristics of the nervous system. Evidence of a genetic basis for the covariance between personality characteristics and biological indices would add to the validity of this assumption.

Traits Are Related to Socially Relevant Outcomes

Measures of intelligence are related to socially significant outcomes such as academic success and intergenerational social mobility (Brody, 1992). Thompson, Detterman, and Plomin (1991) performed a behavioral genetic covariance analysis for the relationship between indices of intelligence and academic achievement. Using data from a sample of 6–12-year-old twins, they found that ability measures were more heritable than achievement measures. They also found that the correlations between ability and achievement for MZ twins ranged between 0.31 and 0.40. The comparable DZ twin correlations were between 0.18 and 0.23. These data are compatible with a model that assigns a heritability value to the covariance between achievement and ability of 0.80 and a value of 0.00 for the influence of shared family environments. Cardon, DiLalla, Plomin, DeFries, and Fulker (1990) reported analogous results for a genetic covariance analysis of the relationship between IQ and reading skills using data from the Colorado adoption project. These analyses indicate that the covariance between intelligence and academic achievement may be substantially mediated by genetic influences.

Personality traits are also related to socially important outcomes. For example, Caspi, Elder, and Bem (1987) related childhood ratings of ill-temperedness to indices of occupational status and occupational stability in a 20-year longitudinal study. Similarly, Huesmann, Eron, Lefkowitz, and Walder (1984) obtained relationships between peer ratings of aggressive behavior in childhood and indices of adult criminal and antisocial behavior in a longitudinal study. These and other analogous studies indicate that personality dispositions assessed by ratings in childhood are related to socially relevant outcomes.

These studies do not inform us of the reasons for the relationships between childhood characteristics and socially relevant adult outcomes. Behavioral genetic covariance analyses provide a useful initial examination of the covariances of the extended nomological network of trait relationships. Consider, for example, how such analyses might be helpful in understanding the results of the study of aggression. Huesmann et al. (1984) had several indices of aggression in their longitudinal study. They used measures obtained at different times to perform a latent trait analysis, which indicated that the 20-year stability for male aggression had a hypothetical correlation of 0.5. If researchers are to understand the impact of latent dispositions over an individual's life span, then they shall have to understand why dispositions that are initially manifested in early childhood influence socially relevant adult behaviors. Why do children who are described as aggressive by their peers manifest criminal, aggressive, and antisocial behavior as young adults? Why do some individuals who are aggressive as children fail to manifest aggression as young adults? To what extent are the covariances among these manifestations of a latent disposition attributable to genetic covariances? There is evidence from adoption studies that some forms of criminal behavior are heritable (Mednick, Gabrelli, & Hutchings, 1987). It is possible that the relationships between childhood aggressive behavior and adult criminal and antisocial behavior are influenced by genetic influences that are common to both. It is reasonable to assume that aggression, like most personality traits, is influenced by genes and by nonshared environmental events. Mednick et al. found in their study of Danish adoptees that the criminality of an adopted child was influenced by the criminal history of the biological father and the social class of both the biological and adopted family. These data indicate that criminality is influenced by both genetic and shared environmental events. These data do not provide any information about the reasons for covariances between such personality traits as aggression and criminality. Although aggression may not be influenced by shared environmental events, whether an aggressive individual exhibits criminal behavior may be substantially dependent on environmental influences shared by individuals reared together. Dispositional analyses of personality have as a central task an understanding of the mechanisms by which dispositions that are manifested initially in childhood are related to adult personality and socially relevant outcomes. We know that such relationships exist. Virtually all accounts of the reasons for the relationships are speculative. Behavioral genetic analyses of such covariances would provide a vitally important first step in the development of an understanding of these relationships.

Traits and Performance in Laboratory Situations

Measures of intelligence are related to performance on various laboratory tasks that measure elementary information-processing skills, such as reaction time (Jensen, 1987) and pitch discrimination for briefly presented tones (Raz, Willerman, & Yama, 1987). Baker, Vernon, and Ho (1991) reported a genetic covariance analysis of the relationship between speed of information processing on a variety of experimental tasks and performance on tests of intelligence (see Vernon, 1983). The correlation between a speed-of-information-processing composite score and performance IQ was −0.60 for MZ twins and −0.25 for DZ twins. These correlations were compatible with a model that attributed virtually all of the covariance between speed of information processing and IQ to genetic influences.

Personality traits are related to theoretically derived indices of performance in various experimental situations. For example, extraversion is related to psychophysiological arousal. Introverts exhibit relatively high levels of arousal potential, but they are less likely to be hyperaroused in situations that are high in arousal potential (Smith, 1983). Evidence for an interaction between extraversion and arousal in situations that differ in arousal potential is compatible with Eysenck's biological theory of extraversion (Brody, 1988; Eysenck, 1967; Eysenck & Eysenck, 1985).

Behavioral genetic analyses of covariances between personality trait measures and performance in laboratory contexts have not, as far as I know, been reported. Research reporting relationships between extraversion and arousability in laboratory contexts has generally been interpreted as supporting a theory of heritable characteristics of the nervous system that influence the development of personality. Evidence indicating that genetic covariances exist between personality and performance in these laboratory contexts would strengthen the assumption that relationships between personality and performance in these situations are attributable to heritable characteristics of the nervous system.

Review

This brief review of covariance analyses suggests that there is more evidence for a genetic basis for the covariance between intelligence and its extended nomological network than there is for a genetic basis for the covariance between personality traits and their extended nomological network. Behavioral genetic analyses of covariances among the manifestations of personality traits would provide a useful first step for analyzing the basis of these relationships.

Martin and Jardine (1986) reported the results of a study that demonstrates the potential of genetic covariance analysis of personality characteristics. They studied relationships among self-report measures of neuroticism and state-dependent measures of depression and anxiety using a large sample of Australian twins. They reported the results of a behavioral genetic analysis of the covariances among these measures. The mean heritability for the three individual differences measures was 0.39. Their behavioral genetic analysis partitioned the additive genetic variance into two components: a genetic component that was common to each of the three measures and specific additive genetic components for each of the measures. Approximately 79% of the additive genetic influence on the three measures was estimated to be common to all of them. The remaining variance on the three measures was split in roughly equal proportions between nonshared environmental influences that were common to each of the measures and nonshared environmental influences that were specific to each of the measures. The analysis, in common with many other behavioral genetic analyses, was compatible with the assumption that there were no shared environmental influences on these measures that lead individuals reared in the same family to resemble each other. This analysis supports the assumption that the genotypes that predispose individuals to develop neurotic tendencies are largely the same as the genotypes that predispose individuals to be depressed and anxious. Martin and Jardine obtained their measures of neuroticism, depression, and anxiety on a single occasion using self-report data. This may have inflated the covariances among the measures, although it is not clear whether this methodology would lead to differences in the magnitude of the correlations of MZ and DZ twin pairs. It is this latter difference that represents the foundation of the analysis indicating that most of the genetic variance that influences responses to each of these questionnaire measures is common to all of them.

Conclusion

General intelligence is a broad disposition of a person that is influenced by genotypes. Not only is it the case that scores on a measure of the disposition (IQ) are heritable, but the degree to which the components of the aggregate measure are related to the aggregate score is related to the heritability of components. Longitudinal continuities in intelligence are attributable in part to genetic continuities and a continuing genetic influence on growth and change in intelligence. Furthermore, relationships between intelligence and its biological substrate, as well as the more extended socially significant outcomes that are related to intelligence, are probably partially mediated by genetic characteristics. These results enable one to conceive of intelligence as a biologically based trait whose structure, continuity and discontinuity, biological correlates, and social and behavioral influences are linked by genetic covariances. There is no personality trait for which a comparable claim can be made. Personality traits are heritable. Their structure, known biological correlates, and longitudinal continuities and discontinuities and the diverse relationships between traits and socially relevant outcomes of traits may or may not be influenced by genetic covariances. In principle, what is known about the behavioral genetics of intelligence provides a model for the analysis of personality traits that may help researchers to decide whether the structure of any personality trait may properly be construed as being derivable from genetic relationships among its components and manifestations. The understanding of personality traits will be enhanced by behavioral genetic studies of personality traits that are longitudinal and use heteromethod methods to measure personality. Such research may help to broaden the understanding of personality traits. Research demonstrating continuities between childhood manifestations of a trait and socially relevant outcomes supports a broad conception of personality dispositions. Twin and adoption studies that analyze genetic and environmental contributions to change and continuity in personality, as well as the covariances among the diverse manifestations of personality characteristics, should enable researchers to discover the breadth of influence of genotypes on personality.

References

Baker, L. A., Vernon, P. A., & Ho, H. Z. (1991). The genetic correlation between intelligence and speed of information processing. *Behavior Genetics*, *21*, 351–367.

Brody, N. (1988). *Personality: In search of individuality*. San Diego, CA: Academic Press.

Brody, N. (1992). *Intelligence* (2nd ed.). San Diego, CA: Academic Press.

Buss, D. M., & Craik, K. H. (1980). The frequency concept of disposition: Dominance and prototypically dominant acts. *Journal of Personality, 48*, 379–392.

Campbell, D. T., & Fiske, D. W. (1959). Covergent and discriminant validation by the multitrait multimethod matrix. *Psychological Bulletin, 56*, 81–105.

Cardon, L. R., DiLalla, L. F., Plomin, R., DeFries, J. C., & Fulker, D. W. (1990). Genetic correlations between reading performance and IQ in the Colorado Adoption Project. *Intelligence, 14*, 245–257.

Cardon, L. R., Fulker, D. W., DeFries, J. C., & Plomin, R. (1992). Multivariate genetic analysis of specific cognitive abilities in the Colorado Adoption Project at age 7. *Intelligence, 16*, 383–400.

Caspi, A., Elder, G. H., Jr., & Bem, D. J. (1987). Moving against the world: Life-course patterns of explosive children. *Developmental Psychology, 23*, 308–313.

Caspi, A., & Moffitt, T. E. (1991). Individual differences are accentuated during periods of social change: The sample case of girls at puberty. *Journal of Personality and Social Psychology, 61*, 157–168.

Caspi, A., & Moffitt, T. E. (1993). *Continuity amidst change: A paradoxical theory of personality coherence.* Manuscript submitted for publication.

Cohn, S. J., Cohn, C. M. G., & Jensen, A. R. (1988). Myopia and intelligence: A pleiotropic relationship? *Human Genetics, 80*, 53–58.

Detterman, D. K., Thompson, L. A., & Plomin, R. (1990). Differences in heritability across groups differing in intelligence. *Behavior Genetics, 20*, 369–384.

Dworkin, R. H. (1979). Genetic–environmental influences on person–situation interactions. *Journal of Research in Personality, 13*, 279–293.

Eaves, L. J., & Eysenck, H. J. (1976). Genetic and environmental components of inconsistency and unrepeatability in twins' responses to a neuroticism questionnaire. *Behavior Genetics, 6*, 145–160.

Eysenck, H. J. (1967). *The biological basis of personality.* Springfield, IL: Charles C Thomas.

Eysenck, H. J., & Eysenck, M. W. (1985). *Personality and individual differences: A natural science approach.* New York: Plenum.

Eysenck, H. J., & Eysenck, S. B. G. (1968). *Personality structure and measurement.* San Diego, CA: EDITS.

Goldsmith, H. H., & Gottesman, I. I. (1981). Origins of variations in behavioral style: A longitudinal study of temperament in young twins. *Child Development, 52*, 91–103.

Heath, A. C., Eaves, L. J., & Martin, N. G. (1989). The genetic structure of personality: III. *Personality and Individual Differences, 10*, 877–888.

Heath, A. C., Neale, M. C., Kessler, R. C., Eaves, L. J., & Kendler, K. S. (1992). Evidence for genetic influences on personality from self-reports and informant ratings. *Journal of Personality and Social Psychology, 63*, 85–96.

Hershberger, S. I., Plomin, R., Pedersen, N. L., & McLearn, G. E. (1993). *Traits and metatraits: Their reliability, stability and shared genetic influence.* Manuscript submitted for publication.

Huesmann, L. R., Eron, L. D., Lefkowitz, M. M., & Walder, L. O. (1984). Stability of aggression over time and generations. *Developmental Psychology, 20*, 1120–1134.

Jackson, D. N., & Paunonen, S. V. (1985). Construct validity and the predictability of behavior. *Journal of Personality and Social Psychology, 49*, 544–570.

Jensen, A. R. (1987). Individual differences in the Hick paradigm. In P. A. Vernon (Ed.), *Speed of information-processing and intelligence* (pp. 101–175). Norwood, NJ: Ablex.

Jensen, A. R., & Sinha, S. N. (1992). Physical correlates of human intelligence. In P. A. Vernon (Ed.), *Biological approaches to human intelligence* (pp. 139–242). Norwood, NJ: Ablex.

Loehlin, J .C. (1992). *Genes and environment in personality development.* Newbury Park, CA: Sage.

Loehlin, J. C., Willerman, L., & Horn, J. M. (1987). Personality resemblances in adoptive families: A 10-year follow-up. *Journal of Personality and Social Psychology, 53*, 961–969.

Lynn, R. (1990). New evidence on brain size and intelligence: A comment on Rushton and Cain and Vanderwolf. *Personality and Individual Differences, 11*, 755–756.

Martin, N., & Jardine, R. (1986). Eysenck's contributions to behaviour genetics. In S. Modgil & C. Modgil (Eds.), *Hans Eysenck: Consensus and controversy* (pp. 13–47). Philadelphia: Falmer.

McCartney, K., Harris, M. J., & Bernieri, F. (1990). Growing up and growing apart: A developmental meta-analysis of twin studies. *Psychological Bulletin, 107*, 226–237.

Mednick, S. A., Gabrelli, W. F., Jr., & Hutchings, B. (1987). Genetic factors in the etiology of criminal behavior. In S. A. Mednick, T. E., Moffitt, & S. A. Stack (Eds.), *The causes of crime: New biological approaches* (pp. 74–91). London: Cambridge University Press.

Pedersen, N. L., Plomin, R., Nesselroade, J. R., & McClearn, G. E. (1992). A quantitative genetic analysis of cognitive abilities during the second half of the life span. *Psychological Science, 3*, 346–353.

Plomin, R., Chipuer, H. M., & Loehlin, J. C. (1990). Behavioral genetics and personality. In L. Pervin (Ed.), *Handbook of personality: Theory and research* (pp. 225–243). New York: Guilford.

Raven, J. C. (1938). *Guide to the standard progressive matrices.* London: H. K. Lewis.

Raz, N., Willerman, L., & Yama, M. (1987). On sense and senses: Intelligence and auditory information processing. *Personality and Individual Differences, 8*, 201–210.

Segal, N. L. (1985). Monozygotic and dizygotic twins: A comparative analysis of mental ability profiles. *Child Development, 56*, 1051–1058.

Smith, B. D. (1983). Extraversion and electrodermal activity: Arousability and the inverted-U. *Personality and Individual Differences, 4*, 411–419.

Thompson, L. A., Detterman, D. K., & Plomin, R. A. (1991). Associations between cognitive abilities and scholastic achievement: Genetic overlap but environmental differences. *Psychological Science*, *2*, 158–165.

Vernon, P. A. (1983). Speed of information processing and general intelligence. *Intelligence*, *7*, 53–70.

Wilson, R. S. (1983). The Louisville twin study: Developmental synchronies in behavior. *Child Development*, 54, 298–316.

Wilson, R. S. (1986). Continuity and change in cognitive ability profile. *Behavior Genetics*, *16*, 45–60.

Zuckerman, M. (1991). *Psychobiology of personality*. New York: Cambridge University Press.

CHAPTER 9

Genetic Perspectives on Personality

David C. Rowe

The consensus among behavioral scientists is that personality is the outcome of both genetic and environmental influences. Even in the controversial domain of IQ testing, most experts recognize the presence of genetic influence (Synderman & Rothman, 1987). The topic for this section is how to go beyond the simple recognition of genetic influences to use knowledge about genetic influences to develop more sophisticated and complete theories of personality.

To begin, it should be understood that the behavioral genetics approach is a theory of personality. This theory attributes variation in personality, at least partly, to variation in the structure and functioning of physiological systems. Genes contain the information that codes for various structural and regulatory proteins (and RNAs) that set pathways of physiological development. Although individual differences may be expressed in any physiological system, those in the brain and other parts of the nervous system are probably most relevant for behavioral traits.

With thousands of genes expressed in the nervous system, ample room exists for genetic variability in its function. Behavioral genetics theory further attributes the transmission of personality between generations, at least partly, to the transmission of genes; and it attributes differences in population means from one human group to another, at least partly, to accumulated individual differences.

With these features in mind, this theory does little to explain the total sociocultural context. According to findings on inherited personality traits, some people should be outgoing and gregarious, and some people should be shy. But a behavioral genetics approach cannot say whether these individuals will be riding horses or driving cars, or whether their economy will be based on farming or on industry. In Gestalt psychology's metaphor, behavioral genetics may miss the "ground" of cultural averages against which the "figure" of individual differences stands out.

Furthermore, the etiology of a trait does not indicate exactly *how* a trait works its effect. The transition from *how much* variation is explained by a particular source (i.e., genes or environment) to how that effect is produced is a long and arduous one. This statement is true regardless of whether one is considering the physiological or psychological realms of explanation. The discovery of a heritability of 50% doesn't indicate in which brain location these physiological differences reside or how they do their work. In the psychological realm, traits may modify behavior because people become consciously aware of their internal (biological trait) capabilities as well as their external (environmental) constraints and opportunities. The role of self-awareness in completing the circuit between traits and behavior is not well explained in behavioral genetics. If the great enthusiasm for cognitive models of behavior is to be followed, then the explanation of behavior must involve some kind of computer simulation of process mechanisms (provided that the simulation is not so complex as to be incomprehensible).

None of these limitations is unique to behavioral genetics. Each theory of personality has its "range of convenience" (i.e., tasks it does well and tasks it does poorly). Behavioral genetics is very good at accounting for trait variation; it is poorer at accounting for mechanism, especially when the "mechanism" and outcome may be in different realms of analysis. But certain theoretical formulations within behavioral genetics allow for greater illumination of mechanistic questions. For example, statistical techniques exist to assign composite variance components to specific environmental influences or to specific genes. Those personality researchers and teachers who want more than a "textbook knowledge" of behavioral genetics should seek these conceptual formulations and the related findings. This chapter addresses three different areas of potential interest to personality researchers and teachers: (a) new developments in the field, (b) evidence regarding environmental influences, and (c) possible linkages between evolutionary and behavioral genetics approaches to personality. I end with a word on genetic determinism.

New Developments in Behavioral Genetics

Personality researchers and teachers should be aware of the behavioral genetics literature on (a) the covariance of traits and (b) the theoretical assumptions of behavioral genetics research designs.

Analysis of Covariance

The analysis of covariance is a tool for understanding mechanism (see also chapter 21 by Hewitt, who discusses the value of this approach in relation to psychopathology). The correlation of two traits, X and Y, or between a trait and some outcome, can be apportioned to genetic and environmental influences according to the equation:

$$r_{XY} = h_X h_Y r_g + c_X c_Y r_c + e_X e_Y r_e,$$

where the path coefficients h, c, and e represent genetic, shared environmental, and nonshared environmental influences, respectively. The heritability of trait X is h_X^2, its shared environmentability is c_X^2, and its nonshared environmentability is e_X^2 (and similarly for trait Y). Heritability is an index of the proportion of trait total variation that is due to genetic variation among members of some population. This variation results from the substitution of one allele for another at different genetic loci relevant to a trait. Shared environmental influences are experienced similarly by family members and increase their similarity on a trait. Nonshared environmental influences produce behavioral differences among family members on a trait and are experienced uniquely by each individual. The correlation coefficients on the right side of the equation represent the association of the respective environmental and genetic influences on a trait within individuals. For instance, r_g depends on the extent to which the same genetic loci affect both traits simultaneously; r_e depends on the extent to which an individual who is exposed to a nonshared environmental influence on trait X is also exposed to the same one on trait Y. Comparably, r_c is the correlation of the shared environmental influence affecting traits X and Y, respectively.

Covariance analysis can be applied to learn more about the factor structure of traits. Ordinary factor analysis can reveal which traits cluster together and may share common determinants. Covariance analysis can break the *phenotypic* correlation matrix into *genetic*, *shared environmental*, and *nonshared environmental* matrices. The factor structure of the genetic matrix may reveal which traits are influenced by a common set of genes and therefore share common physiological influences. For example, speed of information processing as assessed by elementary cognitive tests shares genetic variation with general IQ. (Ho, Baker, & Decker, 1988). Factor analysis of the shared environmental matrix can reveal traits that are correlated for sharing environmental determinants. The nature of the specific environmental influence may be better understood in the context of several mutually correlated traits than with regard to a single trait. The analytic methods are just beginning to be widely applied.

Covariance analysis also elucidates the correlation of traits with developmental outcomes. For example, infant–mother attachment is associated with different temperamental traits. Process interpretations of this association differ, however. Environmentally, it may reflect either general child-rearing style as a shared environmental influence or the specific learning history of infant–mother interaction. Genetically, the kind of attachment children form may depend on their genetic trait dispositions. Given these different process models, a twin study could be conducted to determine whether the phenotypic association of attachment and temperament is mediated genetically or environmentally, and if the latter, then whether shared or nonshared environmental influences predominate. The unique learning history process model would suggest that the attachment–temperament association is dominated by nonshared environment. Thus, behavioral genetics analysis may reveal how two variables are associated in terms of causal genetic or environmental mechanisms.

Assumptions of Research Designs

Personality researchers and teachers should also be aware of the many attempts by behavioral geneticists to test assumptions of their methods. For example, Loehlin and Nichols (1976) pioneered efforts to study biases in twin studies in their book *Heredity, Environment and Personality.* Correlations were calculated between parental ratings of treatment similarity and within–twin pairs differences in personality. Those monozygotic (MZ) twins who were (according to parental reports) treated more alike failed to be more similar in personality traits. Thus, these findings support the equal environments assumption of the twin method, at least for these measures of parental treatment. In another approach to biases in twin studies, misclassified twins were used (i.e., MZ twins who were thought to be dizygotic [DZ] twins, and vice versa). Surprisingly, many parents incorrectly classify their twins, primarily because their doctors had misclassified the twins at birth based on placentation information. Misclassified twins present a nice opportunity to test for the existence of preferential treatment of MZ twins. If DZ twins who were thought to be MZ twins were more alike than other DZ twins, then one might interpret the greater similarity generally of MZ twins in terms of social influence. Contrary to this expectation, personality resemblance followed twins' true biological relatedness, not their perceived classifications (Scarr & Carter-Saltzman, 1979). Other studies have tested other assumptions of twin and adoption studies, such as effects of selective placement on IQ similarity in reared-apart twins (Bouchard, Lykken, McGue, Segal, & Tellegen, 1990) and effects of perceived similarity on actual similarity in adoption studies (Scarr, Scarf, & Weinberg, 1980).

Personality researchers and teachers, then, should be aware that behavioral geneticists have actively sought to test assumptions of their models; indeed, critics often complain that the methods are biased, without offering empirical evidence. The main point is that a violation of assumptions requires several facts to hold: (a) some form of unequal treatment according to twin type or adoptive child versus biological child, (b) some form of nonrandom placement, and (c) a relationship of treatment or placement to trait variation. In studies of reared-together twins, it is the last point in this reasoning that is the weakest. MZ twins do receive more similar treatments than do DZ twins in some domains. But these treatments may lack influence on personality development, so that twins are effectively alike in the environments that matter for personality development, despite the surface appearance of differential treatment.

Personality researchers and teachers should be aware, as well, of new empirical studies. In this category, I would include a large Swedish study and an American study of reared-apart twins (Bouchard et al., 1990; Pedersen, Plomin, McClearn, & Friberg, 1988), large English and Australian studies of reared-together twins (Eaves, Eysenck, & Martin, 1989), a follow-up of transracial adoptees (Weinberg, Scarr, & Waldman, 1992), a masterful survey of genetic and environmental influences on the "Big Five" personality traits (Loehlin, 1992), and first attempts to locate genes that affect behavior (Fulker et al., 1991).

Evidence Regarding Environmental Influences

Psychological environments differ among families: Some parents impose relatively strict rules on their children, whereas others are permissive; some parents display warmth and affection overtly, whereas others are cold and unemotional; some parents are cultured and interested in politics and foreign places, whereas others are more parochial. Economically, even excluding the poorest families, American families differ tremendously in their economic security and accumulated wealth.

But from working- to professional-class families, these psychological environments may fail to influence children's personality development. Personality teachers and researchers should be aware that rearing environments that appear to be psychologically different may be functionally equivalent for children's development (Scarr, 1992). That is, these families' psychological environments fall into a species-typical range that will support the development of a broad range of individual differences in personality traits—from conscientiousness to impulsiveness, or between any other trait extremes. Of course, opportunities present in a particular family environment may influence how a trait will be expressed: A conscientious middle-class child might have a newspaper route and private music lessons that are unavailable to many poor families, but the middle-class family environment does not determine whether a child is conscientious or impulsive.

In behavioral genetics, evidence for the functional equivalence of family environments comes from a lack of between-families environmental influence on many traits. Children's average phenotype on personality traits differs among families. But average differences in phenotype can arise from differences in parental genotypes as well as from differences among families in the specific environmental influences that siblings share.

Consider the evidence on extraversion and body weight, as shown in Table 1. In adoptive families, it makes little difference whether children

TABLE 1

Familial Correlations for Weight and Extraversion

Relationship	Weight	Extraversion
MZ twins reared together[a]	0.80	0.55
MZ twins reared apart[a]	0.72	0.38
DZ twins reared together[b]	0.43	0.11
Biological parent-child[b]	0.26	0.16
Biological siblings[b]	0.34	0.20
Adoptive parent-adopted child[c]	0.04	0.01
Unrelated siblings reared together[c]	0.01	−0.06

Note. MZ = monzygotic; DZ = dizygotic. The weight correlations are from "The Nature of Environmental Influences on Weight and Obesity: A Behavior Genetic Analysis" by C. M. Grilo and M. F. Pogue-Geile, 1991, *Psychological Bulletin, 110,* pp. 520–537; copyright 1991 by the American Psychological Association. The extraversion correlations (weighted average *r*s in the original) are from *Modern Personality Psychology: Critical Reviews and New Directions* (pp. 355 and 356) edited by G. Caprara and G. L. Van Heck, 1992, Hertfordshire, UK: Harvester-Wheatsheaf; copyright 1992 by Harvester-Wheatsheaf. The *unrelated siblings* weight correlation omits step-siblings because of parental assortative mating.

[a]Genetic relatedness = 1.0. [b]Genetic relatedness = 0.5. [c]Genetic relatedness = 0.0.

are raised by overweight or normal-weight adoptive parents: The specific environmental influence of parental weight cannot make children alike in weight because the sibling correlations on the weights of biologically unrelated children were about zero ($rs = 0.01$). A lack of shared rearing influence on children's weights undermines the expectation of many environmentally oriented theories that parental eating habits are an influence on their children's obesity. But sibling correlations do increase with an increase in their genetic similarity, from unrelated children, to siblings, and then to MZ twins; and thus, genotype influences variation in weight. Heredity, but not specific influences of shared environment, then can account for children in some families being heavier than those in other families. As shown in Table 1, correlations were decidedly lower for extraversion than for weight, which suggests more environmental (unshared) influence on extraversion than on weight; and DZ twin correlations were less than half of those for MZ twins, which suggests nonadditive genetic influence or some special environmental effect for extraversion not present in the case of weight. One finding was similar for both traits, however: In the adoptive groups, the weight and extraversion correlations gave equally little evidence of shared family rearing influence on their development.[1]

Thus, the functional equivalence of treatments can be expressed in an important, practical way as an absence of effects from shifting the group mean on many family treatments. If 100 sociable, gregarious fathers adopt infants and another 100 shy, reticent fathers adopt infants, then the proportions of children in both groups who become extraverted adolescents will be about the same.

Genetic inheritance does produce "reliable" effects. The biological children of parents who are extreme on a personality trait will be more extreme on the same trait than children in general. The consistency of genetic influence has led to the design of high-risk studies of schizophrenia, alcoholism, and manic–depressive psychology, all by identifying

[1]Although the difference in correlations between MZ twins who are raised together and those who are raised apart may appear to suggest the influence of a shared family environment, these differences may reflect special twin environments or sampling biases.

affected biological parents. Notice that, because of the lack of specifiable environmental effects, high-risk studies have not been designed to select on particular child-rearing treatments in families in which both parents are phenotypically unaffected.

Personality researchers and teachers should be aware that associations between psychological environments and developmental outcomes may be genetically mediated. Genetic variation can cause variation in psychological environments through its influence on behavioral phenotypes. The psychological environment, after all, is a product of some person's behavior—a parent, a teacher, or a peer. Insofar as heritable traits and dispositions affect that person's behavior as an environmental stimulus to another person, variation in what we label as *environment* can be etiologically genetic.

In some situations, the association between the psychological environment and developmental outcomes is genetically mediated. The passive genetic correlation between parent and child is a prime example. The same cluster of genes may produce two disparate behavioral effects: In a parent, they may influence variation in child-rearing styles; whereas in a child, they may influence variation in outcome traits. This pathway of common causality is possible because a child shares half that parent's genes through either the egg or sperm cell. For example, more punitive discipline in parents of aggressive children may reflect a cluster of genes manifested as greater emotionality in the parent and physical aggression in a young child. This pathway of genetic mediation of environmental association renders much of the research on childhood socialization ambiguous as to the relative influence of genetic and family environmental influences because family members are related genetically as well as environmentally.

The Family Environments Scale's (FES) Personal Growth Factor can illustrate genetic mediation of "environmental" associations (Plomin, Loehlin, & DeFries, 1985). Children who had fewer behavioral problems lived in families that were more supportive and expressive (FES Personal Growth Factor). The mean correlation of this family measure with the children's behavioral problems was 0.23 in nonadoptive families in which parent and child were biologically related. But in adoptive families, it dropped to only 0.07. Under the expectation of environmental influence, of course, the two correlation coefficients should be about equal. The difference between them (0.23–0.07) estimates the part of the correlation due to the passive association of all aspects of family environment with genotypes that are transmissible to offspring. Thus, 0.16, or 69% of the original association in nonadoptive families, was due to genetic mediation of the family environment–child outcome association.

One of the most frequently cited environmental measures is social class. Although measures of parental education, occupation, and income are associated with many broad environmental differences among families, it is clearly wrong to view them as uncontaminated by genetic influences. The idea that "social class" may contain genetic variation is not new (Herrnstein, 1973). Social class statuses are attained by individuals through their own efforts in education, job selection, and climbing up corporate and institutional hierarchies, all behaviors that may be directly and indirectly influenced by a broad range of heritable intellectual and personality traits (Waller, 1971). "Social mobility" tables will show one that, even if every poor adult today were immediately assured a higher income, a lower class would be recreated in the next generation because a good proportion of a its population arises in the children of middle class parents, who fall in social status relative to that of their natal families.

In America today, excluding the extremes of inherited wealth and isolated pockets of poverty in inner city and rural areas, about 40–50% of the variation in earnings is due to genetic variation (Taubman, 1976).[2] In practical terms, this means that a process of genetic self-selection occurs whereby heritable traits lead to increases or decreases in social statuses until individuals with different occupations and income levels differ in their genotypes as well as in their phenotypes. Scarr and McCartney (1983) used the metaphor "niche picking," in analogy with different organisms adopting a preferred environmental habitat best suited for their means

[2]Based on Taubman's (1976) models that used assumptions closest to those of the classical twin method. The subjects were White male twins who had been in the U.S. armed forces during World War II. The earnings' data were for 1973.

of survival and reproduction. With these facts in mind, no research on social class should be interpreted as unambiguously informative about environmental influences on developmental outcomes.

There is clearly a need to find ways to examine environmental influences that are not scientifically ambiguous because they merely capitalize on existing genetic variation. In Rowe and Waldman (chapter 19), a number of approaches to examining environmental influences in the context of heritable individual differences are discussed. Briefly, family environmental influences may become more important at environmental extremes. Thus, one may find evidence for greater shared environmental influences in particular social contexts. Models can test this by examining whether parameter estimates of shared environmental influence vary with environmental context. Furthermore, environmental influences may be nonshared and experienced uniquely by each family member. Behavioral genetics models can also examine the nonshared influence of specific environmental measures.

Another point, sometimes missed in developmental psychology's focus on families, is that environmental influence can occur through a host of nonfamilial routes. Most broadly, a reproduction of environmental context occurs through existing social institutions that structure the education of the young and the means of economic production. More narrowly, environmental influence occurs through same-age peer groups and through adults other than parents, including teachers, other biological relatives, and acquaintances. Dramatic environmental changes may spread through nonfamilial influence routes to produce changes in developmental outcomes. In one example, consider the progressively more negative attitudes toward drug use from 1980 to 1992 and the concomitant decrease in the use of illegal drugs; or consider the institutional change of increasing the legal age of drinking, with its concomitant effect of reducing car accident rates among adolescents. Thus, interventions need not proceed merely through local family environment but can be communitywide. Furthermore, a moderate heritability does not mean an absence of social influence on how traits are expressed as different behavioral outcomes. These ideas receive less attention in developmental psychology because of the discipline's primary focus on development in the family context, but they are part of a broader social science perspective.

Combining Evolutionary and Quantitative Genetic Approaches

Many environmentally oriented researchers may be unaware of the intellectual separation of two different approaches to the study of behavior with a genetic basis: sociobiology and behavioral genetics. Wilson (1975) defined *sociobiology* as the "systematic study of the biological basis of all social behavior" (p. 4). The term sociobiology has acquired a somewhat negative connotation in the social sciences because Wilson boldly argued that this new discipline would recast the foundations of the established social sciences and humanities—an invasion of intellectual territory that was not greeted warmly. Buss (1991) used the term *evolutionary psychology* rather than sociobiology because the former focuses attention on psychological mechanisms shaped by evolution. Whatever the terminology, these new disciplines are all intellectual heirs of Darwinism in their common identity in formulating an understanding of behavior based on the theoretical concepts of a reenergized evolutionary biology.

One of the major causes of intellectual estrangement between behavioral genetics and sociobiology is the emphasis on individual differences in the former discipline and on universal behavioral mechanisms in the latter. Behavioral geneticists study individual differences in IQ, childhood temperament, adult personality, physical traits, and a host of other variables of scientific or social importance. Sociobiologists study universal or species-typical behavioral mechanisms in the areas of altruism, sexual competition and jealousy, other sex differences, aggression, nepotism, xenophobia, and similarly important domains of social behavior. In their studies, the sociobiologists suggest that just one, or some small number, of behavioral patterns, are adaptive outcomes of humans' evolutionary heritage. Patterns for groups may differ, particularly between males and females who are subject to somewhat different evolutionary pressures because of the biological constraints of reproduction. But individual differences are regarded as only "noise," and they are given no more respect than error variance in classical experimental research designs (see Buss, 1991, for an exception).

Tooby and Cosmides (1990) articulated a basis for the view that adaptations are pan-specific (at least within one gender). According to their view, adaptations are mechanisms or systems "designed" by natural selection to answer the problems posed by physical, ecological, and social selection pressures. Adaptive design is complex because "each part must present a uniform, regular, and predictable set of properties to the system" (Tooby & Cosmides, 1990, p. 28). Given interdependent parts, adaptations can only be produced by alleles that everyone in a population shares, that is, by alleles that do not vary among individuals in a population so that everyone has exactly the same genotypes. If not, they argue, then an adaptation would be broken apart by sexual recombination in each generation. In other words, for alleles that vary in a population, a child is different from both parents: Any carefully crafted adaptation would fall apart in just one generation. Taken to an extreme conclusion, adaptations become Platonic ideal types, and genetic variation may be an evil that detracts from the possibility of behavioral perfection (although it may serve other functions, such as protection from microbes).

From the individual difference perspective, people seem hardly to pursue one, or even a small number, of life histories. Lives are tremendously variegated, with many possible life histories of reproduction, social interaction, economic activity or inactivity, and so on. People pick niches in which their particular genetic propensities do best; they avoid niches that conflict with and do not support their propensities. Tooby and Cosmides (1990) offered some concepts to integrate this kind of genetic variation into sociobiological theory. Specifically, they noted that genetic variation may modulate thresholds of response to particular social situations of evolutionary importance. Provocatively, they also suggested that the self-awareness of heritable characteristics may direct the choice of particular behavioral strategies. A man who is aware that he is small and easily made anxious does not try to physically intimidate other men because he lacks both the psychological and physical trait prerequisites. These ideas, and others (e.g., Crawford & Anderson, 1989), may help to integrate individual difference concepts into sociobiology.

Although the issue is certainly undecided, I hold the view that individual variation in personality and intellectual traits may be of some adaptive importance. I argue that a possible role of individual differences needs to be given greater consideration in (a) reconstructions of human evolutionary history; (b) comparisons of humans with nonhuman primates; and (c) theories of variation in reproductive success in modern societies, despite their distance from the Pleistocene environments of humans' evolutionary lineage. As a caution against avoiding adaptive "niche" interpretations of genetic variation, consider that variant genetic alleles in mollusks, which were previously thought to be purely neutral mutations, may be in reality adaptive ones. That is, mollusk populations manage to maintain similar levels of genetic variability at allozyme loci, despite genetic isolation from one another (as determined from molecular genetic markers; see Kari & Avise, 1992).[3]

If sociobiologically oriented scholars can gain from a greater consideration of individual differences, then behavioral geneticists can learn from sociobiology's focus on phenotypes of evolutionary significance. Most personality traits map poorly onto behavioral domains of direct, evolutionary consequence. Sexual and reproductive life histories are not typically assessed. Few behavioral genetics studies have focused on individual differences in prejudice, altruism directed toward relatives versus nonrelatives, or restrained-versus-unrestrained sexual strategies. The merger of behavioral genetics and sociobiology will depend on behavioral geneticists integrating evolutionary concerns into their studies. Personality teachers and researchers should explore both the behavioral genetic and sociobiological approaches to the genetic underpinnings of human social behavior.

Resistance to "Biological Determinism"

Personality teachers and researchers should introduce students to the concept of probabilistic causation (Mulaik, 1987). Few practicing social

[3]Allozymes are alleles coding the proteins for metabolic enzymes.

scientists would endorse perfect determination of any behavior; in complex systems, outcomes are probabilistically related to initial conditions. As Mulaik specifically stated: "[T]he values of an independent or causal variable do not determine the specific *outcomes* of a dependent variable but rather the specific (conditional) *probability distributions* with which the values of the outcome variable occur" (p. 24; italics in original).

Nonetheless, the idea of probabilistic determination may inadvertently lead to another misconception: that no underlying causal process is at work. In Einhorn and Hogarth's (1986) creative illustration, they imagine that the causes of birth were unknown. A contingency table relating sexual intercourse to pregnancy could find a statistical association of less than unity if 20 of 100 women who experienced sexual intercourse were pregnant and if 5 of 100 not experiencing intercourse were pregnant due to either lying or data-entry errors. But this correlation of 0.34 conceals a deterministic causal process: the fertilization of the egg cell by the sperm and its subsequent growth in utero.

To apply these ideas to "genetic determinism," the relationships between particular genotypes and phenotypes is clearly probabilistic. About 13% of the children of a schizophrenic parent become schizophrenic (Gottesman, 1991). With any one child, one would be highly uncertain about whether that child would be schizophrenic. But in a group of 1,000 children of schizophrenic mothers, one could be highly certain that about 130 children would be schizophrenic. The probability of finding no schizophrenic children is so low as to make some cases of schizophrenia appear almost inevitable; thus, probabilistic causation does not imply that all outcomes are improbable. Moreover, the physiological processes linking particular genetic alleles to increased risk presumably occur through deterministic systems. Thus, although they may seem so on the surface, probabilistic causation and deterministic process are not incompatible ideas: Causation can be investigated at lower, more reductionistic levels in which deterministic processes may be clearer; but at each level of analysis, probabilistic processes will also enter. This logic applies as well to the influence of specific environmental mechanisms on trait phenotypes. No reason exists to suppose that behavioral geneticists and environmentally oriented researchers cannot adopt a common philosophical perspective on causal process.

References

Bouchard, T. J., Lykken, D. T., McGue, M., Segal, N. L., & Tellegen, A. (1990). Sources of human psychological differences: The Minnesota study of twins reared apart. *Science, 250*, 223–228.

Buss, D. M. (1991). Evolutionary personality psychology. *Annual Review of Psychology, 42*, 459–491.

Caprara, G., & Van Heck, G. L. (1992). *Modern personality psychology: Critical reviews and new directions*. Hertfordshire, UK: Harvester-Wheatsheaf.

Crawford, C. B., & Anderson, J. L. (1989). Sociobiology: An environmentalist discipline? *American Psychologist, 44*, 1449–1459.

Eaves, L. J., Eysenck, H. J., & Martin, N. G. (1989). *Genes, culture and personality: An empirical approach*. San Diego, CA: Academic Press.

Einhorn, H. J., & Hogarth, R. M. (1986). Judging probable cause. *Psychological Bulletin, 99*, 3–19.

Fulker, D. W., Cardon, L. R., DeFries, J. C., Kimberling, W. J., Pennington, B. F., & Smith, S. D. (1991). Multiple regression analysis of sib-pair data on reading to detect quantitative trait loci. *Reading and Writing: An Interdisciplinary Journal, 3*, 299–313.

Gottesman, I. J. (1991). *Schizophrenia genesis: The origins of madness*. New York: Freeman.

Grilo, C. M., & Pogue-Geile, M. F. (1991). The nature of environmental influences on weight and obesity: A behavior genetic analysis. *Psychological Bulletin, 110*, 520–537.

Herrnstein, R. J. (1973). *IQ and the meritocracy*. Boston: Atlantic-Little Brown.

Ho, H., Baker, L., & Decker, S. N. (1988). Covariation between intelligence and speed of cognitive processing: Genetic and environmental influences. *Behavior Genetics, 18*, 247–261.

Kari, S. A., & Avise, J. C. (1992). Balancing selection at allozyme loci in oysters: Implications from nuclear RFLPs. *Science, 256*, 100–102.

Loehlin, J. C. (1992). *Genes and environment in personality development*. Newbury Park, CA: Sage.

Loehlin, J. C., & Nichols, R. C. (1976). *Heredity, environment, and personality*. Austin: University of Texas Press.

Loehlin, J. C., & Rowe, D. C. (1992). Genes, environment, and personality. In G. Caprara & G. L. Van Heck (Eds.), *Modern personality psychology: Critical reviews and new directions* (pp. 352–370). Hertfordshire, UK: Harvester-Wheatsheaf.

Mulaik, S. A. (1987). Toward a conception of causality applicable to experimentation and causal modeling. *Child Development, 58*, 18–32.

Pedersen, N. L., Plomin, R., McClearn, G. E., & Friberg, L. (1988). Neuroticism, extraversion, and related traits in adult twins reared apart and together. *Journal of Personality and Social Psychology, 55*, 950–957.

Plomin, R., Loehlin, J. C., & DeFries, J. C. (1985). Genetic and environmental components of "environmental" influences. *Developmental Psychology, 21*, 391–402.

Scarr, S. (1992). Developmental theories for the 1990s: Development and individual differences. *Child Development, 63*, 1–19.

Scarr, S., & Carter-Saltzman, L. (1979). Twin method: Defense of a critical assumption. *Behavior Genetics, 9*, 527–542.

Scarr, S., & McCartney, K. (1983). How people make their own environments: A theory of genotype → environment effects. *Child Development, 54*, 424–435.

Scarr, S., Scarf, E., & Weinberg, R. A. (1980). Perceived and actual similarities in biological and adoptive families: Does perceived similarity bias genetic inferences? *Behavior Genetics, 10*, 445–458.

Synderman, M., & Rothman, S. (1987). Survey of expert opinion on intelligence and aptitude testing. *American Psychologist, 42*, 137–144.

Taubman, P. (1976). The determinants of earnings: Genetics, family and other environments: A study of white male twins. *American Economic Review, 66*, 858–870.

Tooby, J., & Cosmides, L. (1990). On the universality of human nature and the uniqueness of the individual: The role of genetics and adaptation. *Journal of Personality and Social Psychology, 58*, 17–67.

Waller, J. H. (1971). Achievement and social mobility: Relationships among IQ score, education, and occupation in two generations. *Social Biology, 18*, 255–263.

Weinberg, R. A., Scarr, S., & Waldman, I. D. (1992). The Minnesota transracial adoption study: A follow-up of IQ test performance at adolescence. *Intelligence, 16*, 117–135.

Wilson, E. O. (1975). *Sociobiology: The new synthesis*. Cambridge, MA: Belknap Press.

CHAPTER 10

The Idea of Temperament: Where Do We Go From Here?

Jerome Kagan, Doreen Arcus, and Nancy Snidman

Psychology is entering its modern synthesis, which is analogous to the theoretically rich movement that followed the evolutionary biologists' recognition that both mutation and natural selection participate in the evolution of animal forms, by recognizing that the major variation in human profiles of cognition, behavior, and emotion emerge from a combination of biological variation and sequences of environmental experiences.

Temperament and prenatal events are two major sources of early biological variation. The most important environmental encounters during the first decade of human life include experiences within the family and with other children. During adolescence, the values of an individual's culture and subculture ascend in influence. This chapter presents two

The research reported in this chapter was supported, in part, by grants from the John D. and Catherine T. MacArthur Foundation and the Leon Lowenstein Foundation. We thank John Hendler, Wang Yu-feng, and Sheila Greene for their contributions to the research described in this chapter.

examples of this synthesis in psychology based on research from our laboratory at Harvard University. The first example describes the influence of maternal behavior on infants who are born with a high-reactive profile—a temperamental quality that is related to the popular concept of ease of arousal (Rothbart, 1989). The second example considers the implications of differences between Asian and Caucasian infants in this temperamental quality.

The Influence of Maternal Behavior on Reactive Infants

Continued study of the temperamental categories of inhibited and uninhibited children, which are defined by dramatic differences in fear and sociability during the 2nd year of life, has led us to hypothesize that the two groups of children differ in the excitability of the amygdala and its multiple circuits to the striatum, cingulate, frontal cortex, central gray, hypothalamus, and sympathetic chain, with the former group of children being more excitable than the latter (Kagan & Snidman, 1991b). Research with animals indicates that the amygdala and its circuits are particularly important participants in producing states of fear to novel events (Dunn & Everitt, 1988). The research on potentiated startle represents a particularly persuasive source of support for the involvement of the central nucleus of the amygdala in conditioned fear states (Davis, 1992; Davis, Hitchcock, & Rosen, 1987).

If inhibited and uninhibited children differ in the excitability of the amygdala and its circuits, then they should, as infants, also differ in reactivity to stimulation because the amygdala receives sensory information from all modalities. In addition, the amygdala is a source of two important efferent circuits whose targets could produce variation among infants in motor reactivity and crying to unfamiliar stimuli (Dunn & Everitt, 1988; Kelley, Domesick, & Nauta, 1982; Mishkin & Aggleton, 1981). One system, which originates in the basolateral area of the amygdala, projects to the ventromedial striatum and skeletal motor system (Nauta, 1986). When this circuit is activated by stimulation, especially unfamiliar events, infants are likely to show an increase in motor activity that takes the form of increased muscle tension and flexing and extending of the limbs (Rolls & Williams, 1987). A second system, which originates in the central nucleus of the amygdala, projects to the cingulate cortex and central gray. One set of projections to the central gray mediates defensive motor responses that, in infants, can take the form of arching of the back. Another target of the cingulate cortex and central gray is the area of the vocal cords and larynx. Research with animals suggests that distress calls are mediated by this circuit (Jurgens, 1982), which is likely to participate in the distress cries of human infants. Because high levels of both motor activity and crying to unfamiliar stimuli can be mediated by low thresholds in the amygdala and its projections, it follows that the study of these two behaviors might supply early predictors of inhibited and uninhibited temperamental types. Specifically, the combination of high motor activity and frequent crying to novel stimulation should predict the later display of inhibited behavior; the opposite profile should predict uninhibited behavior.

The reader should note that the terms *inhibited* and *uninhibited* refer to the behavioral profiles of timidity or lack of it that are displayed during the 2nd year of life. The profiles of reactivity to stimulation that are observed at 4 months of age index the possession of a physiology that theoretically predisposes the child to become inhibited or uninhibited later. Allergies provide an analogy. The presence of high levels of immunoglobulin E (IgE) in children predisposes that person to develop asthma. But a child with a high level of IgE who does not show any symptoms of wheezing is not classified as asthmatic. That diagnosis is only applied after the characteristic symptoms appear.

We administered a relevant set of procedures to over 600 healthy, term, Caucasian-American, primarily middle-class 4-month-old infants from two different cohorts; these children were from the Boston, Massachusetts, area. Data from the first cohort of 100 infants have been described previously (Kagan & Snidman, 1991a). The second larger cohort of 560 infants was administered a similar 40-minute battery of episodes that included colorful moving mobiles, tape-recorded speech, and application to the nostrils of a cotton swab that was dipped in dilute butyl alcohol. As expected, the infants differed in the frequency and vigor of motor activity, especially flexing and extending of the limbs and arching of the back, as well as fretting and crying. The large sample size of this second cohort permits confidence in the estimates of prevalence of four different reactive types (Kagan & Snidman, 1991a). Table 1 outlines the procedures administered to this cohort.

On the basis of the infants' responses to the test battery, the infants were divided into four groups. The largest group, called *low reactive* and comprising about 35% of the sample, was characterized by infrequent motor activity and minimal crying to the battery. A second group, called *high reactive* and comprising about 20% of the same population, displayed a qualitatively different profile that was characterized by frequent displays of vigorous motor activity—pumping of the legs, extending of the limbs, arching of the back—and frequent fretting and crying, especially during periods of motor arousal. The third group included the 25% of the sample that showed low levels of motor activity but frequent crying, and the final group included the 10% that displayed high levels of motor activity but minimal crying. The small number of remaining infants were difficult to classify because of either their failure to complete the battery or their borderline profile.

The children were observed in the laboratory at 14 months of age as they encountered a series of 17 unfamiliar situations, which included confrontations with people, objects, and situations. The child could display a fear reaction on each of the 17 standardized episodes, where *fear*

TABLE 1
Overview of the Investigation of Relation of Experience to the Actualization of Inhibited and Uninhibited Profiles

Age (in months)	Procedure	Sample size
4	Battery to evaluate reactivity	560
5–13	Home visits	96
14	Battery to assess inhibited and uninhibited profiles	460
21	Battery to assess inhibited and uninhibited profiles	275
21	Q-sort to mothers	253

was narrowly defined as fretting or crying to an unfamiliar event or procedure or reluctance to approach an unfamiliar person or object despite an invitation to do so. The 9 episodes that produced a fear response most often were the placement of electrodes on the child for the recording of heart rate, the placement of a blood pressure cuff for the recording of blood pressure, reluctance to imitate an examiner who requested the child to put his or her hand into a cup containing either water or red- or black-colored liquid, refusal to taste a drop of liquid from an eye dropper, distress to the facial frown and stern voice of the examiner speaking a nonsense phrase, presentation of papier-mâché puppets accompanied by a tape-recorded voice speaking nonsense, refusal to approach a stranger who invited the child to play with her toy, refusal to approach a robot despite an invitation to do so, and finally, refusal to approach an adult with a black cloth over her head and shoulders who invited the child to come and play with her.

Of the 460 children who were seen at 14 months of age, 40% showed low fear (zero or one fear); 32% showed moderate fear (two or three fears); and the remaining 28% showed high fear (four or more fears). There was a striking relation between the earlier reactive styles and fearfulness, with the largest difference occurring between the high- and low-reactive infants (Table 2). These results replicate those found with the first, smaller cohort (Kagan & Snidman, 1991a). Of the high-reactive

TABLE 2

Proportion of Each Reactive Group Showing Low, Moderate, or High Fear at 14 Months of Age

	Fear rating		
Reactive group	Low	Moderate	High
High reactive	10	30	60
Low reactive	62	28	10
High motor–low cry	43	29	28
Low motor–high cry	27	40	33

Note. Reactive groups were determined at 4 months of age. Fear ratings were determined by the number of fears displayed during the battery of confrontations: low = 0–1 fears; moderate = 2–3 fears; and high = >4 fears.

infants, 60% showed high fear and only 10% showed low fear; by contrast, 62% of the low-reactive infants showed low fear and only 10% showed high fear ($p < .0001$). The other two groups showed an intermediate number of fears; 28% of the high motor–low cry children and 33% of the low motor–high cry children showed high fear.

These differences were preserved through 21 months of age for 275 children who were administered 21 age-appropriate episodes that were characterized by unfamiliar events, people, and situations. As at 14 months of age, the children were classified as low, moderate, or high fear depending on the number of episodes on which the child displayed the criterial behaviors. Of the 70 high-reactive infants, 50% showed high fear and only 20% showed low fear. Of the 113 low-reactive infants, 21% showed high fear and 51% showed low fear ($p < .001$). When each child's behavior on the batteries at both 14 and 21 months of age were combined, 34% of the high-reactive infants showed high fear at both ages whereas only 5% showed low fear. Among the low-reactive infants, only 6% showed high fear and 38% were low fear at both ages ($\chi^2 = 36.9$; $p < .0001$).

Arcus and colleagues selected from the originally large sample of 560 infants 48 infants who were high reactive and 48 who were low reactive at 4 months of age and observed them in their homes when they were 5, 7, 9, 11, and 13 months of age. Half of the children were first born, and half were later born; half were boys, and half were girls. The purpose of these visits was to determine whether these investigators could understand the variation in fear score at 14 months of age as a function of both the infant's original temperament and maternal practices (Arcus, Gardner, & Anderson, 1992).

The videotapes from these visits have been analyzed for the first 20 high-reactive and 20 low-reactive first-born children who had completed all visits. The results of that analysis suggest an interaction between temperament and home experience on fearful behavior. The maternal behaviors that predicted the fear scores of the high-reactive infants appear to be the result of different maternal philosophies. Some mothers believe they should be sensitive and protective of their infants; others believe that the child must adapt and learn to cope with minor stresses. The ways each mother balances these two imperatives in contemporary American culture is a determinant of her behavior with her infant, and this variation influences how fearful the high-reactive child will be at 14 months of age. One variable was the proportion of time the mother held the infant while it was fretting or crying, especially during the first two visits at 5 and 7 months of age when mothers hold children a great deal. The variable was defined as the ratio of the time the mother held the infant while it was fretting or crying minus the time the mother held the infant while it was calm, divided by the sum of the total time the child was held. A similar variable was created for physical affection toward the infant. A composite variable combined proportion of time holding and proportion of time displaying physical affection.

A second relevant behavior, called *limit setting*, was derived primarily from the visits to the home at 9, 11, and 13 months after the child became mobile. The mother's reaction was coded whenever the infant was engaged in any one of a delimited number of transgressions: behaviors defined as dangerous to the infant or to others, a violation of standards on cleanliness, or an act that was not healthy for the infant. Some examples include reaching for a knife or mouthing an object that might cause the baby to choke. The limit-setting variable was defined as the proportion of all transgressions on which a mother issued a firm, direct command or prohibition or directly blocked the child's access to a forbidden object. A mother who set high limits was characterized by firm, direct strategies, although none of the maternal behaviors was harsh. A mother with a low limit-setting score was characterized by very indirect strategies.

These two maternal behaviors—holding and affection when crying and limit setting—contributed approximately 35% of the variance to the fear score at 14 months of age for high-reactive infants but very little variance for the low-reactive infants. High-reactive infants with mothers who used firm and direct limit setting had significantly lower fear scores at 14 months of age than did high-reactive infants with mothers who were less firm. The mother's response to the child's fretting and crying made a significant, but smaller, contribution to the child's fear score at 14 months of age with high values for maternal holding associated with more fear among high-reactive infants.

Arcus interprets these data as indicating that mothers of high-reactive children who are unusually responsive to their infant's fretting and, in addition, indirect in their limit setting later in the 1st year facilitate the actualization of a highly fearful profile. This relation holds only for infants who possess a high-reactive temperament. This finding may be due, in part, to the fact that infants with mothers who set limits are less likely to have their crying reinforced by the parent and, in addition, are given an opportunity to develop coping strategies to deal with the minor frustrations that occur every day. As a result, when the child encountered the unfamiliar events in the laboratory that generated uncertainty, the child was more likely to tolerate the temporary intrusion and novelty and was less likely to cry.

These results are supported by data from a Q-sort that was administered to the mothers of all infants when their children were seen in the laboratory at 21 months of age. The Q-sort consisted of 14 items describing positive qualities of children, and the mothers were asked to rank the 14 items with respect to their relative desirability when their child was 5 years old. One of the items stated, "I would like my child to obey me most of the time." The infants whose mothers placed that item in Ranks 1–6 (i.e., as desirable) were less fearful at 14 and 21 months than infants whose mothers placed that item as less desirable (Ranks 7–14). Six of the 63 high-reactive infants were unusual because they showed low fear at both 14 and 21 months of age. The mothers of all 6 infants placed the *obey* item as highly desirable; not one high-reactive infant whose mother ranked this item as undesirable showed consistently low fear.

The Q-sort was administered to 253 mothers of children who have been observed at both 14 and 21 months of age. Of the infants with low-fear scores at both 14 and 21 months of age, 61% had mothers who ranked the obey item as desirable. Among the remaining children with moderate- or high-fear scores, 45% of the mothers ranked the obey item as desirable (χ^2 square $= 5.4$; $p < .05$).

These data provide a relatively clear example of how a combination of a specific familial environment and a particular temperamental bias can produce a high- or low-fear profile in the 2nd year of life. The biology of the child and the social environment both contribute to the evolution of the psychological phenotype, which the 19th century called *character*.

Population Differences in Ease of Arousal

Over 20 years ago, Freedman and Freedman (1969) suggested that Asian-American infants, who were only a few days old and born in California, were lower in their ease of arousal compared with Caucasian-American infants. Caudill and Weinstein (1967) and, more recently, Lewis (1989) reported that Japanese infants differ from Caucasians in ease of arousal during the 1st year of life. Although the Japanese data are based on older infants, the Freedman and Freedman evidence, which was gathered during the first few days of life, imply the operation of genetically based differences between Asian and Caucasian infants in ease of arousal to stimulation.

We recently administered the 4-month battery described earlier to 106 Irish infants who were born in Dublin and 80 Chinese infants who were born in Beijing. The Irish data were gathered by John Hendler; the Chinese data were gathered by Wang Yu-feng of Beijing Medical College. The administration of the battery to both the Irish and Chinese infants was identical to that described earlier for the Boston infants.

The videotapes of these sessions were coded for motor activity, vocalization, fretting, crying, and smiling. A total motor score was calculated based on the frequency of movements of both arms, both legs, bursts of movement of either arms or legs, extensions of both arms or legs, or arches of the back. Vocalization, smiling, and fretting were coded as number of trials on which each of those behaviors occurred. Crying was coded as the number of seconds the child cried. Because the examiner or mother terminated the trial on which a child cried for more than a few seconds, the duration of crying rarely exceeded 4 or 5 seconds on any particular trial. The reliabilities of the scoring of these variables, which were evaluated with independent coders, were $r = 0.83$ for motor activity, $r = 0.73$ for crying, $r = 0.94$ for fretting, $r = 0.77$ for vocalization, and $r = 0.55$ for smiling.

Analysis of the videotapes of the infants' behavior revealed a dramatic difference between the Caucasian and Chinese infants. The Chinese infants were significantly lower in motor activity, irritability, and vocalization compared with the Irish infants and the Caucasian-American infants from the previously described Boston area study (Table 3). Smiling was the only behavior for which the Chinese infants did not differ from the other two groups.

These results suggest population differences in ease of arousal that might be a function of the reproductive isolation of the two groups. As noted earlier, motor activity and crying are mediated, in part, by the excitability of the amygdala and its circuits to the corpus striatum, cingulate, central gray, and hypothalamus. The differences in ease of motor activity and crying suggest a muting or modulation of these circuits in the Chinese infants. It is relevant that Lin, Poland, and Lesser (1986) found that Asian-American patients with anxious symptoms require lowered concentrations of psychotropic medication than do Caucasian-American adults with the same symptoms.

These data invite a speculation regarding the difference in the classic philosophies of Asians and Europeans. Postreformation Christian philosophy, which is more clearly an intellectual product of northern rather than southern Europe, emphasizes the inherently dysphoric mood of human beings. The commentaries on human nature that were written by Martin Luther and John Calvin emphasized the anxiety, fear, and guilt that is endemic to the human condition and the extraordinary effort that is necessary to control these unpleasant feelings. Calvin believed that

TABLE 3

Mean Behavioral Scores for Motor Activity, Crying, Fretting, Vocalizing, and Smiling for Caucasian-American, Irish, and Chinese 4-Month-Old Infants

Behavior	American	Irish	Chinese
Motor activity	48.6	36.7	11.2
Crying (in seconds)	7.0	2.9	1.1
Fretting (% trials)	10.0	6.0	1.9
Vocalizing (% trials)	31.4	31.1	8.1
Smiling (% trials)	4.1	2.6	3.6

humans must struggle continually with the fear of impending danger and that total freedom from anxiety is impossible because of the inability of humans to obliterate dark thoughts of the future (Bouwsma, 1988).

In all of our cohorts, inhibited children were more likely than uninhibited children to be blue eyed and have an ectomorphic body build and, therefore, more likely to trace their genetic ancestry to northern European populations. The uninhibited children more often had brown eyes and a mesomorphic build characteristic of southern Europeans (Arcus & Kagan, 1992; Rosenberg & Kagan, 1987, 1990). Thus, there is some basis for a modest association between a northern European ancestry and a disposition toward dysphoric affect and an inhibited temperament.

The Buddhist philosophy, which is attractive to Asians, emphasizes the serenity of the human condition and makes the goal of existence the elimination of desire and the attainment of the state of nirvana in which conscious awareness of the world is temporarily obliterated (Fung, 1983). Most scholars who have commented on these differences in philosophical goals have emphasized only their cultural determinants. Many social scientists have suggested that the differences in modal personality between Asians and Europeans are a derivative of the adoption of one or the other of these two philosophies.

We suggest at a speculative level that perhaps temperamental differences between Asian and European populations make a small, but nonetheless real, contribution to the differential attractiveness of these two philosophical positions. Stated more boldly than is warranted, if a large number of adults in a society are experiencing high levels of tension, uncertainty, and dysphoria because of temperamental factors, then a philosophy that urges them to be calm and serene and to cease striving for external symbols of security may meet some resistance because such an imperative does not match their conscious feeling tone. By contrast, a philosophy that accepts chronic levels of tension, guilt, and anxiety as definitive of the human condition may seem less valid to adults who possess a consciousness that is a derivative of a much lower level of internal arousal. These adults will find the Buddhist message in greater accord with their feeling tone. Perhaps nature and nurture come together even at the level of the deepest philosophical assumptions of a society.

Should these speculations prove to have some validity, they have no political, legal, or ethical implications whatsoever. Unfortunately, today's researchers live in a historical moment when ethnic and racial issues are imbued with unusually strong emotion and all differences are evaluated, incorrectly, as good or bad. We forget that Ludwig Wittgenstein (1922) commented on the absolute independence of ethics and objective facts. Some philosophers have even suggested that the reason the *Tractatus Logico-Philosophicus* was written was to make that division explicit. Science has enriched our lives, made labor easier, and contributed to human health and longevity. But science is not to be used as the sole source of our laws or our morality. Human groups differ in an indeterminate number of genetically based characteristics. This diversity is to be regarded as a set of interesting facts about nature and never as an argument for the awarding of differential privilege or power.

Finally, it is important to appreciate that temperamental differences among young children are malleable. There is no fixed determinism between a particular temperamental bias in infancy and a narrowly defined outcome in later childhood or adolescence. The fact of temperamental differences among children is not inconsistent with political egalitarianism. Introverted and extroverted adults are equally entitled to society's prizes. If introverts are less common in the U.S. Senate, as is likely, then it is because of their choice, not because of a prejudice against shy adolescents. If hyperactive boys are more often arrested for crimes, then that fact does not imply that the police dislike youth with this characteristic. Every unbalanced occupational profile is not the result only of prejudice. An appreciation of temperamental variation enriches our understanding of individual development and is in no way inconsistent with the egalitarian hope.

More important, environmental conditions can modulate a temperamental profile. Daily experiences permit some children to control their irritability and later their fear. It is even possible that experiences that reduce levels of uncertainty can alter the excitability of the limbic system through changes in the density of receptors on neurons.

Membership in a temperamental category simply implies a slight bias for certain affects and actions. The physiology only affects the probabilities that certain states and behaviors will occur in particular rearing environments. There is always the opportunity for the child to learn to control the urge to withdraw from a stranger or a large dog. The temperamentally shy child is not chronically helpless; remember, lions can be trained to sit quietly on a chair even though that posture is not typical of their species. Indeed, the role of the environment is more substantial in helping the child to overcome the tendency to withdraw than in making that child timid in the first place. No human quality, psychological or physiological, is free of the contribution of events both within and outside the organism. No behavior is a first-order, direct product of genes. To rephrase Willard Van Orman Quine, every psychological quality is like a pale gray fabric woven from thin black threads, which represent biology, and thin white ones, which represent experience. But it is not possible to detect any quite black or white threads in the gray cloth.

References

Arcus, D., Gardner, S., & Anderson, C. (1992, May). *Infant reactivity and maternal style in the development of inhibited and uninhibited behavioral profiles*. Paper presented at the meeting of the International Society for Infant Studies, Miami, FL.

Arcus, D., & Kagan, J. (1992). *Temperament and craniofacial variation in infants*. Unpublished manuscript.

Bouwsma, W. J. (1988). *John Calvin*. New York: Oxford University Press.

Caudill, W., & Weinstein, H. (1969). Maternal care and infant behavior in Japan and America. *Psychiatry, 32*, 12–43.

Davis, M. (1992). The role of the amygdala in fear and anxiety. *Annual Review of Neuroscience, 15*, 353–375.

Davis, M., Hitchcock, J. M., & Rosen, J. B. (1987). Anxiety and the amygdala. *Psychobiology of Learning and Motivation, 21*, 263–305.

Dunn, L. T., & Everitt, B. J. (1988). Double dissociations of the effects of amygdala and insular cortex lesions on conditioned taste aversion, passive avoidance, and neophobia in the rat using the excitotoxin ibotenic acid. *Behavioral Neuroscience, 102*, 3–9.

Freedman, D. G., & Freedman, M. (1969). Behavioral differences between Chinese-American and American newborns. *Nature, 224*, 1227.

Fung, Y., (1983). *The history of Chinese philosophy: Vol. 2. The period of classical learning* (E. Bodde, Trans.). Princeton, NJ: Princeton University Press.

Jurgens, U. (1982). Amygdalar vocalization pathways in the squirrel monkey. *Brain Research, 241*, 189–196.

Kagan, J., & Snidman, N. (1991a). Infant predictors of inhibited and uninhibited profiles. *Psychological Science, 2*, 40–44.

Kagan, J., & Snidman, N. (1991b). Temperamental factors in human development. *American Psychologist, 46*, 856–862.

Kelley, A. E., Domesick, V. B., & Nauta, W. J. H. (1982). The amygdalostriatal projection in the rat. *Neuroscience, 7*, 615–630.

Lewis, M. (1989). Culture and biology: The role of temperament. In P. R. Zelazo & R. G. Barr (Eds.), *The challenges to developmental paradigms: Implications for theory assessment and treatment* (pp. 203–223). Hillsdale, NJ: Erlbaum.

Lin, K. M., Poland, R. E., & Lesser, I. M. (1986). Ethnicity and psychopharmacology. *Culture, Medicine, and Psychiatry, 10*, 151–165.

Mishkin, M., & Aggleton, J. (1981). Multiple functional contributions of the amygdala in the monkey. In Y. Ben-Ari (Ed.), *The amygdala complex* (INSERM Symposium No. 20, pp. 409–420). Amsterdam: North-Holland.

Nauta, H. J. W. (1986). The relationship of the basal ganglia to the limbic system. In P. J. Vinken, G. W. Bruyn, & H. L. Kluwans (Eds.), *Handbook of clinical neurology* (Vol. 5, pp. 19–31). New York: Elsevier Science.

Rolls, E. T., & Williams, G. V. (1987). Neuronal activity in the ventral striatum of the primate. In M. B. Carpenter & A. Jayaraman (Eds.), *The basal ganglia* (Vol. 2, pp. 349–356). New York: Plenum.

Rosenberg, A. A., & Kagan, J. (1987). Iris pigmentation and behavioral inhibition. *Developmental Psychobiology, 20*, 377–392.

Rosenberg, A. A., & Kagan, J. (1990). Physical and physiological correlates of behavioral inhibition. *Developmental Psychobiology, 22*, 753–770.

Rothbart, M. K. (1989). Temperament in childhood: A framework. In G. A. Kohnstamm, J. E. Bates, & M. K. Rothbart (Eds.), *Temperament and childhood* (pp. 59–76). New York: Wiley.

Wittgenstein, L. (1922). *Tractatus logico-philosophicus* (C. K. Ogden & F. P. Ramsey, Trans.). London: Routledge.

PART FOUR

Psychopathology: Genetic and Experiential Factors

PART FOUR

Introduction

Irving I. Gottesman

Given the constraint of selecting only five chapters to represent the much larger domain of genetics and psychopathology for this section, it is inevitable that many important and well-developed areas of research have been omitted. One purpose of this introduction is to direct the current generation of students of behavioral genetics to useful entry points to that literature. The authors in this section represent two nations (the United States and the United Kingdom), three disciplines (psychology, genetics, and psychiatry), and five areas of discussion (mood disorders, schizophrenia, alcohol dependence, autistic disorder, and attention-deficit hyperactivity disorder). All of the authors have expertise beyond that displayed in their respective chapters, which has resulted in balanced and broader presentations than might otherwise be the case. They are all mindful of the uncertainties surrounding psychiatric nosology and diagnosis; the 10th edition of the *International Classification of Diseases* (*ICD-10*) (World Health Organization, 1992) has just been published to become effective in 1993, and the next version of the conventions to be used in the U.S.—the 4th edition of the *Diagnostic and Statistical Manual of Mental Disorders* (*DSM-IV*) (American Psychiatric Association, in press)—is on the launching pad, awaiting the completion and integration of 12 field trials. The implications of such nosological revisions for psychiatric genetics are profound for they determine the specification of phenotypes for our research; ambiguities, unreliability, and imperfect construct validity of phenotypes frustrate the attempts at genetic analyses. It is noteworthy that even in this "decade of the brain," none of the criteria for the psychopathological phenotypes with which researchers deal are based on biological or laboratory findings. Researchers depend on the signs and symptoms of behavior that they observe or that are reported to them by the subjects and their relatives.

The chapters in this section are best appreciated with a little historical background. That a critical mass of knowledge about the relationships between genetics and psychopathology of various sorts existed became quite clear when, in short order, Rosenthal (1970) published his book *Genetic Theory and Abnormal Behavior* and Slater and Cowie (1971) produced *The Genetics of Mental Disorders*—both books were distributed by major publishers. These three authors were very self-conscious about the need to "distance" their presentations from the outrageous and barbaric applications of genetic theories in the service of the Nazis that led to the extermination or sterilization of mental patients and to the Holocaust (Müller-Hill, 1988; Proctor, 1988; Weindling, 1989). They were successful in legitimizing the development of the reawakening of objective, apolitical, scientific, and humanitarian research into the causes of major mental disorders.

One major event on the way to these two books was an effort to save the "baby from the bathwater" at the First International Congress of Psychiatry held after the war in Paris in 1950. At that meeting (Sjögren, 1950), an entire session (and volume) boldly dealt with "genetics and eugenics" under the umbrella of "social psychiatry." Franz Kallmann, a German–Jewish refugee then residing in New York, presented his findings on the genetics of psychoses from 1,232 twin index families; Lionel Penrose, a British geneticist later to head the Galton Institute, discussed research methods in human genetics; J. A. Fraser Roberts, a British pediatrician and one of the first medical geneticists, reviewed the genetic aspects of mental retardation in a remarkably prescient paper; Eliot Slater, also ahead of his times, presented genetic aspects of personality and neurosis; and Erik Strömgren, a Danish psychiatrist, defined the field that would become psychiatric genetic epidemiology. At another session at the congress, Ornulf Odegaard, a Norwegian social psychiatrist, discussed the role of genetics in psychiatry and introduced the rudiments of multifactorial-polygenic theory for explaining the transmission of schizophrenia, an idea that would lay dormant until revived by Gottesman and Shields (1966, 1967). All members of the 1950 sextet are deceased.

Some of the major advances since these halting first steps are chronicled in the chapters that follow. The certainty of the findings and interpretations about the various roles of genetic and experiential factors (biological, psychological, societal) in each of the areas of discussion that follow vary as a function of the nomological network surrounding each form of psychopathology. For example, in chapter 13, McGue draws on data from numerous family studies of alcoholism as well as 7 twin studies and 5 adoption studies. In chapter 12, I draw on more than 40 family and twin studies of schizophrenia and 5 adoption studies. In studying a rare disorder like autism, Rutter, Bailey, Bolton, and Le Couteur (chapter 14) have a much slimmer database on which to build genetic inferences. Eaves et al., in chapter 15, examine symptoms of attention-deficit hyperactivity disorder in a small but relatively homogeneous sample of young male twin pairs. No doubt researchers will look back on these chapters in a decade and wonder how they could have been so "naive." The advances in psychiatric genetics are proceeding rapidly, unlike the slower advances in mainstream psychology, precisely because this area is under the umbrella of genetics and the neurosciences in general. Established journals such as *Nature* and *Science* as well as newer ones such as *Genetic Epidemiology, Psychiatric Genetics, Neuroscience Facts* (delivered weekly by FAX), and the neuropsychiatric section of the *American Journal of Medical Genetics* herald the rapid advances and genuine breakthroughs that are occurring as you read these words. One instance is provided by the announcement in the *New York Times* (March 24, 1993, p. 1) of the discovery of the gene for Huntington's disease and another is by Hardy (1992) in regard to the genetics of Alzheimer's disease. Entrée to areas not covered here may be found in the books edited by McGuffin and Murray (1991) and Mendlewicz and Hippius (1992); in the articles to be derived from the National Institute of Mental Health's molecular genetics initiative and the 10-center consortium for schizophrenia, bipolar disorder, and Alzheimer's disease (cf. *Schizophrenia Bulletin*); and for German readers, in the recent book by Propping (1991).

References

American Psychiatric Association (in press). *Diagnostic and statistical manual of mental disorders* (4th ed., *DSM-IV*). Washington, DC: Author.

Gottesman, I. I., & Shields, J. (1966). Schizophrenia in twins: 16 years' consecutive admissions to a psychiatric clinic. *British Journal of Psychiatry, 112*, 809–818.

Gottesman, I. I., & Shields, J. (1967). A polygenic theory of schizophrenia. *Proceedings of the National Academy of Sciences, 58*, 199–205.

Hardy, J. (1992). The genetics of Alzheimer's disease. *Neuroscience Facts, 3*, 65.

McGuffin, P., & Murray, R. (Eds.). (1991). *The new genetics of mental illness.* Oxford, UK: Butterworth-Heinemann.

Mendlewicz, J., & Hippius, H. (Eds.). (1992). *Genetic research in psychiatry.* Berlin: Springer-Verlag.

Müller-Hill, B. (1988). *Murderous science: Elimination by scientific selection of Jews, Gypsies, and others in Germany 1933–1945.* London: Oxford University Press.

Proctor, R. N. (1988). *Racial hygiene: Medicine under the Nazis.* Cambridge, MA: Harvard University Press.

Propping, P. (1991). *Psychiatrische Genetik* [*Genetic psychiatry*]. Berlin: Springer-Verlag.

Rosenthal, D. (1970). *Genetic theory and abnormal behavior.* New York: McGraw-Hill.

Sjögren, T. (Ed.). (1950). Social psychiatry: Genetics and eugenics. In *Proceedings of the International Congress of Psychiatry* (Vol. VI). Paris: Hermann & Cie.

Slater, E., & Cowie, V. (1971). *The genetics of mental disorders.* London: Oxford University Press.

Weindling, P. (1989). *Health, race and German politics between national unification and Nazism 1870–1945.* Cambridge, UK: Cambridge University Press.

World Health Organization. (1992). *The ICD-10 classification of mental and behavioral disorders: Clinical descriptions and diagnostic guidelines* (10th rev., *ICD-10*). Geneva: Author.

CHAPTER 11

Genes, Adversity, and Depression

Peter McGuffin and Randy Katz

One of the most consistent observations about the likely causes of depressive disorders is that they are more frequently found among the relatives of depressed patients than in the population at large. However, familial aggregation could imply environmental causes, a genetic etiology, or a combination of the two. Although it is true that some (mainly uncommon) disorders for which abnormal behavior is prominent are transmitted as simple mendelian traits, for example, Huntington's disease (Harper, 1991) and probably some subforms of Alzheimer's disease (Goate et al., 1991), a more complicated pattern of inheritance is usual. Indeed, as discussed elsewhere in this book, there is a large range of behaviors for which familial resemblance can be demonstrated. This includes scores on IQ tests and paper-and-pencil tests of personality for which genes seem to play a part in family resemblance, as well as characteristics that are likely to reflect cultural transmission such as religious beliefs or political attitudes.

Although it might seem quite straightforward to decide whether it is genes or family culture that is mainly responsible for the transmission of a particular trait, the evidence, based on family data alone, can be quite misleading. The idea that certain traits that are mainly nongenetic can "simulate mendelism" (Edwards, 1960) is not new, but it is something that researchers constantly need to remind themselves of in an era in which dramatic advances are being made in molecular genetics and for which it is tempting to speculate that this branch of science might hold all the answers for biological or behavioral phenomena. For example, McGuffin and Huckle (1990) recently studied attendance at medical school among the relatives of first- and second-year students at the University of Wales College of Medicine. The "risk" of attending medical school in the first-degree relatives of medical students was about 60 times that of the general population and, on carrying out a complex segregational analysis, McGuffin and Huckle found that the trait showed a pattern within families that was closely similar to that of autosomal recessive inheritance. This study thus replicated the finding of Lilienfeld (1959), who 3 decades earlier obtained the same result using a simpler method of analysis. It is clear that if researchers are to proceed from observing familial aggregation to making inferences about the role of genes rather than family environment, then other sources of information will be needed. Fortunately, such information exists for depression.

Natural Experiments

Studies of twins and individuals separated from their biological relatives early in life provide the classical methods of teasing apart the effects of genes and family environment. For depression, the results are mainly consistent in showing an important genetic contribution to at least the more severe forms of affective disturbance (for reviews see McGuffin & Katz, 1989; Tsuang & Faroane, 1990). Thus, for example, the concordance in monozygotic (MZ) twins for manic depression was 67% compared with 20% in dizygotic (DZ) twins in a carefully conducted study for which the index cases were systematically ascertained using the Danish National Twin Register (Bertelsen, Harvard, & Hauge, 1977). This was very similar to the averaged results of earlier studies for which the overall MZ concordance was 69% compared with the DZ concordance of 13% (Gershon, Bunney, Leckman, Van Eerdweegh, & De Bauche, 1976). The adoption data are less extensive and show some inconsistencies; but overall, there is only evidence for an increased rate of affective disorder in biological relatives and not in adopted relatives (McGuffin & Katz, 1989).

Thus, taken together, the family, twin, and adoption data provide a suggestive body of evidence for an important genetic contribution to affective disorder. So far, however, we have used the terms *affective disorder*, *manic depression*, and *depressive disorder* in a general and, more or less, interchangeable way as if these terms were all-embracing and all descriptive of the same entity. This was indeed a prevailing view from the time of Kraepelin's (1909) first description of manic–depressive insanity at the end of the 19th century through most of the 20th century. However, influential family studies coincidentally published in the same year by Angst (1966) and Perris (1966) helped bring about a change. This was a move toward the view first proposed by Leonhard (1959) that manic depression could usefully be divided into *bipolar disorder*, which presents as episodes of both mania and depression, and *unipolar disorder*, which presents as episodes of depression alone. Although Perris' study indicated that the two disorders tended to "breed true," Angst's findings were more complicated. Nearly all subsequent research has found the same pattern as that found by Angst: Family members of unipolar index cases tend toward an excess of only unipolar disorder, whereas the family members of patients with bipolar disorder have increased risk of both bipolar and unipolar disorders. Another fairly consistent finding is that, although both types of disorder are familial, the overall risk of affective disorder is greater in the relatives of bipolar cases (McGuffin & Katz, 1989). The findings of the twin study of Bertelsen et al. (1977) also suggest that the genetic influences on bipolar disorder are strong, with the heritability in excess of 80% (McGuffin & Katz, 1989); whereas in unipolar disorder, the genetic influences may be more modest, with greater room for invoking environmental effects. For this reason, and because it is the commoner form of affective disorder, most studies that have examined environmental effects in affective disorder have focused on the unipolar form. For the remainder of this chapter, we follow this convention in dealing with only unipolar disorder. However, we will also break somewhat with convention by attempting to simultaneously consider both genes and environment and how they coact and interact to cause depression.

Genetic Diathesis, Environmental Stress?

A long-established and common view of depression among psychiatrists in Europe (although less prevalent in the United States) is that there are broadly two types of unipolar disorder with different etiologies. On the one hand, there is endogenous depression, which has prominent "biological" features such as early morning waking, diurnal variation of mood, and loss of appetite and also has a constitutional basis; whereas on the other hand, there is neurotic depression, which lacks biological features and is mainly reactive to stress. For example, Stenstedt (1966) noted that there were psychogenic factors in 90% of his neurotically depressed subjects but other exogenous factors in the remainder. In an earlier study (Stenstedt, 1952), he found that those patients in whom there were obvious precipitants tended to have less family loading than those who did not. Such results then would support the idea that some patients with a marked biological/genetic diathesis develop their illness "out of the blue," whereas other patients in whom there is a smaller and nonspecific constitutional predisposition develop their illness as a result of obvious stress. Unfortunately, either because such dichotomy was taken for granted or because until recently researchers with interests in both factors were uncommon, most studies considered potential stresses and genetic diathesis quite separately. One exception was the investigation by Pollitt (1972), who found a morbid risk of depression among relatives that was particularly high (at around 21%) when precipitants for the proband's illness were doubtful or absent. The morbid risk in relatives fell to between 6% and 12% when the proband's illness was "justifiable" in the sense that it followed severe physical stress, infection, or psychological trauma.

Since then, there have been methodological advances both in the ways that "stress" and psychiatric diagnosis are assessed and in the ways that family studies are conducted. Therefore, in collaboration with colleagues at the Medical Research Council's Social Psychiatry Unit in London, we mounted a study to investigate the relationship between adversity, in the form of life events or chronic difficulties (Brown & Harris, 1978), and familial aggregation of depression in a consecutive series of depressed patients and their families (Bebbington et al., 1988). The three main hypotheses with which we set out were as follows:

1. Depression with a neurotic pattern of symptoms is more often associated with a preceding threatening life event than is depression with an endogenous pattern.
2. Depressed patients whose disorder occurs in the absence of life events more often have depression among their relatives than do patients whose onset of disorder is associated with adversity.
3. Depression associated with adversity and depression not associated with any detectable stress are definable as two different forms of disorder, with each showing a tendency to breed true within families.

All of these hypotheses proved to be incorrect. Threatening life events and chronic difficulties were as often found before the onset of endogenous depression as before the onset of neurotic disorder (Bebbington et al., 1988), which contradicts the traditional view. The frequency of depression, regardless of how the disorder was defined, was higher among the relatives of the depressed subjects than in the general population, but there was no difference between the morbid risk in relatives of index cases who had experienced life events and that in relatives of those who had not (McGuffin, Katz, Aldrich, & Bebbington, 1988). Furthermore, among the currently depressed family members, there was no tendency for subtypes of depression (one life event associated and the other not) to breed true (McGuffin et al., 1988).

We did, however, have one strikingly positive finding that was completely unexpected. Not only did depression aggregate in families but so also did life events. In the 3-month period before the interview (or before the onset of depression if the relative was depressed), 42% of the relatives of depressed patients reported one or more threatening life events compared with only 7% of a community sample studied earlier by Bebbington,

Hurry, and Tennant (1981). One possible explanation of these results might just have been that the recent onset of depression in the probands (the index cases) and their subsequent referral to a hospital caused turmoil throughout the family. However, even when all events that were in any way related to the probands were omitted, there was still a marked excess of reported adversity among family members, with 29% of them having had one or more threatening life events. Furthermore, the timing of life events of family members was unrelated to the proband's onset of depression.

Another intriguing finding was of a surprisingly weak association between recent life events and current depression among family members. Of the first-degree relatives who had recent life events, 21% were found to be current "cases" of depression on interview with the present state examination (PSE) (Wing, Cooper, & Sartorius, 1974) compared with 15% among the relatives who had not experienced life events. However, this difference was nonsignificant statistically and contrasts markedly with the results of a study of a community sample, which used the same methods, for which 57% of those who had a recent life event were cases of depression compared with 7.5% of those who had not had a life event (Bebbington et al., 1981). A logistic regression that put the family and community data together showed that there was a highly significant interaction between life events and being the first-degree relative of a depressed patient in an analysis in which the proportion of affected subjects was taken as the response variable (McGuffin et al., 1988).

These results raised the question of whether the reported association between life events and depression that has been repeatedly observed in community studies (e.g., Bebbington et al., 1981; Brown & Harris, 1978; Paykel, 1978) is partly due to the fact that both are familial. Although this has yet to be replicated directly, recent twin study findings (Kendler, Neale, Heath, Kessler, & Eaves, 1992; Plomin & Bergeman, 1991) support the idea that life events show familial aggregation. In conclusion, the relationships between adversity, familiality, and depression turned out to be more complicated than we had originally thought, and we then attempted to explore this further in a twin study of our own.

Different Diagnostic Definitions, Different Environmental Effects

Our aims in carrying out a twin study of depression were to assess the degree of genetic determination of unipolar depression (a question that has sometimes been neglected compared with bipolar disorder), explore further the effects of different phenotypic definitions, and attempt to understand better the interplay between genes and environment in causing depressive disorder. Our sample was derived from the Maudsley Hospital twin register in London, which was established by Eliot Slater in 1948. We found that the register contained 408 probands with a primary diagnosis of depression of whom 215 fulfilled our screening criteria based on the third edition of the *Diagnostic and Statistical Manual of Mental Disorders (DSM-III)* (American Psychiatric Association, 1980) and the syndrome check list of the PSE (Wing et al., 1974). Of these, 34 probands turned out to have had one or more episodes of bipolar disorder, and therefore, this part of the sample was set aside. The remaining 181 probands and their cotwins were closely investigated by personal interviews when possible, as well as examination of hospital case records and information from general practitioners and relatives. Case abstracts were prepared and assessed by clinicians who were unaware of the twins' zygosity and the diagnosis of the cotwin. This resulted in the elimination of 4 probands, which left us with a sample of 68 MZ and 109 DZ proband cotwin pairs.

A variety of definitions of depression were applied, and the preliminary results have been reported previously (McGuffin, Katz, & Rutherford, 1991). Some of the main findings both for the preliminary analysis and a more recent assessment are summarized in Table 1. The constant finding, regardless of the definition that was applied, was of a significantly higher concordance for MZ twins than for DZ twins, which suggests a genetic effect. However, we then went on to apply a simple biometric model under which the observed phenotype could have resulted from additive gene effects, shared (familial) environment, and nonshared environment with no interactions (i.e., no nonadditive effects). Having made certain

TABLE 1

Concordance (Percentage) for Different Definitions of Unipolar Depression in Monozygotic (MZ) and Dizygotic (DZ) Twins

	MZ		DZ	
Definition	C	*r*	C	*r*
Broad	66	0.88[a]	42	0.66[a]
Narrow	46	0.77[b]	20	0.45[b]

Note. For MZ twins, *n* probands = 68; for DZ twins, *n* probands = 109. Broad = PSE-ID-CATEGO/hospital-treated; narrow = blind *DSM-III-R* major depression. *r* = correlation in liability. [a]Assuming a population risk of 8.9%. [b]Assuming a population risk of 4.2%.

assumptions about the morbid risk of depression in the general population based on the study by Sturt, Kumakura, and Der (1984), which was carried out in Camberwell, the old London borough in which the Maudsley Hospital is based, and about the distribution of liability to depression (Reich, James, & Morris, 1972), we obtained two very different results for two different definitions of depression in the cotwins. The first was a narrow definition that used criteria of the revised *DSM-III (DSM-III-R)* (American Psychiatric Association, 1987) for which we estimated the lifetime risk in the population to be 4.2%. The second broader definition used PSE-ID-CATEGO criteria (Wing et al., 1974), which we had somewhat modified so that having received hospital treatment was a necessary component and which has a population lifetime risk of about 8.9% (Sturt et al., 1984). For the narrower definition, 79% of the variance was accounted for by additive genetic effects and the remainder by nonshared environment. Model fitting allowed no room for shared (familial) environment. By contrast, under the broad definition of depression, the additive genetic component explained only 39% of the variance, with shared environment accounting for 46%, and the remainder due to nonshared environment. Attempts to remove either genes or shared environment as explanatory variables of broadly defined depression resulted in a significant worsening of the fit of each model, which led to the conclusion that all three components were necessary.

We speculate that these differences arose both because of the difference in breadth of the definition of depression and because of the incorporation of the receipt of hospital treatment in the broader definition.

Thus, it might be expected that if an individual develops symptoms of depression, the probability that he or she will seek treatment and be referred to a hospital specialist is increased if he or she has a relative (such as a twin) who has already received such treatment. Therefore, our findings can be interpreted as showing that all of the familial aggregation of narrowly defined depression is explained entirely genetically, whereas treatment seeking for depressive symptoms is influenced by family environment (as in our analysis of the broad definition).

So far in our discussion of these twin results, environment—whether shared or nonshared—has been treated as a "latent variable." However, we also used some direct measures of the environment. Past environment, in particular those aspects of it that had been shared by the twins when growing up, was investigated using a questionnaire. As might be predicted, MZ twins were more alike than were DZ twins, including dressing alike during childhood. However, none of these measures appeared to influence their similarity for the later development of depression, nor did other measures of shared environment, which did not differ between MZ and DZ twins, such as the number of years spent together in the family home. But what of more immediate factors, such as recent life events, which may be of more direct relevance to the development of depression? Our next set of analyses concerned these.

Twin Similarities for Adversity and Depression

In our twin study, we used a measure of life events (Brugha, Bebbington, Tennant, & Hurry, 1985) that was much less elaborate than, but derived from, the life events and difficulties schedule of Brown and Harris (1978). This measure consisted of a checklist of the 12 most common categories of events associated with marked or moderate long-term threat. These included categories of events that were likely to be independent of the subject's symptoms, such as death of a first-degree relative, as well as events that might be associated with depressive symptoms, such as separation due to marital difficulties. In addition, we attempted to quantify subjective distress for each category of event on a 5-point scale ranging from 0, which signifies no event occurring in this category, to 4, which indicates that such an event occurred during this period and was severely distressing.

In general, the results supported the principal findings of our family study (McGuffin et al., 1988). The frequency of reported life events was high, with 74% of probands and 68% of cotwins having had one or more events in the 6 months before interview. Although we do not have a reliable population figure based on the checklist of 12 events with which to compare these results, other studies suggest that the frequency of reported life events for the general population is much lower (Brown & Harris, 1978). Also in keeping with the family study was an apparent lack of relationship between report of threatening life events and the presence of disorder. Although, because of our methods of ascertainment, we were less certain about time of onset than for the family material, threatening life events were actually less commonly reported by those cotwins who were current cases of depression according to PSE-ID-CATEGO (Wing & Sturt, 1978) than in cotwins who were not current cases. Thus, among cases, 66% had one or more life events compared with 70% in the cotwins who were not cases. Life events were similarly reported by 61% of cotwins who had a lifetime diagnosis of major depression compared with 73% of those who did not have such a diagnosis. This difference, which again is opposite to expected direction, is nonsignificant; and in fact, there was no relationship between the frequency of reported life events and any lifetime psychiatric diagnosis. In summary, the twin and family data both suggest a high frequency of reported threatening life events in individuals who have had a depressive disorder and in their relatives regardless of whether or not the relatives themselves are, or have been, depressed.

Interestingly, there was no apparent concordance for life events between probands and cotwins when the data were analyzed according to a simple dichotomy of life events reported/not reported in the past 6 months. However, when we quantified life events according to the number of different categories of events reported or by adding up the total subjective distress score, significant correlations emerged. The overall correlation for number of events was 0.35; and the correlation for MZ twins was only slightly higher at 0.37 than that for DZ twins, for which it was 0.33.

The distress scores turned out to be markedly skewed, but a reasonable approximation to normality was obtained by using the transformation $y = (1 + x)^{-1/2}$. The MZ intraclass correlation for the transformed score was 0.37, and the DZ correlation 0.33, which suggests only modest genetic effects. However, we explored this further using formal model fitting similar to that described earlier for the analysis of depression. The results are summarized in Table 2. Although the full model suggests a large common environmental effect and low heritability, it was impossible to reject reduced models with either common environment constrained to be zero or heritability similarly constrained. The only model that could be rejected with $\chi^2 = 54.7$ ($df = 2$) was the null model with no familial effects. We conclude that the subjective distress associated with life events is familial, but we are unable to differentiate whether this is to family environment, genes, or a combination of the two. We suggest that family environment has the larger effect, but the results of the analysis are inconclusive.

As mentioned earlier, two other recent twin studies (Kendler, Neale, Heath, Kessler, & Eaves, 1991; Plomin, 1990) also found positive correlations for life events in twins. In both of these studies, a differentiation was made between *controllable* events (those likely to have been influenced by the subject's own actions) and *uncontrollable* events (those that were not conceivably influenced by the subject). Only controllable events

TABLE 2

Model Fitting Based on Subjective Distress Score Following Life Event

Model	h^2	c^2	χ^2
Additive genes and family environment	0.06	0.30	0.0
Additive genes only	0.37	[0.0]	0.74
Family environment only	[0.0]	0.36	0.03
No familial effects	[0.0]	[0.0]	54.68

Note. Parameter values in brackets (i.e., [0.0]) are fixed.

(which by implication might be related to the subject's personality or mood state) appeared to have a genetic component.

Conclusions

Our findings strongly suggest that the syndrome of major depression when narrowly defined is highly heritable and that common family environment plays little if any part. These results are in keeping with those of other studies such as that of Torgersen (1986), whose research was also based on a clinical sample, and that of Kendler et al. (1992), who studied a sample of female twins drawn from the general population. We found that the pattern of our results was markedly influenced by the definition of phenotype; and in particular, incorporating hospital treatment into the definition of depression resulted in a substantial family environmental contribution to the variance in liability. This may suggest that the main role of family environment in clinical depression is to influence help seeking.

The findings in relation to life events are surprising, but the familial effect is consistent with the results of our twin and family studies as well as those of two other recent twin studies (Kendler et al., 1991; Plomin, 1990). Our own studies are, as far as we are aware, the only ones to attempt to examine the familiality of both adversity and depression within the same sample. In both the family study and the twin study, there was little or no association between life events and depression in the relatives of depressed probands, which is in marked contrast to the often-repeated finding of an association between onsets of depression and preceding life events in samples of unrelated individuals (Bebbington et al., 1988). This may suggest that at least part of this association between life events and depression is due to the fact that both are familial. At present, it is impossible to say whether the familiality of life events reflects a type of behavior that is event prone or a type of thinking that is characterized by an increased awareness or perception of threat.

However, our studies contribute to an intriguing pattern of findings that are emerging in recent behavioral genetics research suggesting that genetic influences have a role in a variety of phenomena that have traditionally been regarded as indicators of the environment (Plomin & Bergaman, 1991).

References

American Psychiatric Association. (1980). *Diagnostic and statistical manual of mental disorders* (3rd ed., *DSM-III*). Washington, DC: Author.

American Psychiatric Association. (1987). *Diagnostic and statistical manual of mental disorders* (3rd ed. revised, *DSM-III-R*). Washington, DC: Author.

Angst, J. (1966). Zur Atiologie unde Nosologie endogener depressiver Psychosen [On the etiology and nosology of the endogenous depressive psychosis]. *Monographen aus der Neurologie und Psychiatrie, 112.*

Bebbington, P., Brugha, T., McCarthy, B., Potter, J., Sturt, E., Wykes, T., Katz, R., & McGuffin, P. (1988). The Camberwell collaborative depression study: I. Depressed probands. Adversity and the form of depression. *British Journal of Psychiatry, 152*, 754–765.

Bebbington, P., Hurry, J., & Tennant, C. (1981). Epidemiology of mental disorders in Camberwell. *Psychological Medicine, 11*, 561–579.

Bertelsen, A., Harvard, B., & Hauge, M. (1977). A Danish twin study of manic-depressive disorders. *British Journal of Psychiatry, 130*, 330–351.

Brown, G. W., & Harris, T. (1978). The social origins of psychiatry. *130*, 330–351.

Brugha, T., Bebbington, P., Tennant, C., & Hurry, J. (1985). The list of threatening experiences: A subset of 12 life event categories with considerable long-term contextual threat. *Psychological Medicine, 15*, 189–194.

Edwards, J. H. (1960). The simulation of mendelism. *Acta Genetica, 10*, 63–70.

Gershon, E., Bunney, W. E., Leckman, J. F., Van Eerdweegh, M., & De Bauche, B. A. (1976). The inheritance of affective disorders: A review of data of hypotheses. *Behaviour Genetics, 6*, 227–261.

Goate, A. M., Chartier-Harlin, M. C., Mullan, M. C., Brown, J., Crawford, F., Fidani, L., Giuffa, L., Haynes, A., Irving, N., James, L., et al. (1991). Segregation of a missense mutation in the amyloid precursor protein gene with familial Alzheimer's disease. *Nature, 349*, 704–706.

Harper, P. S. (1991). *Huntington's disease.* Philadelphia: W. B. Saunders.

Kendler, K. S., Neale, M. C., Heath, A. C., Kessler, R. C., & Eaves, L. J. (1991). Life events and depressive symptoms: A twin study perspective. In P. McGuffin & R. M. Murray (Eds.), *The new genetics of mental illness* (pp. 146–164). Stoneham, MA: Butterworth-Heinemann.

Kendler, K. S., Neale, M. C., Kessler, R. C., Heath, A. C., & Eaves, L. J. (1992). A population-based twin study of major depression in women: The impact of varying definitions of illness. *Archives of General Psychiatry, 49*, 257–266.

Kraepelin, E. (1909). *Psychiatrie Psychiatry* (8th ed.). Leipzig, Germany: Thieme.

Leonhard, K. (1959). Aufteilung der Endogen Psychosen [Classification of the endogenous psychoses]. Berlin: Akademic Verlag.

Lilienfeld, A. M. (1959). A methodological problem in testing a recessive genetic hypothesis in human disease. *American Journal of Public Health, 49*, 199–204.

McGuffin, P., & Huckle, P. (1990). Simulation of mendelism revisited: The recessive gene for attending medical school. *American Journal of Human Genetics, 46*, 994–999.

McGuffin, P., & Katz, R. (1989). The genetics of depression and manic–depressive illness. *British Journal of Psychiatry, 155*, 294–304.

McGuffin, P., Katz, R., Aldrich, J., & Bebbington, P. (1988). The Camberwell collaborative depression study: II. Investigation of family members. *British Journal of Psychiatry, 152*, 766–774.

McGuffin, P., Katz, R., & Rutherford, J. (1991). Nature, nurture and depression: A twin study. *Psychological Medicine, 21*, 329–335.

Paykel, E. S. (1978). Contribution of life events to causation of psychiatric illness. *Psychological Medicine, 8*, 245–253.

Perris, C. (1966). A study of bipolar, manic-depressive and unipolar recurrent depressive psychoses. *Acta Psychiatrica et Neurologica Scandinavica*, Suppl. 42.

Plomin, R. (1990). The role of inheritance in behavior. *Science, 248*, 183–188.

Plomin, R., & Bergeman, C. S. (1991). The nature of nurture: Genetic influences on environmental measures. *Behavioral and Brain Sciences, 14*, 373–427.

Pollitt, J. (1972). The relationship between genetic and precipitating factors in depressive illness. *British Journal of Psychiatry, 121*, 67–70.

Reich, T., James, J. W., & Morris, C. A. (1972). The use of multiple thresholds in determining the mode of transmission of semi-continuous traits. *Annals of Human Genetics (London), 36*, 163–184.

Stenstedt, A. (1952). A study of manic depressive psychosis: Clinical, social and genetic investigations. *Acta Psychiatrica*, Suppl. 79.

Stenstedt, A. (1966). Genetics of neurotic depression. *Acta Psychiatrica Scandinavica, 42*, 392–409.

Sturt, E., Kumakura, N., & Der, G. (1984). How depressing life is—life long morbidity risk for depressive disorder in the general population. *Journal of Affective Disorders, 7*, 109–122.

Torgersen, S. (1986). Genetic factors in moderately severe and mild affective disorders. *Archives of General Psychiatry, 43*, 222–226.

Tsuang, M. T., & Faroane, S. V. (1990). *The genetics of mood disorders*. Baltimore: John Hopkins University Press.

Wing, J. K., Cooper, J. E., & Sartorius, N. (1974). *The measurement and classification of psychiatric symptoms*. London: Cambridge University Press.

Wing, J. K., & Sturt, E. (1978). *The PSE-ID-CATEGO system supplementary manual*. London: Medical Research Council.

CHAPTER 12

Origins of Schizophrenia: Past as Prologue

Irving I. Gottesman

The pendulum of public and scientific opinions about how much we know about the origins and causes of schizophrenia and how secure is such knowledge is too often perturbed by alternations between exaggerated pessimism and exaggerated optimism. Almost 100 years have elapsed since that brilliant synthesizer of the database on psychopathology, Emil Kraepelin (1896), gave us the concept of dementia praecox that is easily recognized today as 4th edition *Diagnostic and Statistical Manual of Mental Disorders* (*DSM-IV*) (American Psychiatric Association, in press) or 10th edition *International Classification of Disease* (*ICD-10*) (World Health Organization, 1992) schizophrenia. It may have been reasonable for him to conclude at that time that "[t]he real nature

Portions of this chapter were presented at the Ninth Weissenauer Schizophrenia Symposium at the University of Bonn, Federal Republic of Germany, March 13, 1992, on the occasion of receiving the Kurt Schneider Prize. Other portions are adapted from Schizophrenia Genesis—The Origins of Madness (*Gottesman, 1991*).

of dementia praecox is totally obscure." Only a pessimist would agree that his opinion should be repeated today without qualifications, but only an optimist would agree with recent newspaper headlines to the effect that the gene causing schizophrenia has been found on Chromosome 5 or 11 or X or whatever it may be tomorrow (Owen, 1992). One of my favorite psychologists–curmudgeons, the late Joseph Zubin, fantasied the following metadialogue (Zubin, 1987) between himself and the ghost of Kraepelin:

> J.Z.—Why does it take so long to make progress in the field of schizophrenia? E.K.—Well, it may be the case that we "knew" more in the early part of this century than we "know" now. In the USA I once heard someone say "It ain't ignorance that causes all the trouble. It's knowing things that ain't so!" Perhaps we had to unlearn false knowledge before we could advance to the new, cut down the underbrush before new plants could thrive. (p. 361)

Some Durable Facts About Schizophrenia

No interpretations about the familiality of schizophrenia can proceed very far without a strong link to the data from psychiatric epidemiology. That is, how often does a random person "off the street" or even "out of the Punjab" develop schizophrenia by the end of the risk period for schizophrenia that we can take to be 54 or so years of age (Gottesman, 1991; Sartorius et al., 1986). When Luxenburger (1928), a leading figure in Ernst Rüdin's Munich Institute for Psychiatric Genetics, estimated the lifetime risk for developing schizophrenia as it was defined in the 1920s, he came up with a value of 0.85% based on a numerator of 5 and an age-corrected denominator of 590.5. Apparently, he did very accurate field work; the magnitude of that risk has varied little over time despite the increasing sophistication that has occurred with respect to structured interviews and sampling; of course, the standard errors have decreased, and some interesting but unexplained variation still exists (Robins & Regier, 1991; Slater & Cowie, 1971; cf. Torrey, 1989).

Given a benchmark value for the lifetime risk of developing schizophrenia of about 1%, it becomes possible to quantify the observed familiality of schizophrenia and to engage in the exercise of genetic model fitting (McGue & Gottesman, 1989). The obvious familiality of schizophrenia does not lend itself to instant conclusions about the causes of the disorder; naive theorists of a psychodynamic persuasion would use the observations to implicate one or another dysfunctional aspects of shared experiences or environments; naive theorists of a sociological or biological-but-not-genetic orientation would use the far-from-perfect familiality to implicate cultural and unshared factors, including viruses; and naive genetically oriented theorists would use the same observations to confirm their belief that schizophrenia runs in families because genotypes run in families.

Just how familial is schizophrenia, and have the risks changed over the course of this century? In a large, well-done study reported by Schulz in 1932 (cf. Rosenthal, 1970) from Germany with 660 probands, 3.7% of the parents were also schizophrenic, as were 8.3% of the siblings. All family studies are consistent in finding much higher rates in siblings than in parents, although both classes of relatives are of the first degree, that is, sharing 50% of their genes in common. I return to this fact later. In a representative early study of the adult children of 109 schizophrenics reported by Oppler in 1932 (cf. Rosenthal, 1970), the risk for developing schizophrenia was 9.7%.

Pooling the more than 40 systematic family and twin studies conducted in western Europe for definite plus probable diagnoses of schizophrenia in relatives of probands between 1920 and 1987 (Gottesman, 1991) results in the most stable estimates of risks, and they are comfortably close to the individual studies just cited. The pattern of risks, age-corrected except for the twin studies, in Table 1 leads to a number of major inferences by a hypothetical, impartial arbitrator: The straw-person-theorists portrayed earlier are each partially correct. The data as a whole do not appear to accord with simple mendelizing inheritance associated with dominant or recessive genes (McGue & Gottesman, 1989; O'Rourke, Gottesman, Suarez, Rice, & Reich, 1982); and the disorder does not appear to be infectious, either psychologically or by bacterial/viral vectors, because the spouses have such low risks (cf. the results of adoption strategies) and the half-siblings, despite contemporary ecology/

experience-sharing, have only half the risk seen in full siblings. Newer family studies (Kendler & Diehl, in press) that have been sensitive to the added strengths of blind, structured interviews and operationalized criteria confirm the overall picture depicted in Table 1.

Kendler et al. (in press-a, in press-b), using structured interviews that focused on schizophrenia-spectrum disorders and criteria of the third, revised edition of the *Diagnostic and Statistical Manual of Mental Disorders* (*DSM-III-R*) (American Psychiatric Association, 1987) with 285 "chart diagnosed" schizophrenics in a western Irish catchment area, retained 123 probands (44%) for a study of their parents and siblings. Some 2.6% ± 1.8% of the parents and 10.1% ± 2.7% of the siblings met *DSM-III-R* criteria (as did 0.5% ± 0.3% of the relatives of normal controls) for schizophrenia. It is important to note that by adding in schizoaffective disorders and other nonaffective psychoses, the risk to the combined

TABLE 1

Morbid Risk (Percentage) of Schizophrenia (Definite and Probable) for Relatives of Schizophrenics

Relationship	Risk
General population	1
Spouses of patients	2
Third-degree relatives	
First cousins	2
Second-degree relatives	
Uncles/aunts	2
Nephews/nieces	4
Grandchildren	5
Half-siblings	6
First-degree relatives	
Parents	6
Siblings	9
Children	13
Siblings with 1 schizophrenic parent	17
Dizygotic twins	17
Monozygotic twins	48
Children of 2 schizophrenic parents	46

Note. Adapted from *Schizophrenia Genesis—The Origins of Madness* (p. 96) by I. I. Gottesman, 1991, New York: Freeman; copyright 1991 by I. I. Gottesman; adapted by permission.

sample of parents and siblings rose to 12.9%. Important and credible grist for the modeler's mill is provided by the estimates of the lifetime prevalence of schizotypal or paranoid personality disorders in the parents and siblings interviewed specifically to detect such symptoms. The process of parenthood selects out individuals of above-average intellect and mental and physical health; the lifetime risk for schizophrenia among parents in the general population is half that of a random sample (Gottesman, 1991) or 0.5%. This is reflected in the much lower risk for parents than for siblings in the Irish sample, but Kendler et al. (in press-a, in press-b) reported a prevalence of schizotypal/paranoid personality disorders of 13.9% in parents and 6.8% in siblings (2.1% and 1.8%, respectively, in controls). It is worth recording ("past as prologue") that Schulz (in 1932; cf. Zerbin-Rüdin, 1967), using less refined methods, found 11.9% of parents and 9.2% of siblings to have "schizoid personalities." The recent findings would appear to strengthen the clinical hunches since the time of Kraepelin that certain eccentricities of personality characterized the untreated first-degree relatives of schizophrenic patients; some of this psychopathology may turn out to be reactive to having to live with a stress-making person.

In another new family study that used blind structured interviews and research diagnostic criteria for subjects from Mainz, Germany (Maier et al., 1992), the lifetime risks for schizophrenia plus schizoaffective disorder were 8.4% for the first-degree relatives of schizophrenics and 9.4% for the relatives of schizoaffectives; it is interesting that the risk of unipolar major depression was equally high across all proband groups' relatives, including relatives of bipolars and unipolars. Apparently, the current definition of unipolar major depression is too heterogeneous both clinically and etiologically to discriminate between schizophrenia-spectrum disorders and affective-spectrum disorders. Depression as a symptom versus a syndrome is ubiquitous among most sufferers of psychiatric disorders, and it may occur among their relatives as a consequence of care-taking stress.

The classical prewar twin studies of schizophrenia, which have received more than their share of negative criticisms (cf. Gottesman & Shields, 1982), were generally conducted, as they had to be then, by one man with little funding; and the one man was always a genetical partisan. In the first study, reported by Luxenburger in 1928 (cf. Fischer, 1973; Gottesman & Shields, 1966), the identical (monozygotic [MZ]) twin concordance rate ranged from 50–76% depending on the method of calculation by pairs versus probands (McGue, 1992) and whether probable cases were kept or not; none of Luxenburger's 13 same-sex fraternal (dizygotic [DZ]) twin pairs were concordant for schizophrenia. Kallmann's (1946) reported pairwise, age-corrected MZ rate of 86% had few "takers" (Gottesman & Shields, 1966, 1972).

Table 2 summarizes the probandwise concordance rates without age correction for the modern twin studies of schizophrenia that have corrected for most of the alleged faults of the earlier studies. Retrospective application of various operationalized criteria for diagnosing schizophrenia to the United Kingdom study left the original results intact (McGuffin et al., in press). The median MZ rate of 46% and the median DZ rate of 14% are respectably close to the prewar studies once provision is made for the differences in severity between a standing state hospital population and consecutive admissions to modern in- and outpatient departments of

TABLE 2

Total Probandwise Concordance Rates (Percentage) for Schizophrenia in Newer Monozygotic (MZ) and Dizygotic (DZ) Twin Studies

	MZ		DZ	
Study	Pairs	Rate	Pairs	Rate
Finland 1963, 1971	17	35	20	13
Norway 1967	55	45	90	15
Denmark 1973	21	56	41	27
United Kingdom 1968, 1987	22	58	33	15
Norway 1991	31	48	28	4
United States 1969, 1983	164	31	268	6
Pooled concordance (excluding U.S.)				
Median	146	48	212	15
Weighted mean		48		16
Pooled concordance (all studies)				
Median	310	46	480	14
Weighted mean		39		10

Note. Adapted from *Schizophrenia Genesis—The Origins of Madness* (p. 110) by I. I. Gottesman, 1991, New York: Freeman; copyright 1991 by I. I. Gottesman; adapted by permission.

psychiatry. The results of the new study from Norway by Onstad, Skre, Torgersen, and Kringlen (1991) that used blind structured interviews and *DSM-III-R* criteria are not distinguishable from those of other studies in the table that interviewed their twins. Furthermore, when the concept of a "hit" is extended to include schizotypal personality, the MZ rate rises to 68%, and the DZ rate rises to 29%. In an ongoing twin study of *DSM-III-R* schizophrenia from Nagasaki, Japan, by Okazaki (1992), the preliminary probandwise concordance rate in 17 MZ pairs is 53%, and in 6 DZ pairs it is zero. These rates rise to 57% in 21 MZ pairs and 33% in the DZ pairs when both schizophreniform and schizotypal disorders are counted as hits.

Interim Fitting of Genetic Models

Although most researchers in this field favor some kind of multifactorial-polygenic model for the transmission of schizophrenia, it is far too early in the game to exclude other possibilities. A simple guide to the sources of familial resemblance under the provisions of the six most likely single models (cf. Moldin & Gottesman, in press) is given in Table 3. A further "heterogeneity" model for disease transmission is specified, and such a model is consistent with multiple instances of single-major-locus causes such as is seen for mendelizing forms of mental retardation, blindness, and deafness, as well as for sporadic (i.e., nonfamilial) cases.

The pattern and magnitude of the results from the just-surveyed population, family, and twin studies of schizophrenia, especially when confirmed by the American and Scandinavian adoption studies (Gottes-

TABLE 3

Six Models of Genetic Transmission of Mental Disorders

	Source of familial resemblance		
Model	Major locus	Polygenes	CE
Single major locus	yes	no	no
With intraallelic heterogeneity	yes	no	no
With interlocus heterogeneity	>1	no	no
Mixed	yes	yes	yes
Multifactorial oligogenic	>1	yes	yes
Multifactorial polygenic	no	yes	yes

man, 1991; Tienari, 1990), provide a solid foundation for launching schizophrenia research into the world of molecular genetics (McGuffin & Murray, 1991). That foundation will stand despite the fits and starts of applying, unsuccessfully so far, the star wars technologies of the new genetics (cf. Kendler & Diehl, in press); the latter are most suited to confirming single-major-locus models. The bow-and-arrows approach of population genetics model fitting using the data points from interviewed and family history-diagnosed cases from Table 1 informs our current views about the causes of the liability to developing schizophrenia (cf. Meehl, 1990).

The multifactorial model is not monolithic, and it does suggest a diathesis–stressor (Fowles, 1992) framework of some kind. The genetic component of the multifactorial model may involve a limited number of loci, say two to five or so, and therefore only 4–10 genes of "large/oligogenic effect" make for feasible searches in the current research environment. Such genes should, in principle, be identifiable with the linkage and association methods now being applied, including those for detecting quantitative trait loci (QTLs) (cf. Owen, 1992; Plomin, 1990).

Efforts to quantify the proportion of liability to developing schizophrenia result in a heritability of liability of 0.63 and a role for shared cultural and familial effects of 0.29, and deleting all twin data from the modeling to explore for the possibility of them distorting the results does yield a lower genetic heritability of 0.42 and a larger cultural/familial effect of 0.53 (McGue, Gottesman, & Rao, 1983, 1985). Modeling that uses an accepted schizophrenia-spectrum definition as an indicator of the relevant genotype has not yet been accomplished (Moldin, Rice, Gottesman, & Erlenmeyer-Kimling, 1990; Prescott & Gottesman, in press).

The Future Is Already Here

The broadest and most useful perspective about the panorama of the causes of schizophrenia comes from embracing what I have called the combined or ecumenical model (Gottesman, 1991). Such a framework permits many flowers to bloom and helps to identify weeds. Disagreements occur among scientists when the specific amount of explanatory "territory" allocated to each model in Table 3 requires specification. Turf wars bring out the true feelings of all those who subscribe to some kind of "biopsychosocial model" for explaining schizophrenia. At this time, based on my reading of the literature and the various modeling efforts that I have conducted with McGue (McGue & Gottesman, 1989), McGuffin (McGuffin, Farmer, & Gottesman, 1987), O'Rourke (O'Rourke et al., 1982), Rao (Rao, Morton, Gottesman, & Lew, 1981), Shields (Gottesman & Shields, 1967), Vogler (Vogler, Gottesman, McGue, & Rao, 1990), and Moldin (Moldin, Rice, Van Eerdewegh, Gottesman, & Erlenmeyer-Kimling, 1990), I have a "guesstimate" for such allocations as shown in Figure 1. I would allocate the lion's share of causes to various multifactorial–polygenic (oligogenic) models that require a dynamic and epigenetic interplay of various moderate-to-high-risk genetic combinations and a number of "toxic environmental" factors/experiences (Day, 1986). As noted by Fowles (1992), the greater the genetic loading or liability, the more likely it is that the environmental factors operating on a particular person to trigger a schizophrenic episode are subtle. The lower the genetic liability, the more likely it is that the environmental factors will be obvious.

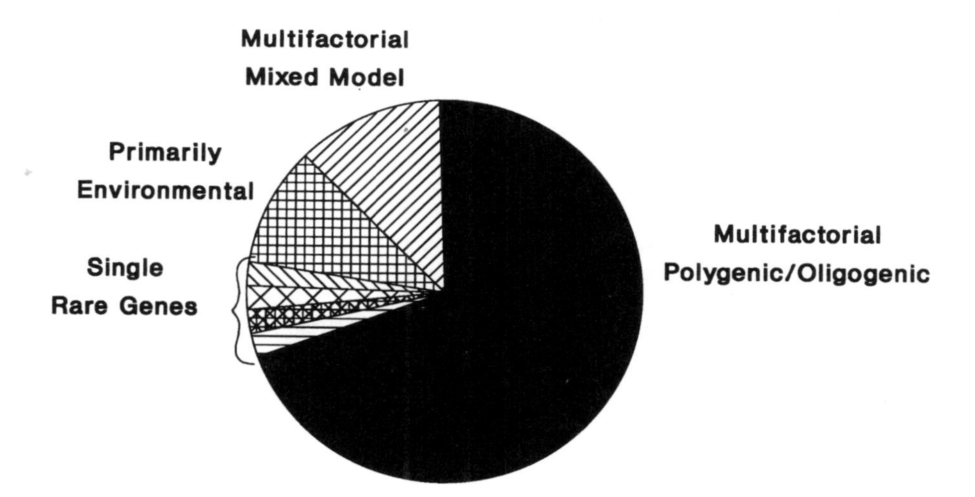

FIGURE 1. "Guesstimates" of the proportional etiologies of schizophrenia inferred from empirical studies, model fitting, and computer simulation. Adapted from *Schizophrenia Genesis—The Origins of Madness* (p. 231) by I. I. Gottesman, 1991, New York: Freeman; copyright 1991 by I. I. Gottesman; adapted by permission.

The remaining territory of causes could be allocated to multifactorial mixed models, to an unknown number of individually rare single major loci, and to apparently sufficient environmental causes (cf. Propping, 1991). Note that phenylketonuria (PKU), one well-known single-major-locus recessive cause of mental retardation, accounts for only 3 in 1,000 cases of the broader category of mental retardation. Rare single-major-locus causes of schizophrenia are to be expected, but they should not be allowed to distort the bigger picture of causes.

Pessimism resulting from the so-far unreplicated demonstrations of linkage between schizophrenia and a genetic marker, despite more than 40 attempts on Chromosomes 5, 11, and X (Kendler & Diehl, in press), must be tempered by our still-modest knowledge about how the brain works and by our still-modest knowledge about molecular genetics for psychopathological research (Mendlewicz & Hippius, 1992). PKU again provides a useful example: Although it occurs with a frequency of 1 in 10,000 Caucasians and was long thought to be just a simple recessive genetic disorder with no apparent heterogeneity, it is as common as 1 in 2,600 in Turkey and as rare as 1 in 120,000 in Japan. At the molecular level, no fewer than 50 different mutations interfere with the proper functioning of the phenylalanine hydroxylase gene and lead to varying degrees of mental retardation (Eisensmith & Woo, 1992). It is clear that negative-linkage findings can be false negatives and that positive-linkage findings can be false positives. Even when the genotype is known to be present, it may not be expressed as evidenced by the equally high risks for developing schizophrenia in the offspring of identical twins discordant for schizophrenia (Gottesman & Bertelsen, 1989).

It has been estimated that 2,900 genes occupy Chromosome 5, but only 82 have been mapped, and only 22 of these are known to be disease related; 2,200 genes occupy Chromosome 11, but only 142 have been mapped, and only 33 of them are known to be disease related (Jasny, 1991). Not too long ago, it was thought there was one receptor and one gene for dopamine neurotransmission, obviously important to any theorizing about the causes of schizophrenia. At last count (Sokoloff et al., 1992), there were five or six different dopamine receptor genes, and they are on at least four different chromosomes—a multifactorialist's dream, or nightmare, come true. Genes are now being discovered at the rate of one a day, and genetic markers are being discovered at the rate of two a day thanks to the amazing automation in the gene industry. Even the most skeptical of psychopathologists must be transduced to optimism in the face of the opportunities (Kidd, 1992; McGuffin, Owen, & Gill, 1992) for discovering the causes of schizophrenia before the "decade of the brain" is over.

References

American Psychiatric Association. (1987). *Diagnostic and statistical manual of mental disorders* (3rd ed., rev., *DSM-III-R*). Washington, DC: Author.

American Psychiatric Association. (in press). *Diagnostic and statistical manual of mental disorders* (4th ed., *DSM-IV*). Washington, DC: Author.

Day, R. (1986). Social stress and schizophrenia: From the concept of recent life events to the notion of toxic environments. In G. D. Burrows & T. R. Norman (Eds.), *Handbook of studies on schizophrenia* (pp. 71–82). Amsterdam: Elsevier.

Eisensmith, R. C., & Woo, S. L. C. (1992). Molecular basis of phenylketonuria and related hyperphenylalaninemias: Mutations and polymorphisms in the human phenylalanine hydroxylase gene. *Human Mutation, 1*, 13–23.

Fischer, M. (1973). Genetic and environmental factors in schizophrenia. *Acta Psychiatrica Scandinavica*, Suppl. 238.

Fowles, D. C. (1992). Schizophrenia: Diathesis-stress revisited. *Annual Review of Psychology, 43*, 303–336.

Gottesman, I. I. (1991). *Schizophrenia genesis: The origins of madness*. New York: Freeman.

Gottesman, I. I., & Bertelsen, A. (1989). Confirming unexpressed genotypes for schizophrenia: Risks in the offspring of Fischer's Danish identical and fraternal discordant twins. *Archives of General Psychiatry, 46*, 867–872.

Gottesman, I. I., & Shields, J. (1966). Contributions of twin studies to perspectives on schizophrenia. *Progress in Experimental Personality Research, 3*, 1–84.

Gottesman, I. I., & Shields, J. (1967). A polygenic theory of schizophrenia. *Proceedings of the National Academy of Sciences, 58*, 199–205.

Gottesman, I. I., & Shields, J. (1972). *Schizophrenia and genetics—A twin study vantage point*. New York: Academic Press.

Gottesman, I. I., & Shields, J. (with the assistance of Hanson, D. R.). (1982). *Schizophrenia: The epigenetic puzzle*. Cambridge, UK: Cambridge University Press.

Jasny, B. R. (Coordinator). (1991). *Genome maps 1991*. Washington, DC: American Association for the Advancement of Science.

Kallmann, F. J. (1946). The genetic theory of schizophrenia: An analysis of 691 schizophrenic twin index families. *American Journal of Psychiatry, 103*, 309–322.

Kendler, K. S., & Diehl, S. R. (in press). The genetics of schizophrenia: A current, genetic-epidemiologic perspective. *Schizophrenia Bulletin, 19.*

Kendler, K. S., McGuire, M., Gruenberg, A. M., O'Hare, A., Spellman, M., & Walsh, D. (in press-a). The Roscommon family study: I. Methods, diagnosis of probands and risk of schizophrenia in relatives. *Archives of General Psychiatry.*

Kendler, K. S., McGuire, M., Gruenberg, A. M., O'Hare, A., Spellman, M., & Walsh, D. (in press-b). The Roscommon family study: III. Schizophrenia-related personality disorders in relatives. *Archives of General Psychiatry.*

Kidd, K. K. (1992). The complexities of linkage studies for neuropsychiatric disorders. In J. Mendlewicz & H. Hippius (Eds.), *Genetic research in psychiatry.* (pp. 61–74). New York: Springer-Verlag.

Kraepelin, E. (1896). *Psychiatrie [Psychiatry]* (5th ed.). Leipzig, Germany: Barth.

Luxenburger, H. (1928). Vorlaufiger Bericht Uber psychiatrische Serienuntersuchungen an Zwillingen [Preliminary report on a serial psychiatric investigation of twins]. *Zeitschrift fur die gesamte Neurologie und Psychiatrie, 116*, 297–326.

Maier, W., Lichtermann, D., Hallmayer, J., Heun, R., Marx, A., & Benkert, O. (1992). *The transmission of affective and schizophrenic disorders in families: Evidence for a unitary concept of major psychiatric disorders?* Unpublished manuscript.

McGue, M. (1992). When assessing twin concordance, use the probandwise not the pairwise rate. *Schizophrenia Bulletin, 18*, 171–176.

McGue, M., & Gottesman, I. I. (1989). Genetic linkage in schizophrenia: Perspectives from genetic epidemiology. *Schizophrenia Bulletin, 15*, 282–292.

McGue, M., Gottesman, I.I., & Rao, D.C. (1983). The transmission of schizophrenia under a multifactorial threshold model. *American Journal of Human Genetics, 35*, 1161–1171.

McGue, M., Gottesman, I. I., & Rao, D. C. (1985). Resolving genetic models for the transmission of schizophrenia. *Genetic Epidemiology, 2*, 99–110.

McGuffin, P., Farmer, A., & Gottesman, I. I. (1987). Modern diagnostic criteria and genetic studies of schizophrenia. In H. Häfner, W. F. Gattaz, & W. Janzarik (Eds.), *Search for the causes of schizophrenia* (pp. 143–156). New York: Springer-Verlag.

McGuffin, P., & Murray, R. (Eds.). (1991). *The new genetics of mental illness.* Stoneham, MA: Butterworth-Heinemann.

McGuffin, P., Owen, M., & Gill, M. (1992). Molecular genetics of schizophrenia. In J. Mendlewicz & H. Hippius (Eds.), *Genetic research in psychiatry* (pp. 25–48). New York: Springer-Verlag.

McGuffin, P., Katz, R., Rutherford, J., Watkins, S., Farmer, A. E., & Gottesman, I. I. (in press). Twin studies as vital indicators of phenotypes in molecular genetic research. In T. Bouchard & P. Propping (Eds.), *Twins as a tool in behavior genetics.* New York: Wiley.

Meehl, P. E. (1990). Toward an integrated theory of schizotaxia, schizotypy, and schizophrenia. *Journal of Personality Disorders, 4*, 1–99.

Mendlewicz, J., & Hippius, H. (Eds.). (1992). *Genetic research in psychiatry*. New York: Springer-Verlag.

Moldin, S. O., & Gottesman, I. I. (in press). Population genetics in psychiatry. In H. I. Kaplan & B. J. Sadock (Eds.), *Comprehensive textbook of psychiatry* (6th ed.). Baltimore: Williams & Wilkins.

Moldin, S. O., Rice, J. P., Gottesman, I. I., & Erlenmeyer-Kimling, L. (1990). Transmission of a psychometric indicator for liability to schizophrenia in normal families. *Genetic Epidemiology, 7*, 163–176.

Moldin, S. O., Rice, J. P., Van Eerdewegh, P., Gottesman, I. I., & Erlenmeyer-Kimling, L. (1990). Estimation of disease risk under bivariate models of multifactorial inheritance. *Genetic Epidemiology*, 7, 371–386.

Okazaki, Y. (1992, June). *A recent schizophrenic twin study using a Japanese sample* [Abstract]. Presented at the Seventh International Congress on Twin Studies, Tokyo, Japan.

Onstad, S., Skre, I., Torgersen, S., & Kringlen, E. (1991). Subtypes of schizophrenia—Evidence from a twin–family study. *Acta Psychiatrica Scandinavica, 84*, 203–206.

O'Rourke, D. H., Gottesman, I. I., Suarez, B. K., Rice, J., & Reich, T. (1982). Refutation of the general single-locus model for the etiology of schizophrenia. *American Journal of Human Genetics*, 34, 630–649.

Owen, M. J. (1992). Will schizophrenia become a graveyard for molecular geneticists? *Psychological Medicine, 22*, 289–293.

Plomin, R. (1990). The role of inheritance in behavior. *Science, 248*, 183–188.

Prescott, C. A., & Gottesman, I. I. (in press). Genetically mediated vulnerability to schizophrenia. *Psychiatric Clinics in North America, 16*.

Propping, P. (1991). *Psychiatrische Genetik* [*Genetic psychiatry*]. Berlin: Springer-Verlag.

Rao, D. C., Morton, N. E., Gottesman, I. I., & Lew, R. (1981). Path analysis of qualitative data on pairs of relatives: Application to schizophrenia. *Human Heredity, 31*, 325–333.

Robins, L., & Regier, D. (1991). *Psychiatric disorders in America*. New York: Free Press.

Rosenthal, D. (1970). *Genetic theory and abnormal behavior*. New York: McGraw-Hill.

Sartorius, N., Jablensky, A., Korten, A., Ernberg, G., Anker, M., Cooper, J. E., & Day, R. (1986). Early manifestations and first-contact incidence of schizophrenia in different cultures. *Psychological Medicine, 16*, 909–928.

Slater, E., & Cowie, V. (1971). *The genetics of mental disorders*. London: Oxford University Press.

Sokoloff, P., Lannfelt, L., Martres, M. P., Giros, B., Bouthenet, M. L., Schwartz, J. C., & Leckerman, J. F. (1992). The D3 dopamine receptor gene as a candidate gene for genetic linkage studies. In J. Mendlewicz & H. Hippius (Eds.), *Genetic research in psychiatry* (pp. 135–149). New York: Springer-Verlag.

Tienari, P. (1990). Gene–environment interaction in adoptive families. In H. Häfner & W. F. Gattaz (Eds.), *Search for the causes of schizophrenia* (Vol. 2, pp. 126–143). New York: Springer-Verlag.

Torrey, E. F. (1989). Schizophrenia: Fixed incidence or fixed thinking? *Psychological Medicine, 19*, 285–287.

Vogler, G. P., Gottesman, I. I., McGue, M., & Rao, D. C. (1990). Mixed model segregation analysis of schizophrenia in the Lindelius Swedish pedigrees. *Behavior Genetics, 20*, 461–472.

World Health Organization. (1992). *The ICD-10 classification of mental and behavioral disorders: Clinical descriptions and diagnostic guidelines* (10th rev., *ICD-10*). Geneva: Author.

Zerbin-Rüdin, E. (1967). Endogene Psychosen [Endogenous psychoses]. In P. E. Becker (Ed.), *Human genetik* [*Human genetics*] (Vol. 2, pp. 446–577). Stuttgart, FRG: Thieme.

Zubin, J. (1987). Closing comments. In H. Häfner, W. F. Gattaz, & W. Janzarik (Eds.), *Search for the causes of schizophrenia* (pp. 359–365). New York: Springer-Verlag.

CHAPTER 13

From Proteins to Cognitions: The Behavioral Genetics of Alcoholism

Matt McGue

A positive family history is one of the most consistent and robust predictors of the risk of alcoholism. Children of alcoholics are approximately four to five times more likely to develop alcoholism sometime during their lifetime than are children of nonalcoholics (Cotton, 1979). But familial resemblance may owe to a shared environment, shared genes, or both, and the nature–nurture debate has nowhere been more heated than in the alcohol research field. Biological explanations of alcoholism gained prominence in the early part of this century only to be dashed by Roe's (1944) finding that the reared-away offspring of problem drinkers experienced no apparent problems with alcohol in adulthood. The reemergence of genetic models of alcoholism began a generation after Roe with publication of the influential Danish adoption studies of Goodwin and colleagues (e.g., Goodwin, Schulsinger, Hermansen, Guze, & Winokur,

The preparation of this chapter was supported, in part, by U.S. Public Health Service grants AA09367 and DA05147.

1973) and has culminated in a current zeitgeist for which genetic models of alcoholism etiology enjoy clear prominence over alternative explanations.

Classical behavioral genetics methods like twin and adoption studies have been instrumental to the fall and rise of genetic theorizing on alcoholism. Modern molecular genetics techniques are likely to play an increasingly important role in future studies aimed at identifying and characterizing genetic influences. In this chapter, a review of behavioral genetics research on alcoholism is presented and organized around four broad conclusions. This review is necessarily brief; interested readers are directed to the more comprehensive reviews provided by McGue (in press) and Heath (in press). It is argued that behavioral genetics research to date has convincingly established the existence of a genetic influence on alcoholism, at least in males. Nonetheless, little is known about either the nature of that influence or how genetic factors combine and/or interact with environmental factors to influence the development of alcoholism. It is concluded that future progress in understanding the behavioral genetics of alcoholism will require moving beyond the artificial dichotomy implicit in the nature–nurture debate to seek linkages among genetic, psychosocial, and developmental models of alcoholism.

Behavioral Genetics Research on Alcoholism

As with most behavioral disorders, there is no single agreed-upon definition for alcoholism (Tarter, Moss, Arria, Mezzich, & Vanyukov, 1992). Different behavioral genetics studies have used different diagnostic standards, giving the appearance of inconsistent findings across studies. A discussion of the different diagnostic approaches to alcoholism is far beyond the scope of the present review. Suffice it to say that in selecting studies for review, only those with treatment-ascertained samples or that applied generally accepted diagnostic criteria for alcoholism were included. Thus, the well-known twin study by Partanen, Bruun, and Markkanen (1966) was not included in the review because it considered symptoms of problem drinking, rather than alcoholism per se, in a normative twin sample. In addition, the twin studies by Allgulander, Nowak, and

Rice (1991) and Romanov, Kaprio, Rose, and Koskenvuo (1991) were not considered because in both cases the assessment of alcoholism (based entirely on treatment seeking) was so insensitive as to lead to a substantial underestimation of twin resemblance. Table 1 provides a brief summary of the salient characteristics of the twin and adoption studies included in the review. For studies that used more than one system for diagnosing alcoholism, the most exclusive form of the diagnosis was generally used in compiling results for the present summary.

Although the total number of relevant twin and adoption studies is relatively small, several conclusions, of varying certainty, appear warranted at this time. Given the relatively early stage of inquiry into the genetics of alcoholism, however, it is recognized that some of these conclusions will need to be amended in the light of future research. Nonetheless, these general results can serve as a guide to future behavioral genetics research in the area.

Conclusion 1: Alcoholism Is Moderately Heritable in Males but Only Modestly Heritable in Females

Both twin and adoption studies support the conclusion that genetic factors influence risk of alcoholism. Nonetheless, the evidence implicating a genetic influence on alcoholism risk is not nearly as consistent as that for schizophrenia or the other major psychiatric disorders. Gender appears to be one factor that contributes to the apparent inconsistency of findings. Figure 1 gives the probandwise concordance rates for alcoholism (i.e., the conditional probabilities that a twin is alcoholic given that his or her cotwin is alcoholic) (McGue, 1992) published in studies of male (Figure 1a) and female (Figure 1b) monozygotic (MZ) and dizygotic (DZ) twins. Despite using different ascertainment schemes, assessment methods, and diagnostic standards, the six studies of male twins are consistent in reporting higher MZ than DZ concordance for alcoholism (the zygosity difference being statistically significant in every study except that by Gurling, Oppenheim, & Murray, 1984). In contrast, in only three of the five twin studies of female alcoholics is the MZ concordance greater than the DZ concordance, with the difference attaining statistical significance in only two of the studies.

TABLE 1

Characteristics of Twin and Adoption Studies of Alcoholism

			No. male		No. female	
Twin Studies	Country	Diagnosis	MZ	DZ	MZ	DZ
Kaij, 1960	Sweden	Chronic alcoholism	14	31	0	0
Hrubec & Omenn, 1981	USA	ICD-8	271	444	0	0
Gurling et al., 1984	UK	WHO alcohol dependence	15	20	13	8
Pickens et al., 1991	USA	DSM-III alcohol dependence	39	47	24	20
McGue et al., 1992	USA	DSM-III alcohol abuse/dependence	85	96	44	43
Caldwell & Gottesman, 1991	USA	DSM-III alcohol dependence	20	15	7	12
Kendler et al., 1992	USA	DSM-III-R alcohol dependence	0	0	81	79
			No. male		No. female	
Adoption studies	Country	Diagnosis	FHP	FHN	FHP	FHN
Roe, 1944	USA	Unspecified	21	11	11	14
Goodwin et al., 1973, 1977	Denmark	Feighner	55	78	49	47
Cloninger et al., 1981, & Bohman et al., 1981	Sweden	Temperance board registration	291	571	336	577
Cadoret et al., 1985	USA	DSM-III alcohol abuse/dependence	18	109	12	75
Cadoret et al., 1987	USA	DSM-III alcohol abuse/dependence	8	152	0	0

Note. Samples overlap in the Pickens et al. (1991) and McGue et al. (1992) studies. MZ = monozygotic; DZ = same-sex dizygotic; FHP = adoptees with positive biological background for alcoholism; FHN = adoptees with negative biological for alcoholism.

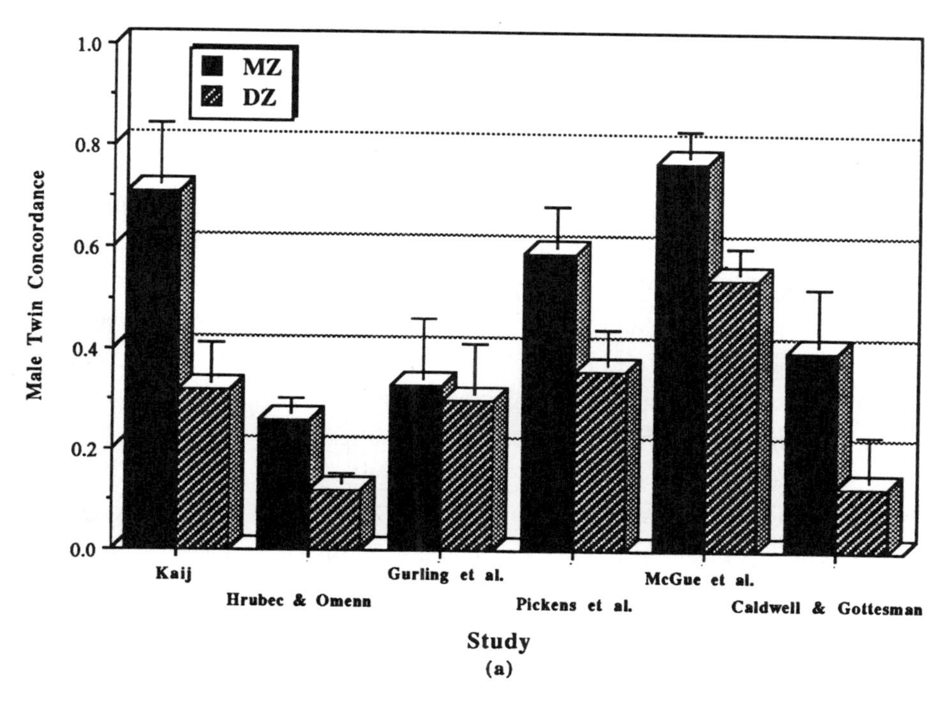

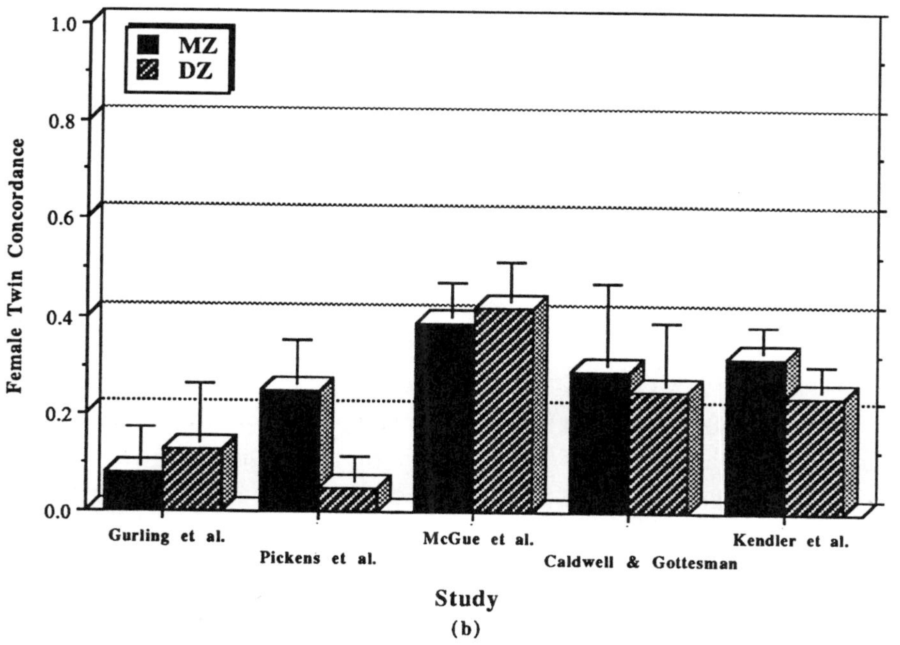

FIGURE 1. Monozygotic (MZ) and dizygotic (DZ) probandwise concordance rates for alcoholism in studies of male (a) and female (b) same-sex twins. Vertical bars mark one standard error. Study characteristics are summarized in Table 1.

Findings from adoption studies provide additional support for the hypothesis that gender moderates the inheritance of alcoholism (Figure 2). With the exception of Roe (1944), studies of male adoptees are consistent in reporting a significantly greater risk of alcoholism among the reared-away biological sons of alcoholics than among the reared-away biological sons of nonalcoholics. In contrast, in only two of four studies of female adoptees is the rate of alcoholism significantly greater among the reared-away daughters of alcoholics than among the reared-away daughters of nonalcoholics.

The comparisons summarized in Figures 1 and 2 can be used to establish the existence of a genetic influence; they do not directly estimate the strength of that influence, especially given the varying diagnostic standards and the consequent differences in alcoholism prevalence that exist across studies. The heritability coefficient gives the proportion of variance in a given trait that is associated with genetic factors and is consequently used by geneticists to indicate the strength of genetic influence. Although the analytical procedures for heritability estimation are more complex when the phenotype is qualitative rather than quantitative, the conceptual formulation is similar (Rice & Reich, 1985). Figure 3 summarizes heritability estimates from those twin studies of alcoholism that reported estimates of heritability[1] (given the generally atypical circumstances that surround adoption, it is relatively difficult to derive generalizable heritability estimates from adoption studies). The heritability estimates support the general impressions gained from Figures 1 and 2: Estimates for the heritability of alcoholism liability are consistently moderate among males but vary markedly across studies of females. The relatively small size of the female twin samples certainly contributes to the marked variation in estimates and suggests the need for caution against overstating the significance of any single estimate. Although, given the existing literature, one cannot entirely rule out the possibility that the

[1]The prevalence of alcoholism in MZ and DZ cotwins as well as in the general population are needed to derive an estimate of heritability. As no general population estimate of prevalence was available from Gurling et al. (1984), a heritability estimate is not reported for this study. Nonetheless, given the lack of a significant zygosity effect, the most likely estimate for heritability for this study would be zero.

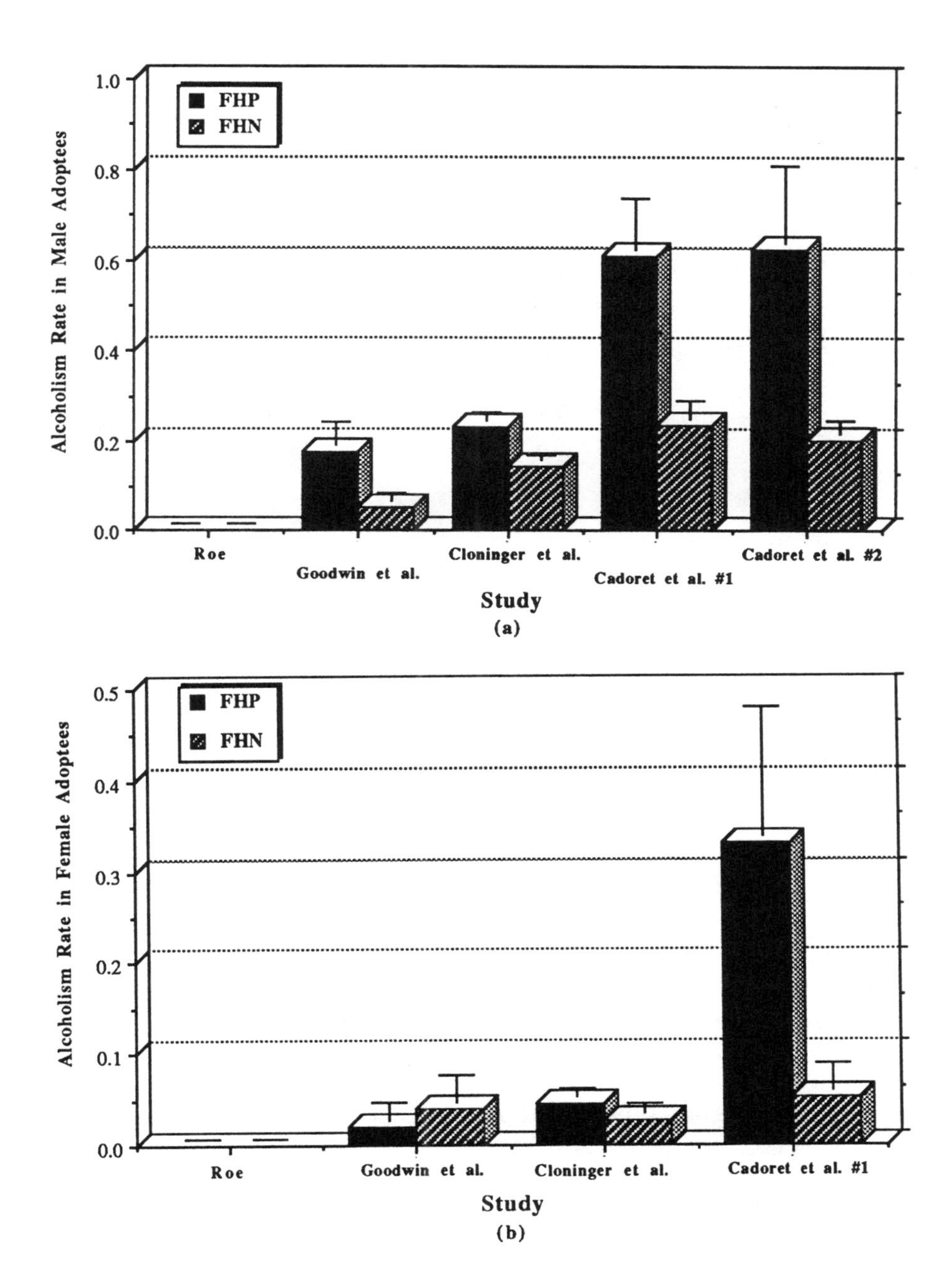

FIGURE 2. Risk of alcoholism among male (a) and female (b) adoptees with either a positive (FHP) or negative (FHN) biological background of alcoholism. Vertical bars mark one standard error. Study characteristics are summarized in Table 1.

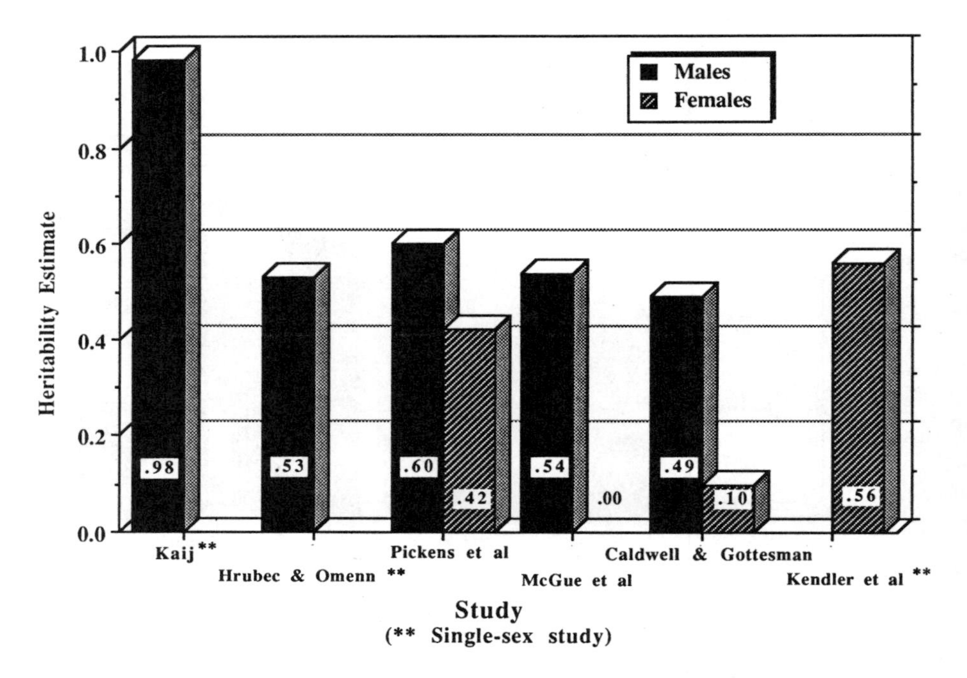

FIGURE 3. Estimates of the heritability of alcoholism liability derived from studies of male and female same-sex twins. Not every study included samples of both male and female twins. Study characteristics are summarized in Table 1.

heritability of alcoholism is equal in the two sexes (e.g., Kendler, Heath, Neale, Kessler, & Eaves, 1992), the aggregate data do suggest greater heritability among men than among women. If so, then identifying the mechanism by which gender moderates the inheritance of alcoholism would constitute a major question to be addressed in future research.

Conclusion 2: The Genetic Influences That Underlie Alcoholism, Like the Disorder Itself, Are Heterogeneous

It has long been recognized that alcoholics represent a clinically diverse group (Knight, 1937). In the many attempts at classifying alcoholics, one distinction has consistently emerged (Sher, 1991): For some alcoholics, drinking appears to be a means of coping with psychological distress (i.e., those who are neurotic and have a relatively late age of alcoholism onset);

whereas for others, drinking appears to be a manifestation of an underlying personality disorder (i.e., those who are undersocialized and have a relatively early age of problem drinking onset). Underlying this distinction may be different biological pathways to alcoholism.

Cloninger, Bohman, and Sigvardsson (1981) reported that genetic influences appeared to be stronger among Swedish male adoptees with moderate alcoholism compared with those with mild or severe alcoholism. In addition, compared with mild or severe alcoholics, moderate alcoholics were more likely to have biological fathers who were criminal and had teenage onset of problem drinking. From these empirical observations arose Cloninger's neurobiological model of alcoholism (Cloninger, 1987) and what has rapidly become the most prominent classification model of alcoholism—the distinction between Type I and Type II alcoholics. Type I alcoholism (posited to account for the mild and severe alcohol abusers in the Swedish study) is hypothesized to be moderately heritable, to afflict both males and females, and to be characterized by a relatively late age of onset and by anxiety surrounding drinking. In contrast, Type II alcoholism (posited to account for moderate forms of alcohol abuse) is hypothesized to be strongly heritable, afflict males primarily, and be characterized by a relatively early age of onset, alcohol-related aggression, and legal complications resulting from alcohol abuse. Although some of the specifics of Cloninger's neurobiological model and clinical predictions remain controversial, without controversy is the centrality accorded age of alcoholism onset in Cloninger's etiological model. Age of onset does appear to be an important moderator of alcoholism inheritance, and early onset alcoholics do appear to differ from late onset alcoholics on a host of psychological and psychiatric characteristics (Helzer & Pryzbeck, 1988).

A recent twin study from the University of Minnesota illustrates the moderating influence of age of onset. McGue, Pickens, and Svikis (1992) estimated the heritability of alcoholism liability to be large and significant among males with a problem drinking onset at or before age 20 years (heritability [h^2] $= 0.73 \pm 0.18$) but to be modest and nonsignificant among males with a relatively late age of onset ($h^2 = 0.30 \pm 0.26$) (Figure 4). Moreover, compared with the late onset cases, males with an early

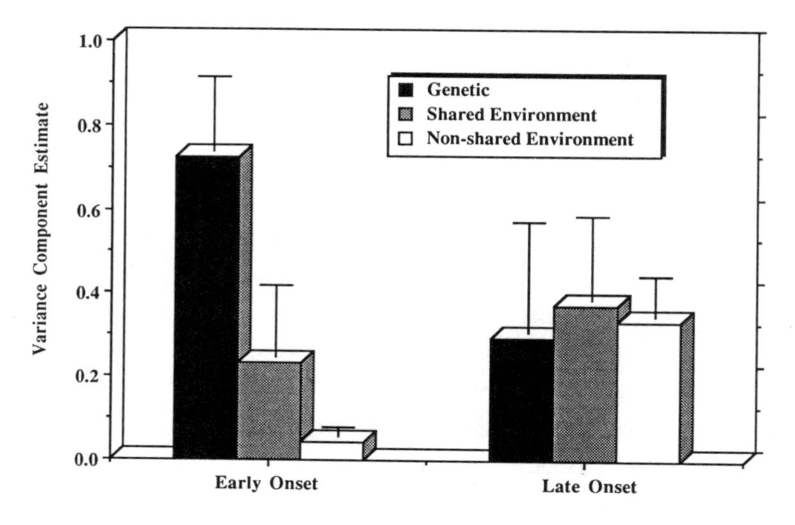

FIGURE 4. Biometrical decomposition of alcoholism liability variance derived from a sample of male twins according to whether the proband had either an early (≤ 20 years) versus late (> 20 years) onset of alcohol-related problems. Adapted from "Sex and Age Effects on the Inheritance of Alcohol Problems: A Twin Study" by McGue, Pickens, and Svikis, 1992, *Journal of Abnormal Psychology, 101,* p. 11; copyright 1992 by the American Psychological Association; adapted by permission.

onset of problem drinking were more likely to exhibit other signs of undersocialized behavior including conduct disorder, other drug use and abuse, and precocious sexuality. In contrast, female twins with early versus late onset of problem drinking showed no differences for these variables.

Conclusion 3: The Genetic Diathesis Underlying Alcoholism Has Not Yet Been Characterized but Is Likely to Be Polygenic

The genetic diathesis that underlies alcoholism, like that for most behavioral disorders, is likely to represent the combined effect of many rather than a single or even a few genes. These "alcoholism polygenes" are among the 50,000 to 100,000 functional genes distributed among the 22 pairs of autosomal and single pair of sex chromosomes that comprise the human genome. Human geneticists use primarily two methods to locate and identify the specific genes contributing to individual differences in a given human characteristic: linkage and association studies. The power of both methods has been substantially increased by the discovery of DNA markers (Botstein, White, Skolnick, & Davis, 1980), which are highly polymorphic, noncoding segments of DNA. The use of DNA polymorphisms has allowed human geneticists to successfully map much of the human genome; more than 2,000 structural genes have been mapped using the nearly 3,000 available polymorphic markers. The day is rapidly approaching when the only obstacles to localizing the specific genes influencing disease expression will be ascertainment of informative families, possession of a rich library of polymorphic markers, and perseverance.

It is beyond the scope of the present review to provide either a critical description of the methodology used in linkage and association studies or a systematic review of findings from application of these methodologies to alcoholism. The search for single-gene effects on alcoholism is an important, emerging area of research, and the nature of findings in this area is such that specific conclusions are subject to rapid revision. The interested reader is again referred to alternative sources including McGue (in press) and Merikangas (1990) for review of linkage and association studies of alcoholism. The focus here is to broadly summarize findings from attempts to identify single-gene effects on alcoholism and discuss what these studies tell us about the nature of alcoholism.

Although linkage studies of alcoholism, as with most behavioral traits, have not yet yielded replicable findings, two separate genetic loci have been implicated in alcoholism etiology by association studies. The first, a gene affecting the alcohol metabolizing liver enzyme, aldehyde dehydrogenase (ALDH), may help account for some of the ethnic variation in alcoholism rates. Goedde, Harada, and Agarwal (1979) reported that approximately 50% of Japanese and Chinese, but 0% of Caucasian, donors had an inherited allelic variant of the low-K_m ALDH isozyme (ALDH2) resulting in deficient ALDH activity. Moreover, Harada, Agarwal, Goedde, Tagaki, and Ishikawa (1982) found that, although 44% of Japanese nonalcoholics were ALDH2 deficient, only 2.3% of Japanese alcoholics were. By inhibiting the conversion of acetaldehyde to acetic acid, the ALDH2 variant provides a natural, protective analog to disulfiram. In short, individuals who inherit ALDH2 deficiency are protected against alcoholism because drinking alcohol is likely to make them sick.

A second and more controversial association is that between alcoholism and the D2 dopamine receptor (DRD2) locus. Animal research has implicated the dopaminergic system in alcohol self-administration (Wise & Rompre, 1989), thereby providing a rationale for exploring the relationship between alcoholism and polymorphisms affecting dopamine neurotransmission. Blum et al. (1990) reported that 24 (69%) of 35 alcoholics, but only 7 (20%) of 35 nonalcoholics, carried the A1 allele at DRD2. Since publication of Blum et al.'s findings, two additional studies have replicated the association between alcoholism and DRD2 (Comings et al., 1991; Parsian et al., 1991), whereas three others reported failures to replicate (Bolos, Dean, Lucas-Derse, Ramsburg, & Brown, 1990; Gelertner et al., 1991; Turner et al., 1992). Moreover, despite reporting a positive association, Parsian et al. (1991) failed to find evidence of linkage between alcoholism and DRD2 in several large informative families, which suggests that the alcoholism–DRD2 association is more complex than a simple direct affect of DRD2 on alcoholism risk, as had been suggested in previous studies. Several hypotheses have been advanced to account for the inconsistency of results including (a) that the association is a statistical artifact of failure to match alcoholic and nonalcoholic samples for ethnic composition and (b) that the DRD2 locus influences susceptibility to alcohol-induced tissue damage rather than alcoholism per se. In any case, the significance of the DRD2–alcoholism association remains unresolved at this time.

The ALDH and DRD2 associations with alcoholism, however uncertain, do illustrate two factors that are likely to characterize the search for single-gene effects on multifactorial disorders like alcoholism. First, the specific genetic mechanisms that underlie genetic associations are likely to be diverse. The association between ALDH2 and alcoholism is due to protective metabolic processes; the putative association between DRD2 and alcoholism is thought to reflect genetically influenced susceptibility to alcohol-induced tissue damage. When the genes contributing to the development of alcoholism are finally identified, they likely will involve specific effects on traits ranging from metabolic reactions and neurophysiological processes to personality and psychopathological disorder. Second, why and how people become alcoholic is not likely to be answered entirely through reductionistic appeals to single-gene effects. It is a truism that the gene-to-behavior pathway is long, which thereby provides multiple opportunities for modulation by other gene systems and environmental factors. The apparently straightforward association between ALDH2 deficiency and alcoholism illustrates the point. Although the conventional wisdom is that ALDH2-deficient individuals do not become alcoholic because drinking makes them sick, it is clear that this cannot be the whole of the story. Some who inherit ALDH2 deficiency do drink; indeed, they apparently drink a lot because approximately 2% of Japanese alcoholics are ALDH2 deficient (Harada et al., 1982). Understanding why some who inherit ALDH2 deficiency become alcoholic whereas others do not will require characterizing the ALDH2 gene effect within the complex biosocial network of alcoholism determinants. That is, it will require determining how the effect of inheriting ALDH2 is moderated by or moderates the effects of other "alcohol metabolizing" genes, the individual's pharmacological and psychological responses to alcohol ingestion, the individual's attitudes and expectancies concerning alcohol's effects, and the impact of societal values for and sanctions against specific forms of drinking. When viewed from this perspective, findings from molecular genetics research are more likely to broaden than narrow the scope of inquiry.

Conclusion 4: Alcoholic Rearing May Not Be One of the Critical Environmental Factors Influencing the Development of Alcoholism

Twin and adoption studies of alcoholism have not only established the existence of genetic influences but also confirmed the importance of environmental factors: MZ twins are not perfectly concordant for alcoholism, and the estimated heritability of alcoholism liability is far from unity. Behavioral geneticists have distinguished between two types of environmental influence: (a) shared environmental effects that correspond to factors shared by reared-together relatives and thus contribute to their similarity and (b) nonshared environmental effects that correspond to factors that are not shared by reared-together relatives and thus contribute to their differences. One of the more remarkable findings from behavioral genetics research is that, although environmental factors exert a substantial influence on individual differences in all psychological traits, in most cases the relevant environmental factors are nonshared rather than shared (Plomin & Daniels, 1987). However, significant shared environmental effects on alcoholism have been reported in several twin studies (e.g., McGue et al., 1992), which suggests that alcoholism may be one exception to the general rule.

Alcoholic rearing is one shared environmental factor that has received much empirical and theoretical attention. For some social learning theorists, the observed excess rate of alcoholism among the children of alcoholics is due, at least in part, to their modeling the drinking behavior of their parents. But this hypothesis is challenged by the fact that elevated rates of alcoholism are observed among the biological offspring of alcoholics even when they are reared in nonalcoholic adoptive families. Studies of intact nuclear families (i.e., relatives who share both genes and rearing environments) cannot unequivocally implicate shared environmental contributions to familial resemblance any more than they can unequivocally implicate genetic influences. Behavioral genetics methodology is needed to resolve the separate influence of genes and shared environment on familial resemblance.

Figure 5 summarizes findings from adoption studies on the effect of being reared in adoptive families containing alcoholic members. As can be seen, the data are rather inconsistent. Although the two Scandinavian studies suggest no effect associated with being reared with alcoholics, the two U.S. studies by Cadoret and colleagues (Cadoret, O'Gorman, Troughten, & Heywood, 1985; Cadoret, Troughten, & O'Gorman, 1987) are both positive. Moreover, the half-sibling study by Schuckit, Goodwin, and Winokur (1972), which is not included in the figure because it is not formally an adoption study, also suggests that there is no increased risk of alcoholism associated with being reared by a nonbiologically related alcoholic parent. Nonetheless, the findings by Cadoret et al. (1985, 1987) cannot be dismissed as aberrations; in these two separate studies in both males and females, Cadoret et al. (1985, 1987) reported increased

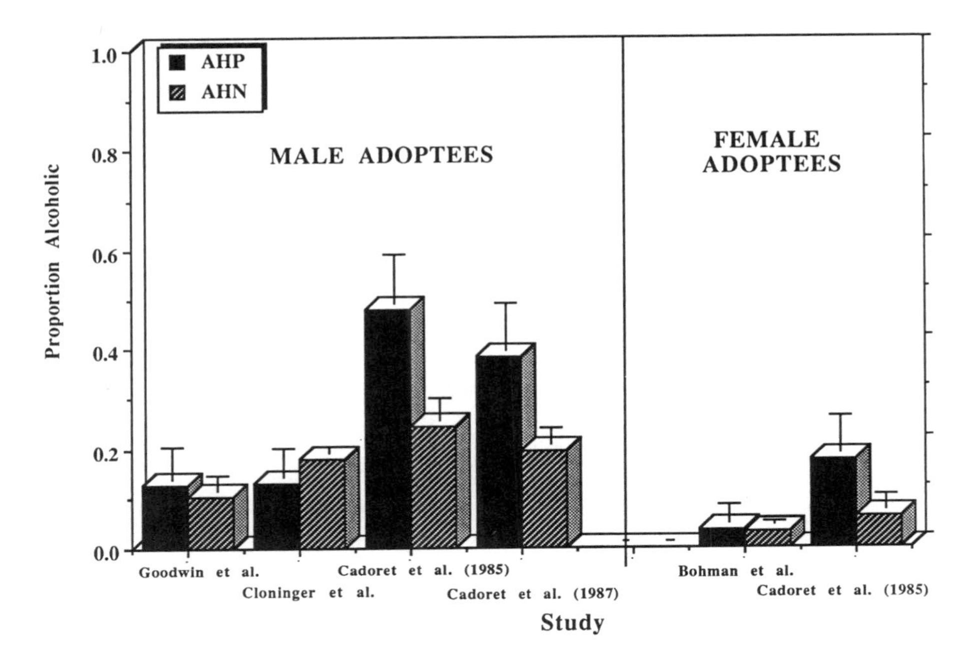

FIGURE 5. Adoptee risk of alcoholism when reared in adoptive home with at least one alcoholic (AHP) or no alcoholics (AHN). In all studies except those by Cadoret and colleagues, rearing designation was based on adoptive parent alcoholism status. For the two Cadoret studies, rearing designation based on history of alcoholism in any alcoholic relative including parent and/or sibling as well as second-degree relatives. Vertical bars mark one standard error. Study characteristics are summarized in Table 1.

rates of alcoholism among adoptees reared by or with other alcoholics. There are two factors that might help account for the apparent inconsistency. First, although the Scandinavian studies and the study by Schuckit et al. (1972) specifically investigated the effect of being reared by an alcoholic parent, Cadoret et al. (1985, 1987) investigated the effect of any alcoholism in the adoptive family (including parents, siblings, and more remote adoptive relatives). Second, the positive association found in at least one of the studies (Cadoret et al., 1985) held in rural (i.e., communities with populations less than 2,500) but not nonrural (population of 2,500 or greater) communities. It may be that alcoholic family members represent a specific environmental liability for developing alcoholism only in those environments where there is limited opportunity to observe problem drinking outside the family. In any case, adoption studies clearly demonstrate the need to consider both the biological and social mechanisms by which relatives come to resemble one another.

Characterizing the Nature of the Genetic Influence on Alcoholism

For behavioral geneticists, the question is no longer whether but how alcoholism is inherited. Mapping the pathway from genes to alcoholic phenotype will involve determining how the inherited diathesis is manifested at the molecular, physiological, behavioral, and social levels. But such characterization is difficult to achieve through traditional standard case-control comparisons. Given the multiple social and medical complications associated with chronic, heavy alcohol consumption, differences between alcoholics and nonalcoholics may reflect consequences as well as causes of the disorder. The high-risk design, and in particular the study of the children of alcoholics, has provided a powerful paradigm for identifying risk factors that predate alcoholism onset. High-risk research on alcoholism has identified three promising candidates that may serve, singly or in combination, as inherited behavioral precursors of alcoholism: alcohol sensitivity, personality–temperament, and cognitive factors.

Alcohol Sensitivity

Pharmacological and psychological responses to alcohol certainly exert powerful influences on individual patterns of drinking. Moreover, in the mouse, alcohol preference is strongly related to alcohol sensitivity, and both phenotypes appear to be partially inherited (McClearn & Kakihana, 1981). Thus, there is much reason to explore in humans the inheritance of alcohol sensitivity and the relationship between alcohol sensitivity and alcohol consumption. Paradoxically, alcohol researchers have hypothesized that both hypo- and hypersensitivity to alcohol's effects may, in part, mediate the inheritance of alcoholism.

Schuckit has been the major proponent of the hypothesis that reduced sensitivity to alcohol constitutes an important inherited risk factor for alcoholism (e.g., Schuckit & Gold, 1988). When asked to report their subjective reactions after a standard dose of alcohol, sons of alcoholics, compared with sons of nonalcoholics matched for drinking history, consistently report lower levels of intoxication. However, alcohol sensitivity differences between sons of alcoholics and nonalcoholics have been less consistently observed when they have been objectively (e.g., body sway or hormonal response) (Sher, 1991) rather than subjectively assessed. If alcohol hyposensitivity is an inherited risk factor for alcoholism, then its mediating role may be due to the inhibition of feedback mechanisms and/or to an encouragement toward alcohol overconsumption to ensure that desired psychological and pharmacological effects are achieved.

Finn and Pihl (1987, 1988) are the major proponents of the complementary hypothesis: Individuals at relatively high risk of developing alcoholism are characterized by heightened sensitivity to the reinforcing properties of alcohol. A venerable hypothesis in the alcohol research field is that alcohol can attenuate response to stress and is thus particularly reinforcing when consumed in stressful situations. In a series of studies, Finn and Pihl (1987, 1988) showed that after alcohol ingestion, sons of alcoholics show greater attenuation of cardiovascular response to stress than do sons of nonalcoholics.

In another important study, Newlin and Thomson (1990) identified many of the methodological limitations of research in this area (e.g., almost exclusive reliance on male offspring, designating 21–25-year-old males with no history of problem drinking high-risk because of a positive family history when their own drinking history suggests that they might be low-risk). They also proposed a resolution to the apparent inconsistency between the results of Schuckit and colleagues (Schuckit et al., 1972; Schuckit & Gold, 1988) on the one hand and those of Finn and Pihl (1987, 1988) on the other. If one plots differences between sons of alcoholics and sons of nonalcoholics on alcohol sensitivity measures according to the time after alcohol ingestion when the measure was taken, sons of alcoholics appear to show heightened sensitivity during the first 30 minutes after alcohol ingestion (i.e., as blood alcohol concentration

[BAC] rises) and reduced sensitivity thereafter (i.e., as BAC drops). That is, in relation to sons of nonalcoholics, individuals at high risk for developing alcoholism were more sensitive to the early, reinforcing effects of alcohol but less sensitive to the late, aversive effects. Because Newlin and Thomson's conclusion is based on an aggregation of studies, many of which they considered methodologically weak, determining whether children of alcoholics are, psychopharmacologically, at double jeopardy for developing alcoholism remains a question to be addressed by future research in this area.

Personality–Temperament

There is a vast research literature relating personality factors to alcoholism. In many ways, the yield from this literature is disappointing: Many of the reported personality differences between alcoholics and nonalcoholics are small and nonreplicable, some of the differences appear to be a consequence rather than a cause of the disorder, and no personality factor has been shown to uniquely characterize "the alcoholic." Nonetheless, due to theoretical studies by Cloninger (1987) and others, as well as the growing body of research documenting the personality characteristics of the children of alcoholics, interest in personality factors has undergone a recent revival. There are inherited personality differences between children of alcoholics and nonalcoholics, and these differences appear to exist well before the children have experienced problems with alcohol (for an excellent review see Sher, 1991). Two personality dimensions appear to be particularly relevant. The first, and most consistently implicated, dimension has been termed *behavioral undercontrol*, or its complement *behavioral constraint*, and roughly corresponds to the individual's ability to inhibit behavioral responses. In relation to children of nonalcoholics, the children of alcoholics are more likely to be diagnosed as hyperactive, oppositional, and conduct disordered; they are also more likely to be rated as impulsive, inattentive, and undersocialized. The second, and less consistently implicated, dimension of personality is *negative emotionality*, or the tendency to experience negative mood states. Although the evidence is less clear than with behavioral undercontrol, in relation to the children of nonalcoholics, the children of alcoholics are more likely to be rated as neurotic and diagnosed with an anxiety disorder. The developmental pathway that can begin in childhood with either neuroticism or impulsivity and end in adulthood with alcoholism remains to be described.

Cognitive Factors

Cognitively oriented alcohol researchers have demonstrated that attitudes and expectancies surrounding alcohol can develop relatively early in life, before direct experience with alcohol, but yet can be powerful predictors of alcohol use and abuse (Brown, 1985; Christiansen, Goldman, & Inn, 1982). Of particular relevance to the present discussion is a small study of 84 twin pairs that reported greater MZ than DZ twin similarity in attitudes about alcohol use (Perry, 1973). It will be important to not only replicate Perry's findings but extend them by determining how cognitive factors moderate or are moderated by inherited risk factors for alcoholism.

Alcoholism and the Environment

Among behavioral disorders, alcoholism is unique in having a necessary environmental determinant: continued exposure to alcohol. As a consequence, environmental factors are likely to be more central to the development of alcoholism than, say, to schizophrenia or autism. Factors that affect an individual's access to alcohol or the perceived utility of drinking are known to affect rates of alcoholism. Thus, rates of problem drinking go down when the relative cost of alcohol increases (e.g., due to tax rate) but go up when taverns remain open longer hours (Smith, 1980). Rates of alcoholism are higher among ethnic groups that condone heavy drinking than among those that value responsible drinking (Vaillant & Milofsky, 1983). Rates of alcoholism also vary across cohorts (Helzer, Burnam, & McEvoy, 1991), over the life span (Helzer et al., 1991), geographically within a given culture (Room, 1983), as a function of religious affiliation (Weissman, Myers, & Harding, 1980), and between the sexes (Helzer et al., 1991)—variation that cannot be accounted for entirely by biological factors. The centrality of environmental influence suggests that alcoholism may provide a prototype for exploring how genetic and environmental factors combine to influence behavior. Indeed, alcoholism may ultimately tell us much about two processes of substantial theoretical interest to behavioral geneticists: genotype–environment correlation and genotype–environment interaction.

The relationship between hyperactivity and alcoholism can be used to illustrate. A 6-year-old hyperactive child carries an increased risk of developing alcoholism. He also, as a consequence of his impulsivity, overactivity, and inattentiveness, experiences a much different world than his nonhyperactive peers. The hyperactive child is much more likely to evoke parental hostility and punitiveness, as well as to engender frustration in his teachers. It is no great leap to suggest that the environment the hyperactive child experiences as a result of inherited behavioral tendencies is instrumental to the development of adult behavioral disorders including alcoholism (i.e., a genotype–environment correlation). In addition, compared with his nonhyperactive peers, the hyperactive child is more likely to be vulnerable to the environmental influences (e.g., the influence of undersocialized peers) that encourage the development of problem drinking (a genotype–environment interaction).

Conclusion

A generation of behavioral genetics research has established the existence of genetic influences on alcoholism. Nonetheless, little is known about either the nature of that influence or how genetic factors combine and/or interact with environmental factors to influence the development of alcoholism. One of the most exciting developments in alcoholism research is the multiplicity of approaches, ranging from the molecular (e.g., individual gene products, receptor complexes) to the molar (e.g., contextual factors, social policy), being used to bring about a better understanding of the etiology of the disorder. The challenge in such diversity is to move beyond insular accounts of specific phenomenon to seek integrated models of alcohol addiction; narrow, unilevel approaches, be they biological or social, will not produce comprehensive accounts of the development of alcoholism. As behavioral genetics moves beyond the narrow confines of the nature–nurture debate, it can offer the integrative framework that alcohol researchers need.

References

Allgulander, C., Nowak, J., & Rice, J. P. (1991). Psychopathology and treatment of 30,344 twins in Sweden: II. Heritability estimates of psychiatric diagnosis and treatment in 12,884 twin pairs. *Acta Psychiatrica Scandinavica, 83*, 12–15.

Blum, K., Noble, E. P., Sheridan, P. J., Montomery, A., Ritchie, T., Jagadeeswaran, P., Nogami, H., Briggs, A. H., & Cohn, J. B. (1990). Allelic association of human dopamine D2 receptor gene and alcoholism. *Journal of American Medical Association, 263*, 2055–2060.

Bohman, M., Sigvardsson, S., & Cloninger, C. R. (1981). Maternal inheritance of alcohol abuse: Cross-fostering analysis of adopted women. *Archives of General Psychiatry, 38*, 965–969.

Bolos, A. M., Dean, M., Luca-Derse, S., Ramsburg, M., & Brown, G. L. (1990). Population and pedigree studies reveal a lack of association between the dopamine D2 receptor gene and alcoholism. *Journal of the American Medical Association, 264*, 3156–3160.

Botstein, D., White, R. L., Skolnick, M., & Davis, R. W. (1980). Construction of a genetic linkage map in man using restriction fragment length polymorphisms. *American Journal of Human Genetics, 32*, 314–331.

Brown, S. A. (1985). Expectancies versus background in the prediction of college drinking patterns. *Journal of Consulting and Clinical Psychology, 53*, 123–130.

Cadoret, R. J., O'Gorman, T., Troughton, E., & Heywood, E. (1985). Alcoholism and antisocial personality: Interrelationships, genetic and environmental factors. *Archives of General Psychiatry, 42*, 161–167.

Cadoret, R. J., Troughton, E., & O'Gorman, T. W. (1987). Genetic and environmental factors in alcohol abuse and antisocial personality. *Journal of Studies on Alcohol, 48*, 1–8.

Caldwell, C. B., & Gottesman, I. I. (1991). Sex differences in the risk for alcoholism: A twin study (Abstract). *Behavior Genetics, 21*, 563–563.

Christiansen, B. A., Goldman, M. S., & Inn, A. (1982). Development of alcohol-related expectancies in adolescents: Separating pharmacological from social-learning influences. *Journal of Consulting and Clinical Psychology, 50*, 336–344.

Cloninger, C. R. (1987). Neurogenetic adaptive mechanisms in alcoholism. *Science, 236*, 410–416.

Cloninger, C. R., Bohman, M., & Sigvardsson, S. (1981). Inheritance of alcohol abuse: Cross-fostering analysis of adopted men. *Archives of General Psychiatry, 38*, 861–868.

Comings, D. E., Comings, B. G., Mubleman, M. S., Dietz, G., Shahbahrami, B., Tast, D., Knell, E., Kocsis, P., Baumgarten, R., Kovacs, B. W., Levy, D. L., Smith, M., Borison, R. L., Evans, D., Klein, D. N., MacMurray, J., Tosk, J. M., Sverd, J., Gysin, R., & Flanagan, S. D. (1991). The dopamine D2 receptor locus as a modifying gene in neuropsychiatric disorders. *Journal of the American Medical Association, 266*, 1793–1800.

Cotton, N. S. (1979). The familial incidence of alcoholism: A review. *Journal of Studies on Alcohol, 40*, 89–116.

Finn, P. R., & Pihl, R. O. (1987). Men at high risk for alcoholism: The effect of alcohol on cardiovascular response to unavoidable shock. *Journal of Abnormal Psychology, 96*, 230–236.

Finn, P. R., & Pihl, R. O. (1988). Risk for alcoholism: A comparison between two different groups of sons of alcoholics on cardiovascular reactivity and sensitivity to alcohol. *Alcoholism: Clinical and Experimental Research, 12*, 742–747.

Gelernter, J., O'Malley, S., Risch, N., Kranzier, H., Krystal, J., Merikangas, K., Kennedy, J., & Kidd, K. K. (1991). No association between an allele at the D2 dopamine receptor gene (DRD2) and alcoholism. *Journal of the American Medical Association, 266*, 1801–1807.

Goedde, H. W., Harada, S., & Agarwal, D. P. (1979). Racial differences in alcohol sensitivity: A new hypothesis. *Human Genetics, 51*, 331–334.

Goodwin, D. W., Schulsinger, F., Hermansen, L., Guze, S. B., & Winokur. G. (1973). Alcohol problems in adoptees raised apart from alcoholic biological parents. *Archives of General Psychiatry, 28*, 238–243.

Goodwin, D. W., Schulsinger, F., Knop, J., Mednick, S., & Guze, S. B. (1977). Alcoholism and depression in adopted-out daughters of alcoholics. *Archives of General Psychiatry, 34*, 751–755.

Gurling, H. M. D., Oppenheim, B. E., & Murray, R. M. (1984). Depression, criminality and psychopathology associated with alcoholism: Evidence from a twin study. *Acta Geneticae Medicae et Gemellologiae, 33*, 333–339.

Harada, S., Agarwal, D. P., Goedde, H. W., Tagaki, S., & Ishikawa, B. (1982). Possible protective role against alcoholism for aldehyde dehydrogenase isozyme deficiency in Japan. *Lancet, ii*, 827.

Heath, A. C. (in press). Genetic influences on drinking behavior in humans. In H. Begleiter & B. Kissin (Eds.), *Alcohol and alcoholism: Vol. 1. Genetic factors*. London: Oxford University Press.

Helzer, J. E., Burnam, A., & McEvoy, L. T. (1991). Alcohol abuse and dependence. In L. N. Robins & D. A. Reiger (Eds.), *Psychiatric disorders in America: The epidemiologic catchment area study* (pp. 81—115). New York: Free Press.

Helzer, J. E., & Pryzbeck, T. R. (1988). The co-occurrence of alcoholism with other psychiatric disorders in the general population and its impact on treatment. *Journal of Studies on Alcohol, 49*, 219–224.

Hrubec, Z., & Omenn, G. S. (1981). Evidence of genetic predisposition to alcoholic cirrhosis and psychosis: Twin concordances for alcoholism and its biological endpoints by zygosity among male veterans. *Alcoholism: Clinical and Experimental Research, 5*, 207–212.

Kaij, L. (1960). *Alcoholism in twins*. Stockholm: Almqvist and Wiksell.

Kendler, K. S., Heath, A. C., Neale, M. C., Kessler, R. C., & Eaves, L. J. (1992). A population based twin study of alcoholism in women. *Journal of American Psychiatric Association, 268*, 1877–1882.

Knight, R. P. (1937). The dynamics and treatment of chronic alcohol addiction. *Bulletin of the Menninger Clinic, 1*, 233–250.

McClearn, G. E., & Kakihana, R. (1981). Selective breeding for ethanol sensitivity: Short-sleep vs. long-sleep mice. In G. E. McClearn, R. A. Deitrich, & V. G. Erwin (Eds.), *Development of animal models as pharmacologic tools* (National Institute on Alcoholism and Alcohol Abuse Research Monograph No. 60, pp. 147–159). Rockville, MD: National Institute on Alcoholism and Alcohol Abuse.

McGue, M. (1992). When assessing twin concordance, use the probandwise not the pairwise rate. *Schizophrenia Bulletin, 18*, 171–176.

McGue, M. (in press). Genes, environment and the etiology of alcoholism. In R. Zucker, J. Howard, & G. Boyd (Eds.), *Development of alcohol-related problems in high-risk youth: Establishing linkages across biogenetic and psychosocial domains* (National Institute on Alcoholism and Alcohol Abuse Research Monograph No. 24). Rockville, MD: National Institute on Alcoholism and Alcohol Abuse.

McGue, M., Pickens, R. W., & Svikis, D. S. (1992). Sex and age effects on the inheritance of alcohol problems: A twin study. *Journal of Abnormal Psychology, 101*, 3–17.

Merikangas, K. R. (1990). The genetic epidemiology of alcoholism. *Psychological Medicine, 20*, 11–22.

Newlin, D. B., & Thomson, J. B. (1990). Alcohol challenge with sons of alcoholics: A critical review and analysis. *Psychological Bulletin, 108*, 383–402.

Parsian, A., Todd, R. D., Devor, E. J., O'Malley, K. L., Suarez, B. K., Reich, T., & Cloninger, C. R. (1991). Alcoholism and alleles of the human D2 dopamine receptor locus: Studies of association and linkage. *Archives of General Psychiatry, 48*, 655–663.

Partanen, J., Bruun, K., & Markkanen, T. (1966). *Inheritance of drinking behavior*. Helsinki, Finland: Finnish Foundation for Alcohol Studies.

Perry, A. (1973). The effect of heredity on attitudes toward alcohol, cigarettes, and coffee. *Journal of Applied Psychology, 58*, 275–277.

Pickens, R. W., Svikis, D. S., McGue, M., Lykken, D. T., Heston, L. L., & Clayton, P. J. (1991). Heterogeneity in the inheritance of alcoholism: A study of male and female twins. *Archives of General Psychiatry, 48*, 19–28.

Plomin, R., & Daniels, D. (1987). Why are children in the same family so different from one another? *Behavioral and Brain Sciences, 10*, 1–60.

Rice, J., & Reich, T. (1985). Familial analysis of qualitative traits under multifactorial inheritance. *Genetic Epidemiology, 2*, 301–315.

Roe, A. (1944). The adult adjustment of children of alcoholic parents raised in foster homes. *Quarterly Journal of Studies on Alcohol, 5*, 378–393.

Romanov, K., Kaprio, J., Rose, R., & Koskenvuo, M. (1991). Genetics of alcoholism: Effects of migration on concordance rates among male twins. *Alcohol and Alcoholism, Suppl. 1*, 137–140.

Room, R. (1983). Region and urbanization as factors in drinking practices and problems. In B. Kissin & H. Begleiter (Eds.), *The pathogenesis of alcoholism: Psychological factors* (pp. 555–604). New York: Plenum.

Schuckit, M. A., & Gold, E. O. (1988). A simultaneous evaluation of multiple markers of ethanol/placebo challenges in sons of alcoholics and controls. *Archives of General Psychiatry, 45*, 211–216.

Schuckit, M. A., Goodwin, D. W., & Winokur, G. (1972). A study of alcoholism in half siblings. *American Journal of Psychiatry, 128*, 1132–1136.

Sher, K. J. (1991). *Children of alcoholics: A critical appraisal of theory and research.* Chicago: University of Chicago Press.

Smith, D. I. (1980). The introduction of Sunday alcohol sales in Perth: Some methodological observations. *Community Health Studies, 4*, 289–293.

Tarter, R. E., Moss, H. B., Arria, A., Mezzich, A. C., & Vanyukov, M. M. (1992). The psychiatric diagnosis of alcoholism: Critique and proposed reformulation. *Alcoholism: Clinical and Experimental Research, 16*, 106–116.

Turner, E., Ewing, J., Shilling, P., Smith, T. L., Irwin, M., Schuckit, M., & Kelsoe, J. R. (1992). Lack of association between an RFLP near the D2 dopamine receptor gene and severe alcoholism. *Biological Psychiatry, 31*, 285–290.

Vaillant, G. E., & Milofsky, E. S. (1982). The etiology of alcoholism: A prospective viewpoint. *American Psychologist, 37*, 494–503.

Weissman, M. M., Myers, J. K., & Harding, P. S. (1980). Prevalence and psychiatric heterogeneity of alcoholism in a United States urban community. *Journal of Studies on Alcohol, 41*, 672–681.

Wise, R. A., & Rompre, P. P. (1989). Brain dopamine and reward. *Annual Review of Psychology, 40*, 191–225.

CHAPTER 14

Autism: Syndrome Definition and Possible Genetic Mechanisms

Michael Rutter, Anthony Bailey, Patrick Bolton, and Ann Le Couteur

In the first description of the syndrome of autism, Kanner (1943) described it as innate and inborn, drawing attention to the presence of abnormalities in infancy and to the fact that most autistic children never show a period of normal development. Research during the 1960s and 1970s showed that it was highly likely that autism arose on the basis of some form of organic brain dysfunction (Rutter, 1979). Yet there was a general reluctance to consider a genetic etiology (Hanson & Gottesman, 1976; Rutter, 1967). That was because of the following: First, there were no reported cases of an autistic child having an autistic parent, and hence no evidence of vertical transmission; second, the rate of autism in siblings was very low (estimates at that time suggested about 2%); and third, there was no evidence of any association with chromosome abnormalities that

We are grateful to the Medical Research Council of the United Kingdom, the Mental Health Foundation, and the John D. and Catherine T. MacArthur Foundation for their support of the research reported in this chapter.

were detectable at that time. These findings seemed to be out of keeping with a strongly genetic etiology, but in fact they were the wrong features to be considered (Bolton & Rutter, 1990; Folstein & Rutter, 1988; Rutter, 1991; Smalley, 1991; Smalley, Asarnow, & Spence, 1988).

The important point about the rate of autism in siblings was not that it was low in absolute terms but, rather, that it was extremely high in relation to the rarity of autism in the general population—an increase in risk of some 50–100 times. Also, because it was known that very few autistic individuals married and had children, vertical transmission was not to be expected. In addition, there were reports of a possibly increased loading for language disorders in families of autistic individuals (Bartak, Rutter, & Cox, 1975).

Twin Studies

It was a recognition of these considerations that stimulated the first systematic twin study of autism (Folstein & Rutter, 1977a, 1977b). This was based on a nationwide search throughout the United Kingdom (UK) for same-sex pairs; 11 monozygotic (MZ) and 10 dizygotic (DZ) pairs were found. Calculations showed that both the absolute number of twins and the MZ–DZ ratio were generally in keeping with the population incidence of autism and the rate of twinning. Zygosity was determined by blood groups except when dizygosity was obvious from genetically determined physical characteristics.

Two aspects in the findings require comment. First, there was a 36% concordance for autism in the MZ pairs compared with 0% in the DZ pairs—a difference that points to the likelihood of a strong genetic component. The second feature, however, is that most of the MZ pairs, but only 1 in 10 of the DZ pairs, were concordant for some type of cognitive deficit, usually involving language delay. The implication was that it may not be autism as such that is inherited but rather some broader type of cognitive abnormality including, but not restricted to, autism.

That finding raised the query of what are the diagnostic features of this broader pattern. At first, there was a focus on the language delay, but its significance was uncertain because normal children vary considerably in the age at which they acquire language and because twins tend to be somewhat behind singletons in their language development (Rutter & Redshaw, 1991). So what might be special about the variety connected with autism? The beginnings of an answer were provided by a recent follow-up into adult life of that original twin sample that was undertaken by Le Couteur et al. (1993). What was most striking in the results was the extent of continuing problems in social relationships. It was not that social problems got worse but rather that they became more obvious as the social demands went up in terms of the expectation of developing close friendships and love relationships.

Inevitably, this first study relied on a relatively small number of twin pairs, so the next need was to determine how well the findings would hold up in further studies. Two have been undertaken. First, there was a Scandinavian study of a sample that was somewhat atypical with respect to a low male-to-female sex ratio and a rather high rate of mental retardation (Steffenburg et al., 1989). This study showed a 91% concordance for autism in MZ pairs versus 0% in DZ pairs. Second, Bailey and colleagues (Bailey et al., 1991, 1993) undertook a second UK study. The design was the same as in the first UK study, with total population coverage and blood groups for zygosity. But, in addition, there was the use of well-standardized, and more discriminating, diagnostic instruments—the Autism Diagnostic Interview (ADI) and the Autism Diagnostic Observation Schedule (Le Couteur et al., 1989; Lord et al., 1989). The results were strikingly similar to those of the first UK study, providing powerful confirmation of the conclusions on genetic factors. The investigation included examination of all pairs in both the new and original samples for the fragile X anomaly, a chromosomal abnormality that was not known at the time of the first study and that might have accounted for the concordance patterns found. In fact, there were no cases of fragile X anomaly in the new sample and only one in the original twin sample; so, obviously, this could not account for the high concordance in MZ pairs.

The two samples were pooled to reexamine the concordance findings; however, first there was the exclusion of the following: one fragile X pair, one with a genetic form of retinoblastoma, one with hypsarrhythmia, and two cases in the original sample that did not meet 10th edition of the *International Classification of Diseases (ICD-10)* diagnostic criteria for autism (World Health Organization, 1992). The MZ–DZ difference in this pooled sample was striking. There was 60% concordance for autism in MZ pairs versus 0% in DZ pairs. In addition, another third of MZ pairs, but only 1 in 10 of DZ pairs, showed a broader pattern of cognitive and social deficits. Again, the findings suggested that the phenotype extended beyond autism as traditionally diagnosed.

Autism is a rare disorder, about 2–4 per 10,000, so the next question was how to translate these concordance figures into a more quantitative estimate of the strength of the genetic component. This may be done using a multifactorial liability model. For the necessary calculations, probandwise correlations were transformed into tetrachoric correlations. The results provided an estimate of heritability for an underlying liability to autism of 91–93% (the exact figure depending on the assumptions about the base rate). Of course, the model involves a number of assumptions, and the precise figure should not be overinterpreted; however, it is clear that there is a very strong genetic component.

The original Folstein and Rutter (1977a, 1977b) twin study, as well as the more recent study reported by Steffenburg et al. (1989), showed that obstetric complications differentiated twins with autism from their cotwins without autism. In both studies, this was interpreted as indicating the possible role of environmentally induced brain damage. However, this seemed out of line with the heritability findings. Also, most of the obstetric complications were quite minor. In singletons, too, obstetric complications are associated with autism; but again, most complications are minor and not of a kind that are usually associated with a high risk of brain damage (Tsai, 1987). The issue was reexamined in the new twin study, and again, the same pattern was found.

However, this time there was also a systematic assessment of minor congenital anomalies. Strikingly, a strong association was found between such anomalies and obstetric complications (Bailey et al., 1993). Indeed, in those twin pairs (mostly DZ) in which obstetric complications created a biological difference or hazard that affected just the twin with autism, and in all cases on whom there were data, this was accompanied by a difference in the congenital anomalies score. The importance of this finding is that most of the congenital anomalies derive from something going wrong in the early part of pregnancy. The implication is that the obstetric complications may stem from a genetically abnormal fetus and may not represent an environmental effect at all. It is well known, of course, that genetically abnormal fetuses (e.g., as with Down's syndrome or the fragile X anomaly) are indeed associated with a substantial increase in obstetric complications (Bolton & Holland, in press).

Family Genetic Studies

The twin genetic method is a powerful one, but it is important to complement it with other research strategies. One of those is the family method. The first systematic study was undertaken by August, Stewart, and Tsai (1981), who found a 15% rate of cognitive impairment (assessed by direct testing) in the siblings of autistic probands compared with 3% in the siblings of Down's syndrome probands. Of the 11 affected siblings in the autism group, 6 showed mental retardation. Two later studies without comparison groups (Baird & August, 1985; Minton, Campbell, Green, Jennings, & Samit, 1982) showed much the same but also drew attention to the finding that the familial loading for mental retardation was largely confined to autistic subjects who themselves were severely retarded.

These early family studies were quite limited in their coverage of conditions in relatives, but fuller data are available from more recent investigations. Bolton et al. (1991, 1993) made a detailed systematic standardized study of the first-degree relatives of 99 individuals with autism and 36 individuals with Down's syndrome. The detailed pedigree findings focused on three main domains: (a) cognitive abnormalities such as severe language delay (meaning that the child had no single words until 24 months of age and/or no phrase speech until 33 months of age) or severe reading and spelling difficulties, (b) social abnormalities in terms of features such as impaired social reciprocity and lack of friends, and (c) repetitive stereotyped behaviors such as circumscribed interests. Furthermore, a confirmatory factor analysis was undertaken to determine whether the selected characteristics grouped together in the way expected. The empirical findings did indeed support the concepts.

The next issue was to determine which features differentiated the siblings of autistic individuals from the siblings of individuals with Down's syndrome. It was clear that the rate of autism in the siblings of autistic individuals was raised—the rate being about 3% compared with 0% in the Down's syndrome group. However, there was also a substantial increase in language or communication difficulties, social deficits, and, to a lesser extent, stereotyped behaviors. This difference was most evident when there was a combination of at least two out of these three domains of abnormality. However, there was also some increase when such abnormalities occurred in isolation.

Some 3% of the siblings had clearcut autism, another 3% had a somewhat atypical syndrome of autism, and a further 3% had a combination of cognitive and social abnormalities of a kind that are qualitatively similar to those seen in autism but which fall well outside the diagnostic boundaries of autism as they are usually understood. There is some difficulty in knowing just how far this broader phenotype extends. The just-presented figures (i.e., 9% in sum) provide a minimum estimate. If isolated cognitive and social abnormalities, with or without repetitive behaviors, are included, then the rate of disorder in siblings rises to 20%. This probably represents something like the approximate upper limit of the frequency of the phenotype. Two other family studies have provided systematic data of a comparable kind (although neither included a control group). The findings of Piven et al. (1990), which were based on Kanner's (1943) cases of autism, were broadly similar to those of Bolton et al. (1993), with a 3% rate of autism in siblings, a 4% rate of severe social impairment, and a 15% rate of cognitive abnormalities.

The Utah family study (Jorde et al., 1990, 1991; Mason-Brothers et al., 1987, 1990; Ritvo, Freeman et al., 1989; Ritvo, Jorde et al., 1989; Ritvo et al., 1990) was much larger, being based on 185 families, but the diagnosis was not based on standardized measures, and furthermore, there was no systematic assessment to detect chromosome anomalies. The results reported in the various published articles are somewhat difficult to interpret because the earlier reports did not differentiate cases of autism associated with known medical conditions and because the figures given in different articles do not tally. Thus, Ritvo, Jorde et al. (1989) reported a sibling recurrence risk of autism of 7% if the first autistic child in that family was male and 14.5% if it was female, but Jorde et al. (1991) gave figures of 3.7% versus 7.0%—rates that are half of those previously reported. The most systematic analyses seem to be those of Jorde et al. (1991) who excluded cases with known medical conditions. Apart from the already-noted sex difference, which fell short of statistical significance, the most notable feature of their findings was the marked fall-off in rate of autism in second- and third-degree relatives (0.13% and 0.05%, respectively) compared with first-degree relatives.

In addition, Gillberg, Gillberg, and Steffenburg (1992) reported a much smaller scale study with rather different findings (being essentially negative with respect to siblings). However, a third of their sample had a known medical syndrome, and half were severely retarded.

Mode of Genetic Transmission

If researchers are to understand the meaning of the twin and family findings in terms of possible modes of genetic transmission, then it is necessary that they go beyond rates of abnormalities in first-degree relatives and look at the patterns in more detail. There are two key findings in this connection. First, Bolton et al. (1993) looked to see if the familial loading varied according to the severity of the autism, which was defined in terms of the score of autistic symptoms on the ADI (Le Couteur et al., 1989). Strikingly, the familial loading was much greater in the case of severe autism. A similar, but not so marked, trend was found with respect to verbal IQ, but there was no association with performance IQ. The importance of this finding that the familial loading varied according to the severity of the autism lies in the implication that several genes are involved and not just one major gene as in Mendelian disorders (see Emery, 1986). However, it was also notable that this association between familial loading and severity of autism did not apply within nonverbal subjects, most of whom were also markedly retarded, which suggests that this most profoundly handicapped group may be genetically different.

The second finding was that the familial loading also varied according to obstetric optimality (using the scale developed by Gillberg & Gillberg, 1983); that is, the loading was greater when the autistic individual showed poor optimality—meaning obstetric complications of one sort or another. This finding is incompatible with any hypothesis that the obstetric factors are creating an environmental risk. Rather, the difference in familial loading suggests that the obstetric complications are the result of a genetically abnormal fetus.

As in the twin study, it was important to check that this familial loading was not a consequence of the fragile X chromosomal anomaly. The findings clearly showed that it was not because only one autistic individual in the Bolton et al. study (1993) showed the fragile X anomaly. When the twin and singleton data were pooled, the overall rate of the fragile X anomaly was about 2% (Bailey et al., in press). This is broadly in line with most modern studies, and it is clear that the earlier claims of a much higher rate have not been borne out (Bailey et al., in press; Bolton & Rutter, 1990).

Discussion and Conclusions

In putting together the findings of twin and family studies, several key issues need to be considered. First, there is the quantification of the genetic contribution to autism. As already mentioned, the combined twin data from the UK studies gave rise to a heritability estimate of 91–93% for an underlying liability to autism. Steffenburg et al. (1989) did not calculate heritability, but the pairwise concordance figures from this study (91% in MZ pairs vs. 0% in DZ pairs) are obviously in keeping with an extremely strong genetic component. The family data, showing a 50–100 times increase in the rate of autism in siblings (Bolton et al., 1993; Folstein & Rutter, 1988; Piven et al., 1991; Smalley et al., 1988), point in the same direction. Despite a much lower estimate by Gillberg (1992), it may be concluded that most cases of autism are largely genetic and, in particular, that obstetric complications do not constitute a frequent primary environmental causal factor (although occasionally they may do so). Of course, in spite of the very high heritability figure, there may be contributory environmental factors (perhaps particularly with respect to the difference between the broader phenotype and traditional autism of a more handicapping variety). So far, there is no positive evidence that this is the case. Nevertheless, the available data do indicate that autism proper differs from the broader phenotype with respect to associations with epilepsy, mental retardation, and possibly head circumference (Bailey et al., 1993; Bolton et al., 1993). Multifactorial models in psychiatry have been popular, but to date, little attention has been paid to factors involved in crossing the threshold because there have been no measures of the inferred liability (Plomin, Rende, & Rutter, 1991). The broader phenotype in autism may provide an approach to this issue.

The second issue is whether autism is genetically homogeneous or heterogeneous. Clearly, there must be some heterogeneity, as shown by the replicated associations with both the fragile X anomaly (Bailey et al., in press) and tuberous sclerosis (Hunt & Dennis, 1987; Hunt & Shepherd, in press; Smalley, Tanguay, Smith, & Gutierrez, 1992), as well as the less certain associations with other single-gene disorders (Folstein & Rutter, 1988; Reiss, Feinstein, & Rosenbaum, 1986). Both Steffenburg (1991) and Gillberg (1990) argued that some two fifths of cases of autism are due to some specific diagnosable medical condition, but other studies have produced much lower figures. A rate of 10% is probably a more realistic estimate (Rutter, Bailey, Bolton, & Le Couteur, in press). Nevertheless, that does not mean that the remaining 90% are genetically homogeneous; indeed, the history of medical genetics suggests that that is most unlikely (Folstein & Rutter, 1988). The rather different twin and family findings in cases of autism associated with profound mental retardation raise the possibility that this may include genetically distinct subvarieties. The finding that autism accompanied by profound mental retardation is much more likely to be associated with known medical conditions (Rutter et al., in press) points in the same direction. The matter warrants exploration (Rutter, 1991).

The third question is whether the autism phenotype extends beyond the traditional diagnostic boundaries. The data from twin and family studies by Folstein and Rutter (1977a, 1977b); Bailey et al. (1993); Bolton et al. (1993); Piven et al. (1990); Wolff, Narayan, and Moyes (1988); Landa, Folstein, and Isaacs (1991); and Landa et al. (1992) all suggest that it does. The overall picture from these combined studies indicates a combination of cognitive and social abnormalities in individuals of normal intelligence. However, despite earlier suggestions to the contrary, the phenotype does not appear to include mental retardation when it is unassociated with autism in the same individual. Accordingly, although autism is likely to be genetically heterogeneous, it seems that the genetic contribution is autism specific and not part of undifferentiated mental retardation (Rutter, 1991).

There have been suggestions that the phenotype should be broadened still further to include Tourette's syndrome (Comings & Comings, 1991) and even anorexia nervosa and obsessional disorders (Gillberg, 1992), but the supporting evidence so far is unconvincing. Family studies have reported an apparent excess in the loading for affective and/or anxiety disorders in relatives (De Long & Dwyer, 1988; Piven et al., 1990, 1991), but it is quite uncertain whether this association is genetically mediated. At present, the findings do not justify an extension of the phenotype beyond cognition and social deficits, although the limits have yet to be firmly established.

Fourth, there is the crucial issue of the mode of inheritance. Segregation studies have been contradictory in their findings (Jones & Szatmari, 1988; Jorde et al., 1990; Ritvo et al., 1985), perhaps because of inconsistencies in sampling and diagnosis, as well as a failure to take into account a broader phenotype and/or to test for and exclude cases due to known medical conditions. However, the marked fall-off in rate going from MZ cotwins to DZ cotwins or siblings (Bailey et al., 1993; Folstein & Rutter, 1977a, 1977b; Steffenburg et al., 1989), together with the further marked fall-off going from first-degree to second-degree relatives (Jorde et al., 1990), indicates that multiple, interacting genes are likely to be involved (Risch, 1990). The association between familial loading and severity of autism points to the same conclusion. A multigene model also leads to the expectation that the loading should be higher in the case of females (because they are the less-often affected sex). The data on this point are somewhat contradictory, although several studies suggest that there may be a greater loading in the families of female autistic subjects. However, the statistical power to detect a sex difference was low in all studies, and the matter remains unresolved (August et al., 1981; Bolton et al., 1993; Lord, DiLavore, & Schopler, 1991; Ritvo, Jorde et al., 1989; Tsai & Beisler, 1983; Tsai, Stewart, & August, 1981).

The final point concerns the need to bring together the clinical, genetic, neuropsychological, and neurobiological data to redefine autism. Although there is very good agreement on the key diagnostic criteria, there is continuing discussion on where and how the diagnostic boundaries should be drawn (Rutter & Schopler, 1988, 1992). The genetic data clearly point to the need to widen the diagnostic concept, but the data do not yet provide a precise set of criteria. Neurobiological findings might help, but so far they do not because the results are so inconsistent and inconclusive (Bailey, 1993; Dawson, 1989; Schopler & Mesibov, 1987). Methodological improvements may make this approach the most fruitful in the future. However, at present, neuropsychological approaches are much more promising (Baron-Cohen, Tager-Flusberg, & Cohen, 1993; Frith, 1989). Although there is some disagreement on the inferences to be drawn regarding the precise nature of a possible core cognitive deficit, the findings are reasonably well replicated. Clearly, an important next step will be to determine the extent to which the cognitive deficits that are associated with autism apply similarly to affected relatives of normal intelligence with the broader phenotype.

Of course, molecular genetics strategies also constitute an essential next step, and they will be crucial in determining the genetic mechanisms in the etiology of autism. However, at present, the paucity of candidate genes, the extreme rarity of heavily loaded families, the likelihood of multiple genes, and the uncertainties regarding the phenotype make for considerable practical difficulties.

Autism has been shown to be the most strongly genetic of all psychiatric disorders (apart from Huntington's disease), and further genetic investigations should be highly rewarding. There is some way still to go before the riddle of autism is solved, but it is likely that genetic data will provide a key element in its solution.

References

August, G. J., Stewart, M. A., & Tsai, L. (1981). The incidence of cognitive disabilities in the siblings of autistic children. *British Journal of Psychiatry, 138*, 416–422.

Bailey, A. (1993). The biology of autism. *Psychological Medicine, 23,* 7–11.

Bailey, A., Bolton, P., Butler, L., Le Couteur, A., Murphy, M., Scott, S., Webb, T., & Rutter, M. (in press). Prevalence of the fragile X anomaly amongst autistic twins and singletons. *Journal of Child Psychology and Psychiatry.*

Bailey, A., Le Couteur, A., Gottesman, I. I., Bolton, P., Simonoff, E., Yuzda, E., & Rutter, M. (1993). *Autism as a strongly genetic disorder: Evidence from a British twin study.* Manuscript submitted for publication.

Bailey, A., Le Couteur, A., Rutter, M., Pickles, A., Yuzda, E., Schmidt, D., & Gottesman, I. (1991). *Obstetric and neurodevelopmental data from the British twin study of autism.* Paper presented at the Second World Congress on Psychiatric Genetics. (Abstracted in *Psychiatric Genetics, 2,* 49)

Baird, T. D., & August, G. J. (1985). Familial heterogeneity in infantile autism. *Journal of Autism and Developmental Disorders, 15,* 315–321.

Baron-Cohen, S., Tager-Flusberg, H., & Cohen, D. (Eds.). (1993). *Understanding other minds: Perspectives from autism.* London: Oxford University Press.

Bartak, L., Rutter, M., & Cox, A. (1975). A comparative study of infantile autism and specific developmental receptive language disorder: I. The children. *British Journal of Psychiatry, 126,* 127–145.

Bolton, P., & Holland, A. (in press). Chromosomal abnormalities. In M. Rutter, E. Taylor, & L. Hersov (Eds.), *Child and adolescent psychiatry: Modern approaches* (3rd ed.). Oxford, UK: Blackwell Scientific.

Bolton, P., Macdonald, H., Murphy, M., Scott, S., Yuzda, E., Whitlock, B., Pickles, A., & Rutter, M. (1991). *Genetic findings and heterogeneity in autism.* Paper presented at the Second World Congress on Psychiatric Genetics. (Abstracted in *Psychiatric Genetics, 2,* 49)

Bolton, P., Macdonald, H., Pickles, A., Rios, P., Goode, S., Crowson, M., Bailey, A., & Rutter, M. (1993). *A case-control family history study of autism.* Manuscript submitted for publication.

Bolton, P., & Rutter, M. (1990). Genetic influences in autism. *International Review of Psychiatry, 2,* 67–80.

Comings, D. E., & Comings, B. G. (1991). Clinical and genetic relationships between autism-pervasive developmental disorder and Tourette syndrome: A study of 19 cases. *American Journal of Medical Genetics, 39,* 180–191.

Dawson, G. (Ed.). (1989). *Autism: Nature, diagnosis, and treatment.* New York: Guilford Press.

De Long, R., & Dwyer, J. T. (1988). Correlation of family history with specific autistic subgroups: Asperger's syndrome and bipolar affective disease. *Journal of Autism and Developmental Disorders, 18,* 593–600.

Emery, A. E. H. (1986). *Methodology in medical genetics: An introduction to statistical methods* (2nd ed.). Edinburgh, UK: Churchill Livingstone.

Folstein, S., & Rutter, M. (1977a). Genetic influences and infantile autism. *Nature, 265*, 726–728.

Folstein, S., & Rutter, M. (1977b). Infantile autism: A genetic study of 21 twin pairs. *Journal of Child Psychology and Psychiatry, 18*, 297–321.

Folstein, S., & Rutter, M. (1988). Autism: Familial aggregation and genetic implications. *Journal of Autism and Developmental Disorders, 18*, 3–30.

Frith, U. (1989). *Autism: Explaining the enigma*. Oxford, UK: Basil Blackwell.

Gillberg, C. (1990). Autism and pervasive developmental disorders. *Journal of Child Psychology and Psychiatry, 31*, 99–119.

Gillberg, C. (1992). Autism and autistic-like conditions: Sub-classes among disorders of empathy. *Journal of Child Psychology and Psychiatry, 33*, 813–842.

Gillberg, C., & Gillberg, I. C. (1983) Infantile autism: A total population study of reduced optimality in the pre-, peri- and neonatal periods. *Journal of Autism and Developmental Disorders, 13*, 153–166.

Gillberg, C., Gillberg, I. C., & Steffenburg, S. (1992). Siblings and parents of children with autism: A controlled population-based study. *Developmental Medicine and Child Neurology, 34*, 389–398.

Hanson, D. R., & Gottesman, I. (1976). The genetics, if any, of infantile autism and childhood schizophrenia. *Journal of Autism and Schizophrenia, 6*, 209–233.

Hunt, A., & Dennis, J. (1987). Psychiatric disorder among children with tuberous sclerosis. *Developmental Medicine and Child Neurology, 29*, 190–198.

Hunt, A., & Shepherd, C. (in press). A prevalence study of autism in tuberous sclerosis. *Journal of Autism and Developmental Disorders*.

Jones, M. B., & Szatmari, P. (1988). Stoppage rules and genetic studies of autism. *Journal of Autism and Developmental Disorders, 18*, 31–40.

Jorde, L. B., Hasstedt, S. J., Ritvo, E. R., Mason-Brothers, A., Freeman, B. J., Pingree, C., McMahon, W. M., Petersen, P. B., Jenson, W. R., & Moll, A. (1991). Complex segregation analysis of autism. *American Journal of Human Genetics, 49*, 932–938.

Jorde, L. B., Mason-Brothers, A., Waldmann, R., Ritvo, E. R., Freeman, B. J., Pingree, C., McMahon, W. M., Petersen, P. B., Jenson, W. R., & Mo, A. (1990). The UCLA–University of Utah epidemiologic survey of autism: Genealogical analysis of familial aggregation. *American Journal of Medical Genetics, 36*, 85–88.

Kanner, L. (1943). Autistic disturbances of affective contact. *Nervous Child, 2*, 217–250.

Landa, R., Folstein, S. E., & Isaacs, C. (1991). Spontaneous narrative–discourse performance of parents of autistic individuals. *Journal of Speech and Hearing Research, 34*, 1339–1345.

Landa, R., Piven, J., Wzorek, M. M., Gayle, J. O., Chase, G. A., & Folstein, S. E. (1992). Social language use in parents of autistic individuals. *Psychological Medicine, 22*, 245–254.

Le Couteur, A., Bailey, A., Goode, S., Robertson, S., Gottesman, I. I., Schmidt, D., & Rutter, M. (1993). *A broader phenotype of autism: The clinical spectrum in twins*. Unpublished manuscript.

Le Couteur, A., Rutter, M., Lord, C., Rios, P., Robertson, S. Holdgrafer, M., & McLennan, J. (1989). Autism Diagnostic Interview: A standardized investigator-based instrument. *Journal of Autism and Developmental Disorders, 19*, 363–387.

Lord, C., DiLavore, P., & Schopler, E. (1991, May). *Sex differences in autism.* Paper presented at the annual meeting of the American Psychiatric Association, New Orleans, LA.

Lord, C., Rutter, M., Goode, S., Heemsbergen, J., Jordan, H., Mawhood, L., & Schopler, E. (1989). Autism Diagnostic Observation Schedule: A standardized observation of communicative and social behavior. *Journal of Autism and Developmental Disorders, 19*, 185–212.

Mason-Brothers, A., Ritvo, E. R., Guze, B., Mo, A., Freeman, B. J., Funderburk, S. J., & Schroth, P. C. (1987). Pre-, peri-, and postnatal factors in 181 autistic patients from single and multiple incidence families. *Journal of the American Academy of Child and Adolescent Psychiatry, 26*, 39–42.

Mason-Brothers, A., Ritvo, E. R., Pingree, C., Petersen, P. B., Jenson, W. R., McMahon, W. M., Freeman, B. J., Jorde, L. B., Spencer, M. J., Mo, A., & Ritvo, A. (1990). The UCLA–University of Utah epidemiologic survey of autism: Prenatal, perinatal, and postnatal factors. *Paediatrics, 86*, 514–519.

Minton, J., Campbell, M., Green, W., Jennings, S., & Samit, C. (1982). Cognitive assessment of siblings of autistic children. *Journal of the American Academy of Child Psychiatry, 21*, 256–261.

Piven, J., Chase, G. A., Landa, R., Wzorek, M., Gayle, J., Cloud, D., & Folstein, S. (1991). Psychiatric disorders in the parents of autistic individuals. *Journal of the American Academy of Child and Adolescent Psychiatry, 30*, 471–478.

Piven, J., Gayle, J., Chase, J., Fink, B., Landa, R., Wzorek, M., & Folstein, S. (1990). A family history study of neuropsychiatric disorders in the adult siblings of autistic individuals. *Journal of the American Academy of Child and Adolescent Psychiatry, 29*, 177–183.

Plomin, R., Rende, R., & Rutter, M. (1991). Quantitative genetics and developmental psychopathology. In D. Cicchetti & S. L. Toth (Eds.), *Internalizing and externalizing expressions of dysfunction: Rochester Symposium on Developmental Psychopathology* (Vol. 2, pp. 155–202). Hillsdale, NJ: Erlbaum.

Reiss, A. L., Feinstein, C., & Rosenbaum, K. N. (1986). Autism and genetic disorders. *Schizophrenia Bulletin, 12*, 724–728.

Risch, N. (1990). Linkage strategies for genetically complex traits. *American Journal of Human Genetics, 46*, 222–253.

Ritvo, E. R., Freeman, B. J., Pingree, C., Mason-Brothers, A., Jorde, L., Jenson, W. R., McMahon, W. M., Petersen, P. B., Mo., A., & Ritvo, A. (1989). The UCLA–University of Utah epidemiologic survey of autism: Prevalence. *American Journal of Psychiatry, 146*, 194–199.

Ritvo, E. R., Jorde, L. B., Mason-Brothers, A., Freeman, B. J., Pingree, C., Jones, M. B., McMahon, W. M., Petersen, P. B., Jenson, W. R., & Mo, A. (1989). The UCLA–University of Utah epidemiologic survey of autism: Recurrence risk estimates and genetic counseling. *American Journal of Psychiatry, 146*, 1032–1036.

Ritvo, E. R., Mason-Brothers, A., Freeman, B. J., Pingree, C., Jenson, W. R., McMahon, W. M., Petersen, P. B., Jorde, L. B., Mo, A., & Ritvo, A. (1990). The UCLA–University of Utah epidemiologic survey of autism: The etiologic role of rare diseases. *American Journal of Psychiatry, 147*, 1614–1621.

Ritvo, E. R., Spence, M. A., Freeman, B. J., Mason-Brothers, A., Mo, A., & Marazita, M. L. (1985). Evidence for autosomal recessive inheritance in 46 families with multiple incidences of autism. *American Journal of Psychiatry, 142*, 187–192.

Rutter, M. (1967). Psychotic disorders in early childhood. In A. J. Coppen & A. Walk (Eds.), *Recent developments in schizophrenia. British Journal of Psychiatry* (Special Publication 2, pp. 133–158). Ashford, UK: Headley Bros./RMPA.

Rutter, M. (1979). Language, cognition and autism. In R. Katzman (Ed.), *Congenital and acquired cognitive disorders* (pp. 247–264). New York: Raven Press.

Rutter, M. (1991). Autism as a genetic disorder. In P. McGuffin & R. Murray (Eds.), *The new genetics of mental illness* (pp. 225–244). Oxford, UK: Butterworth-Heinemann.

Rutter, M., Bailey, A., Bolton, P., & Le Couteur, A. (in press). Autism and known medical conditions. *Journal of Child Psychology and Psychiatry.*

Rutter, M., & Redshaw, J. (1991). Annotation: Growing up as a twin. Twin–singleton differences in psychological development. *Journal of Child Psychology and Psychiatry, 32*, 885–895.

Rutter, M., & Schopler, E. (1988). Autism and pervasive developmental disorders. In M. Rutter, A. H. Tuma, & I. S. Lann (Eds.), *Assessment and diagnosis in child psychopathology* (pp. 408–434). New York: Guilford Press.

Rutter, M., & Schopler, E. (1992). Classification of pervasive developmental disorders: Some concepts and practical considerations. *Journal of Autism and Developmental Disorders, 22*, 459–482.

Schopler, E., & Mesibov, G. (Eds.). (1987). *Neurobiological issues in autism.* New York: Plenum.

Smalley, S. (1991). Genetic influences in autism. *Psychiatric Clinics of North America, 14*, 125–139.

Smalley, S. L., Asarnow, R. F., & Spence, M. A. (1988). Autism and genetics: A decade of research. *Archives of General Psychiatry, 45*, 953–961.

Smalley, S. L., Tanguay, P. E., Smith, M., & Gutierrez, G. (1992). Autism and tuberous sclerosis. *Journal of Autism and Developmental Disorders, 22*, 339–355.

Steffenburg, S. (1991). Neuropsychiatric assessment of children with autism: A population-based study. *Developmental Medicine and Child Neurology, 33*, 495–511.

Steffenburg, S., Gillberg, C., Hellgren, L., Anderson, L., Gillberg, I., Jakobsson, G., & Bohman, M. (1989). A twin study of autism in Denmark, Finland, Iceland, Norway and Sweden. *Journal of Child Psychology and Psychiatry, 30*, 405–416.

Tsai, L. Y. (1987). Pre-, peri-, and neonatal factors in autism. In E. Schopler & G. B. Mesibov (Eds.), *Neurobiological issues in autism* (pp. 180–189). New York: Plenum.

Tsai, L. Y., & Beisler, J. M. (1983). The development of sex differences in infantile autism. *British Journal of Psychiatry, 142*, 373–378.

Tsai, L. Y., Stewart, M. A., & August, G. (1981). Implication of sex differences in the familial transmission of infantile autism. *Journal of Autism and Developmental Disorders, 11*, 165–173.

Wolff, S., Narayan, S., & Moyes, B. (1988). Personality characteristics of parents of autistic children. *Journal of Child Psychology and Psychiatry, 29*, 143–154.

World Health Organization. (1992). *ICD-10: Categories F00–F99 mental and behavioural disorders (including disorders of psychological development). Clinical descriptions and diagnostic guidelines*. Geneva: Author.

CHAPTER 15

Genes, Personality, and Psychopathology: A Latent Class Analysis of Liability to Symptoms of Attention-Deficit Hyperactivity Disorder in Twins

Lindon Eaves, Judy Silberg, John K. Hewitt, Joanne Meyer, Michael Rutter, Emily Simonoff, Michael Neale, and Andrew Pickles

The nature of the relationship between normal differences in personality and psychopathology is still unclear. As long ago as 1952, Eysenck formulated a dimensional model of normal personality that was rooted in the assumption that many of the major psychiatric disorders recognized at the time might be better understood not as distinct "categories" of behavior but as extreme manifestations of continuous and normal variations in personality. Thus, the "disease" model for psychopathology was regarded as a special case of a more general "psychometric" model of behavior. The principal dimensions of Eysenck's theory—extraversion and neuroticism—began as constructs postulated

This research was supported by grants MH45268 and MH48604 from the National Institute of Mental Health and by grant AG04945 from the National Institute on Aging.

to account for the differences among various categories of psychiatric disorder. More recently, other researchers, including Gray (1970, 1981) and Cloninger (1986, 1987), have elaborated different dimensional models that were related to those of Eysenck in an attempt to account for the neuropsychological basis of certain common behavioral disorders.

In the end, the dimensional and categorical models for psychopathology are not mutually exclusive. Normal ("dimensional") variations in personality may account for differences in liability, but major environmental or genetic events may superimpose categorical distinctions between individuals who are symptomatic or asymptomatic or who show different patterns of symptomatology.

In an attempt to provide a bridge between dimensional and categorical models for the genetics of multivariate categorical data, we have begun to explore genetic applications of latent class models to multivariate categorical data on twins. A more mathematical treatment of the method, applied to symptoms of conduct disorder in twins, has been given elsewhere (Eaves et al., 1993). In this chapter, we concentrate on the basic ideas and show how they worked out in practice when we applied them in an exploratory way to the symptoms of attention-deficit hyperactivity disorder (ADHD) in a small but, developmentally speaking, relatively homogeneous sample of young male twin pairs who were interviewed with their parents as part of the much larger Virginia Study of Adolescent Behavioral Development (VSABD).

ADHD is one of the most common causes of referral of children to mental health care in the United States (Barkley, 1990). The great variation in the degree of symptoms and pervasiveness across situations suggests a great deal of heterogeneity in the disorder at the phenotypic level. The goal of our genetic analysis was to help resolve some of this heterogeneity by grounding clinical distinctions at the phenotypic level in identifiable etiological differences at the genetic and environmental levels.

Assessing ADHD in the VSABD

The backbone of the assessment of ADHD in this analysis was the detailed semistructured home interview using the Parental form of the Child and

Adolescent Psychiatric Assessment (P-CAPA), which was developed for epidemiological study by Rutter and his colleagues (Angold, Cox, Prendergast, Rutter, & Simonoff, 1989). The P-CAPA is designed to be investigator based rather than subject based, searching for specific descriptions of relevant behavior and using specified criteria for endorsement and impairment. The instrument seeks systematically the rich range of behavioral indexes that are needed to address questions of severity, heterogeneity, and comorbidity in a research setting.

Sample Ascertainment and Selection of Items

With the cooperation of the Virginia Department of Education, more than 6,000 pairs of school-age twins were identified through local public school districts. We have thus far completed home interviews of more than 1,300 families. The current analysis pertains only to male twins who were 8–11 years old at the time of the interview because ADHD is especially common in this group. These subjects comprised 84 monozygotic (MZ) and 63 dizygotic (DZ) pairs on whom zygosity was sufficiently certain for inclusion at this stage.

The P-CAPA was administered in the home by trained interviewers. Interview protocols and tapes were reviewed in detail prior to data entry by trained interviewers–monitors under the supervision of a doctoral-level clinical psychologist. We focused on a selection of 16 items from the rich assessments provided by the ADHD section of the mother's P-CAPA interviews. In separate assessments for each twin, the mother was asked to identify typical activities in which the child chooses his own activity ("self-imposed") or in which the activity is suggested by someone else such as a parent ("imposed by other"), or "passive activities" such as eating a meal or watching TV. For each such situation, the mother was then asked to rate a series of ways of behaving (e.g., fidgeting, running about, etc.) that typically characterized ADHD. The items reflect a series of different ways of behaving under a number of different settings. For the purposes of analysis, we coded the response intensities at three levels: 0 = not present, 1 = present in at least two activities and at least sometimes uncontrollable by the child or by admonition, and 2 = present in

TABLE 1

Summary of Ratings of 294 Male Twins on 16 ADHD Items of the Interview

Item/behavior	Response (%) 0	1	2	Number missing	Factor loading
Activities imposed by another					
Fidgets	87.3	8.2	4.5	2	0.67
Runs	95.9	2.1	2.1	2	0.87
Can't sit	95.6	3.1	1.4	1	0.76
Shifts	97.3	1.4	1.4	2	0.89
No follow-through	92.8	6.1	1.0	1	0.48
No concentration	91.0	8.3	0.7	4	0.63
Passive activities					
Fidgets	93.2	2.7	4.1	1	0.76
Runs	97.3	1.0	1.7	0	0.82
Can't sit	96.6	1.4	2.1	2	0.89
Shifts	98.3	1.0	0.7	2	0.77
Self-imposed activities					
Fidgets	95.2	2.4	2.4	1	0.69
Runs	98.0	1.4	0.7	0	0.85
Can't sit	97.6	1.0	1.4	1	0.87
Shifts	98.0	0.7	0.3	1	0.79
No follow-through	97.9	1.0	1.0	2	0.76
No concentration	98.3	1.4	0.3	1	0.68

Note. Response categories were coded as follows: 0 = not present; 1 = present in at least two activities, sometimes uncontrollable; and 2 = present in most activities, almost never controllable.

most activities and almost never controllable by the child or by an admonition. The items are summarized in Table 1. The response frequencies for all male twins 8–11 years old are also reported. Clearly, in the absence of data on onset, frequency and duration of symptoms, or impairment of normal function (all of which are available in the P-CAPA), we could not arrive at confident clinical diagnoses of ADHD (see, e.g., Rutter & Gould, 1985). However, the basic intensity data illustrate many important features of the genetic analysis of a complex disorder such as ADHD.

Preliminary Data Analysis

The first principal component of the raw Pearson product–moment correlations (see Table 1) explained 59% of the total variance in item responses.

For preliminary purposes, the items were combined into a single unweighted total ADHD score by adding up the responses to the 16 items. The product–moment correlations between the scores of first and second twins for whom there were no missing values were 0.71 and −0.05, respectively, for MZ pairs (n = 79) and DZ pairs (n = 58). Spearman correlations computed on the ranks of the scores were 0.24 and −0.03, respectively. The large difference between MZ and DZ correlations for the raw scores, with the MZ correlation being greatly in excess of twice the DZ correlation, is normally regarded by geneticists as evidence of marked dominance or recessivity in the effects of the genes responsible for the behavior in question (Eaves, 1982). The very low DZ correlation, relative to that of MZs, is consistent with a model of epistatic interactions between duplicate genes at different loci (see Eaves, 1988), that is, a model in which liability to ADHD is only increased markedly when more than one locus carries a defective (high-risk) allele. The discrepancy between the Pearson and Spearman correlations for MZ pairs reflects the marked skewness in the raw scores.

A Latent Class Analysis of the Twins' Behavior

The latent class approach to genetic analysis is best described and illustrated in two stages. First, the data are treated to a more conventional latent class analysis that ignores the resemblance between twins and makes no attempt to test etiological hypotheses. Having obtained some insight into the nature of the problem from the conventional analysis, one then attempts to probe more deeply into the nature of the underlying categories and how they would affect the phenotype.

Stage 1: Latent Class Analysis Ignoring Genetic Effects

The approach of latent class analysis (see, e.g., Goodman, 1974; Lazarsfeld, 1960; Lazarsfeld & Henry, 1968) has some conceptual similarities with and some differences from factor analysis. Both approaches are concerned with taking into account the patterns of association observed in multivariate data. Both postulate latent constructs—factors or classes—to account for the observed associations. The factor model, however, usually assumes that the associations are linear and best characterized by the correlation coefficient. Indeed, even when the raw observations are categorical, the typical factor model usually treats the categories as little more than arbitrary divisions imposed on a continuous latent trait. The summary of a factor model is typically an estimate of the number of latent dimensions or factors; a measure of the relative importance of each dimension as a source of phenotypic variation; a summary of how each dimension affects each measured variable (the factor loadings); and, if desired, estimates of the "scores" of the individual subjects on each of the major factors to emerge from the analysis. Similarly, in a latent class analysis, one tries to determine how many underlying categories of subjects (latent "classes") are needed to explain the pattern of association between the variables; to estimate to proportion of subjects falling into each class; to characterize the members of each class in terms of the probabilities of endorsing each item conditional on class membership; and, if desired, to estimate the relative probability that a subject with a given response profile will belong to each of the classes.

In the latent class analysis, one tries to predict all of the different patterns of responses of subjects to the 16 items. For a given data set, one begins by postulating a number of latent classes. These may correspond to genotypes, but in the initial stages of the analysis, there is no way of knowing. One may start with a small number of classes, say two. Then, for each class, one estimates the population class frequency and the probabilities that someone in a given class may fall into a particular response category for each of the items. In our case, we first estimated for each of the 16 ADHD items the probability that a subject in each class would show a symptom at all and then, if he did show the symptom, the probability that he would show it severely rather than mildly.

The statistical and computational method we used for estimating these frequencies and probabilities was the method of maximum likelihood, which is described in more detail elsewhere (Clogg, 1977; Eaves et al., 1993; Haberman, 1979). This approach makes the best use of the data in that, among other things, the estimates that we generate under a given hypothesis (e.g., the two-class model) give the model the best chance of fitting the data. We can compare different models (e.g., models postulating three or more classes) with one another to see whether it is really necessary to make a model more complicated (by switching from two to three classes, for example, or from three to four classes). To a first approximation, we can compute a chi-square statistic that allows us to judge the relative gain (or loss) from making a model more (or less) complicated.

For the 16 ADHD symptoms, we fitted models that assumed one, two, three, and four latent classes successively. For each new class, the model requires that an additional 33 probabilities be estimated. These are the frequency of the new class (one more parameter), the probabilities that a member of the new class will show each of the 16 symptoms at all (16 additional probabilities), and the probabilities that a member of the new class will, if he shows a symptom at all, express it severely (another 16 probabilities). Under the one-class model, there are only 32 probabilities to be estimated because with only one class, the probability of belonging to it is fixed at unity.

Table 2 shows how adding extra latent classes to the model improved our ability to explain the relationships between the symptoms. The "baseline" model allows for one class. We then added a second class and found that the chi-square that assessed the importance of the change had a value of 582.44 ($df = 33$). Clearly, the improvement is highly significant, so we knew at least two classes were needed. Adding a third class produced a chi-square of 185.93 ($df = 33$), which was again highly significant. This implies that two classes were insufficient and that at least three were needed. When we added a fourth class, however, the change produced only a nonsignificant chi-square of 35.70 ($df = 33$). At this point, we de-

TABLE 2

Summary of Tests of Nonfamilial Latent Class Models for Attention-Deficit Hyperactivity Disorder in Young Male Twins

Number of classes assumed	Number of probabilities estimated	χ^2	*df*	*P* (%)
1	32			
2	65	582.44	33	<0.1
3	98	185.93	33	<0.1
4	131	35.70	33	25.50

cided that the addition of a fourth class was unnecessary with these data and began to examine the detailed results of the three-class model.

The characteristics of the classes are summarized in Table 3. An estimated 89.8% of boys belonged to the most common class. These boys had virtually a zero probability of showing *any* symptoms at all. Because they could show almost no symptoms (Column A in Table 3 under Class 1), they also had no detectable chance of showing "severe" symptoms (probabilities in Column S). This class corresponded to the vast majority of 8–11-year-old boys in our sample. We estimated that the second class would constitute 7.8% of the population, and it typically consisted of boys who had low to intermediate probabilities of displaying the symptoms but virtually no chance of being described as severe for many of the symptoms. Even those symptoms, such as "fidgeting," which had a

TABLE 3

Latent Three-Class Model for Hyperactivity Symptoms in Young Male Twins: Probabilities That a Member of Each Class Will Display a Symptom (A) and, if so, Will Display Severe Expression (S)

	Probability of symptom					
	Class 1		Class 2		Class 3	
Item/behavior	A	S	A	S	A	S
Activities imposed by another						
Fidgets	.030	.096	1.000	1.000	1.000	1.000
Runs	.000	.000	.261	.000	1.000	1.000
Can't sit	.005	.000	.293	.000	.833	.800
Shifts	.000	.000	.130	.000	.833	.800
No follow-through	.038	.099	.300	.000	.571	.500
No concentration	.040	.000	.410	.000	1.000	.333
Passive activities						
Fidgets	.005	1.000	.516	.326	1.000	1.000
Runs	.007	.000	.046	.000	.716	1.000
Can't sit	.004	.000	.130	.333	1.000	.833
Shifts	.000	.000	.045	.000	.571	.500
Self-imposed activities						
Fidgets	.000	.000	.391	.222	.714	1.000
Runs	.000	.000	.087	.000	.571	.500
Can't sit	.000	.000	.087	.000	.714	1.000
Shifts	.004	.000	.000	.000	.714	.800
No follow-through	.008	.000	.000	.000	.429	.333
Class frequency	89.8%		7.8%		2.4%	

22–33% chance of being rated as severe depending on the context of assessment, still had a relatively low chance of being expressed severely by members of the class. The third class, constituting an estimated 2.4% of the population, had much higher probabilities (0.43–1.00) of showing each of the symptoms individually and, typically, very high odds of being described as severely affected. The fact that some of the probabilities under column S for the third class were "round" numbers such as 0.500 and 0.800 was a function of the small numbers of extreme individuals in the sample and did not concern us. Although the model strongly suggests that only three classes are sufficient to account for the observations, we note that the pattern of endorsement probabilities is consistent with an underlying dimensional ordering of the categories from "asymptomatic" through "mildly symptomatic" to "severely symptomatic."

These findings, although based on a small sample at this stage, have some implications for how ADHD is conceived and assessed clinically. If we were to ignore the context of behavior, or the severity of its manifestation, or concentrate only on the "milder" symptom of fidgetiness, we would expect a prevalence of about 10% for ADHD (the sum of Classes 1 and 2). On the other hand, if we were to require that the abnormal behavior be expressed severely in a series of activities (e.g., running about, inability to sit still, etc.) across a variety of contexts apart from activities imposed by others, we would obtain a reduced prevalence (2.4% according to our small-sample estimate). Thus, our preliminary data analysis supports a clinical distinction between general fidgetiness, which may be associated with milder expressions of inattention and impulsivity when the child is performing imposed activity, and the more extreme, "truly" ADHD behavior in which fidgetiness is accompanied by severe inattention and impulsivity across a series of contexts and activities. Analysis of impairment (which is also obtained in the P-CAPA but still has to be integrated into this model) and etiology (to which we now turn) may provide additional clues about how and when to intervene clinically.

Stage 2: Analyzing Twin Resemblance for Latent Classes

The ADHD data on twins provided a unique opportunity to develop new ways of looking at critical genetic issues that rescued us from deciding too early between the categorical and dimensional approaches. In the first stage of the analysis summarized earlier, twins were treated as *individuals*. Therefore, the results we obtained turned out to be exactly what we would obtain if we analyzed the twins as *pairs* but assumed that pair members were not associated for class membership. That is, the models summarized in Tables 2 and 3 assume that there are no genetic effects, or effects of the shared family environment on whether a boy is normal, or whether he belongs to either the milder or more severe of the two "symptomatic" classes. We therefore had to determine whether twins would be correlated for class membership and, if they were, to begin to examine why. We asked the following: "Is there any genetic basis for distinguishing between the normal, mild, and severe forms of the disorder?" We explore six possible models for twin class membership.

Model 1: No Family Resemblance

Our starting point for the next stage of analysis was the three-class model already presented because two classes were not enough and four seemed to be too many. By treating the twins as independent individuals when we fitted the three-class model at the first state, we had already fitted this model to the data (see Table 3).

Model 2: MZ Association Is Greater Than DZ Association

The most general model for twin resemblance allows twins to be partly associated in class membership, but it lets the pattern of association in MZ twins be different from that in DZ twins. A more technical presentation of this model is given by Eaves et al. (1993) but the basic elements can be described simply.

1. Each twin considered as an individual can belong to one of three classes.
2. It does not matter whether a twin is first or second in a pair; the number of possible classes is the same and the chance of an individual twin belonging to a class is the same.
3. Whether an individual twin is MZ or DZ has no effect on the number of possible classes, nor on the chances of belonging to a given class.
4. The chances of showing a symptom depend only on the class to which an individual twin belongs and not on whether he is MZ or DZ or the first or second twin in a pair.

5. Because there are three classes of individual twins, there are nine possible classes of twin pairs according to the possible pairwise combinations for the classes of the individual twins. Thus, under the most general hypothesis, the first twin may belong to the second class, for example, and the second twin may belong to the first class. Some pairs may be concordant for class membership and others discordant.
6. The patterns of pairwise concordance and discordance in class membership may differ between MZ and DZ pairs.

The specific statement for the three-class model can be made more general to encompass more or fewer classes as needed (see Eaves et al., 1993.) Allowing for the constraints on class frequencies for first and second twins and for MZ and DZ twins requires a total of only seven free frequencies to account for the pairwise twin associations in class membership. The full model, allowing for three latent classes that may be associated differently in MZ and DZ twins thus required that we try to estimate a total of 103 probabilities from the responses of the 294 twins to the 16 ADHD items.

Model 3: MZ Association Equals DZ Association

This model embodies the notion that twin pairs may be associated in class membership but that the degree of concordance is identical for MZ and DZ twins. Thus, association is assumed to be familial, caused by aspects of the shared family environment, but nongenetic.

Model 4: MZ Pairs (but Not DZ Pairs) Belong to Identical Classes

Model 4 is a stronger form of Model 2. It assumes that MZ pairs are perfectly correlated for membership of the three latent classes (i.e., there are no MZ pairs discordant for class membership). This model amounts to assuming that latent class membership is completely familial and possibly genetic. Although class membership is completely familial, there may be some item-specific variation in response profiles within identical twin pairs caused by chance.

Model 5: Single Gene With Complete Penetrance

This model is still stronger than the previous one. Not only does it assume that class membership is entirely genetic but that the three classes correspond to three genotypes at a single locus with two alleles. MZ pairs are completely concordant for class membership, and in DZ pairs the frequencies of concordant and discordant pairs for all combinations of classes are known functions of the frequencies, p and q, of the two alleles, A and B, respectively, and Mendel's law of segregation. The frequencies of the three genotypic classes, AA : AB : BB, are expected to follow the Hardy–Weinberg law and occur in the ratios $p^2 : 2pq : q^2$, respectively. As may be the case for the fourth model, there may be chance variations in response profiles among individuals who belong to the same genetic class.

Model 6: Single Gene With Reduced Penetrance

The previous model assumes that class membership is caused by a single gene. There is a 1 : 1 correspondence between genotypes and the latent classes. Typically, with many common disorders, this is not the case, but the expression of some genotypes may be modified by environmental factors so that it is impossible to infer the genotype from the phenotypic classes with perfect reliability. For each genotype, we thus defined a series of "penetrances" that represented the set of probabilities that each genotype would result in each phenotypic class. Thus, under the single-gene three-class model, there are potentially nine penetrances, where f_{ij} denotes the probability that the ith genotype produces the jth phenotypic class. However, for a given genotype, when two of the three penetrances are known, the third is fixed by the fact that the three phenotypic classes represent all of the possible mutually exclusive outcomes for that genotype. Furthermore, it is necessary to fix at least two additional penetrances at zero. In our example, we assume that the low-risk (AA) genotype has no chance of producing the extremely high-risk phenotype and that the high-risk (BB) genotype can never produce the lowest risk phenotype.

Results of Fitting Alternative Models for Twin Resemblance

The method of maximum likelihood may be used to estimate the large numbers of probabilities implied by each of the aforementioned models (see Eaves et al., 1993). The method yields a measure of how likely it is that the model could yield the particular data in hand ($-2\log(L)$, with

TABLE 4

Comparison of Different Explanations of Monozygotic (MZ) and Dizygotic (DZ) Twin Resemblance for Latent Classes of Hyperactivity

Model description	Number of probabilities	$-2\log(L)$	χ^2	*df* *N*	*P* (%)
1. No family resemblance	98	1,059.92			
2. MZ association > DZ	103	1,049.19	10.73[1]	5	5–10
3. MZ association = DZ	101	1,053.18	3.99[2]	2	10–25
4. MZ pairs' classes same	100	1,135.08	85.89[2]	6	<0.01
5. Single-gene/complete penetrance	97	1,162.93	113.74[2]	6	<0.01
6. Single-gene/incomplete penetrance	103	1,053.81	109.12[5]	6	<0.01

Note. A superscript in the chi-square value denotes the model being used for comparison. In all models, individual twins are assumed to belong to one of three classes.

−2 multiplied by the natural log of the likelihood [L]), which can be used to compare certain subsets of models with one another and to assess whether enhancements or simplifications are justified on statistical grounds. Table 4 summarizes the results of comparing the six models we have enumerated as possible explanations of the pattern of twin resemblance for ADHD in young boys from the VSABD.

The statistical comparisons of the various explanations were not very powerful with the small samples currently available. Nevertheless, the findings have considerable heuristic value. The significance tests showed that there was only a borderline difference at best between the two "benchmark" models, which assume that there is no family resemblance (Model 1) or that there is family resemblance greater in MZ than DZ pairs (Model 2). Even deleting the genetic effects on twin resemblance and trying to account for associations purely in terms of the shared environment (Model 3) was neither clearly better than a model that assumes twins are independent nor worse than a model that allows for the effects of genes. Thus, although allowing for genetic effects in the three "weaker" models gave a marginally better fit to the data, we could not really justify excluding the other alternatives in the absence of larger samples or still more detailed measures.

The two "purely genetic" explanations (Models 4 and 5) were much worse. It was clear that identical twins were not perfectly correlated for class membership and the fact that Model 5 gave such a poor fit, relative to Model 2, meant that we could discount a perfectly penetrant single gene as a more parsimonious account of the ADHD data in young boys.

The last model (Model 6) allowed for the effects of a single gene to be modified by the environment so that each genotype could produce more than one phenotypic class and MZ twins were no longer expected to be perfectly concordant for class membership. This model came close to the more general model for differences in MZ and DZ association in terms of fitting the data because the likelihoods were highly similar for Models 2 and 6. However, Model 2 was more agnostic about the kinds of genetic factors operating and also allowed for additional similarity due to the shared environment.

Table 5 provides estimated frequencies of the MZ and DZ twin pairs in each combination of classes for all six models. The major impact of twin resemblance was observed in MZ concordance for membership of the relatively rare extreme "attention-deficit/hyperactive" class (Class 3). The expected frequency under the model that has no twin resemblance (Model 1) was 0.06%. Under Model 2, which allowed MZ and DZ pairs to be correlated, we estimated that 1.16% of MZ pairs would be concordant. This result points to the frustration of our (currently) small random sample and justified the further pursuit of this especially informative class through a strategy of high-risk sampling of twin pairs.

We note two additional aspects of the MZ pattern under the least restrictive Model 2, compared with more restricted Models 4 and 5. The proportion of MZ pairs belonging to discordant classes was clearly not zero, as predicted under the two purely genetic models. The effects of the environment cannot be relegated to chance vagaries of individual symptomatology but have to be taken seriously as part of the ontogenetic process intervening between the genotype and its expression in the major phenotypic classes emerging from this analysis. Second, when we compared the single-gene/complete penetrance model (Model 6) with all of the others, we found that the proportion of MZ pairs in the middle (fidgety) class of putative heterozygotes under the Mendelian model was far higher (20.15%) than was estimated under Model 2, which allows for genetic *and* environmental effects on class membership. Briefly stated, if there is a

TABLE 5

Estimated Frequencies for Pairwise Membership of Latent Classes in Monozygotic (MZ) and Dizygotic (DZ) Twins Under Six Etiological Hypotheses

	Frequencies (%)					
	MZ Twin 1's class			DZ Twin 2's class		
Model/twin 2's class	1	2	3	1	2	3
1. No family resemblance						
1	79.21	6.94	2.14	79.21	6.94	2.14
2	6.94	0.61	0.19	6.94	0.61	0.19
3	2.14	0.19	0.06	2.14	0.19	0.06
2. MZ association > DZ association						
1	81.98	7.11	0.58	79.34	7.85	2.48
2	7.11	0.00	0.74	7.85	0.00	0.00
3	0.58	0.74	1.16	2.48	0.00	0.00
3. MZ association = DZ association						
1	78.93	8.49	1.35	78.93	8.49	1.35
2	8.49	0.00	0.35	8.49	0.00	0.35
3	1.35	0.35	1.35	1.35	0.35	1.35
4. MZ pairs' classes indentical						
1	89.00	0.00	0.00	78.00	9.56	1.44
2	0.00	9.56	0.00	9.56	0.00	0.00
3	0.00	0.00	1.44	1.44	0.00	0.00
5. Single-gene/complete penetrance						
1	78.56	0.00	0.00	69.88	8.42	0.25
2	0.00	20.15	0.00	8.42	11.09	0.64
3	0.00	0.00	1.29	0.25	0.64	0.40
6. Single-gene/reduced penetrance						
1	80.52	7.49	0.72	79.28	7.74	1.71
2	7.49	0.86	0.46	7.74	0.80	0.27
3	0.72	0.46	1.28	1.71	0.27	0.48

major gene increasing the risk for ADHD, it is surely not fully penetrant. Rather, the environment causes considerable "scrambling" between the latent genotypic classes and the manifest categories of behavior.

Discussion

The danger of presenting numbers to three significant figures is that they may be taken as proving too much. We reiterate the fact that our sample was small, random, and hitherto incomplete, comprising only 147 pairs of young male twins, and that a proportion of pairs was excluded because we had not interviewed them yet or because we did not have final zygosity diagnoses as of this writing. Our power is low at this point, and our data may not be truly representative of the final outcome. Furthermore, the data analyzed here only included maternal ratings, which may have their own unique perceptual biases. A more exhaustive analysis will require that we address the consistency of ratings over fathers, teachers, and the children themselves and integrate the additional data on frequency of symptoms and impairment of normal function essential for a complete picture of ADHD.

If there is a major gene affecting the risk for ADHD, it is not fully penetrant. Figure 1 summarizes the single-gene/reduced penetrance model (Model 6). The frequency of the high-risk (B) allele is estimated to be around 15%. This means that approximately 2% of the population are homozygous for elevated risk. The "penetrances" of the three putative genotypes are the odds that a person having a particular genotype will fall into each of the three phenotypic categories. We note that an estimated 75% of the hypothetical high-risk (BB) genotype are expected to be phenotypically ADHD. The penetrances of the other two classes—AA and AB—show that approximately 90% of these individuals will be phenotypically normal. That is, the low-risk allele shows virtually complete dom-

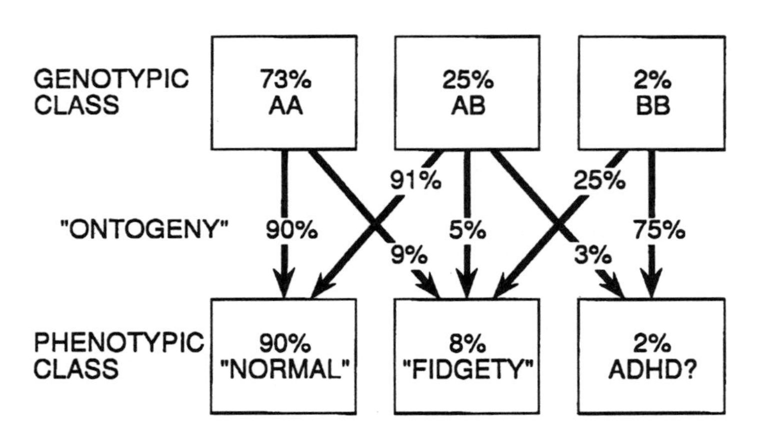

FIGURE 1. Provisional model for the etiology of attention-deficit hyperactivity disorder (ADHD) in young boys. The percentages are approximate.

inance under the model because the AA homozygote and the AB heterozygote are typically indistinguishable given the behavioral items included in this analysis. This provisional interpretation of the latent class analysis, of an apparently recessive allele with relatively low frequency, is consistent with the distributional and correlational data for the raw scale scores summarized earlier. The current model suggests that environmental factors alone are sufficient to account for the reduced penetrance of the primary locus. The low DZ correlation, relative to the MZ correlation, however, suggests that a second locus, interacting epistatically with the first, could turn out to be a significant factor in the genetic architecture underlying ADHD, but it is too early to tell. Although polygenic or oligogenic inheritance cannot be excluded, the large excess of very low scores and the very low correlation of DZ twins compared with MZ twins are both consistent with dominance or epistasis in the direction of relatively common low-risk alleles (Eaves, 1988).

From a clinical perspective, severity of behavioral expression, especially inattention and impulsivity across activities, which are not simply imposed on the child by others but are self-generated or merely occurring passively, is a crucial facet of the correct identification of the high-risk genotype. All in all, however, approximately 25% of the children defined as being possibly ADHD with the current 16 "intensity" items are putatively heterozygotes and not the high-risk recessive homozygotes. Correspondingly, about 25% of the high-risk BB genotypes would be assessed as merely fidgety if classification were based only on these 16 items. Errors of behavioral assessment in both directions can lead geneticists to the wrong place as they try to detect linkages between a putative high-risk allele and markers of known genomic location.

Our analysis suggests that elements of both dimensional and categorical models may be necessary for a full understanding of some forms of psychopathology. We see that even if there were a single gene of large effect contributing to the risk for ADHD, there is evidence of a dimensional ordering of severity of symptomatology at the phenotypic level. Furthermore, even when one assumes that there is such a gene, the pattern of penetrances is consistent with a dimensional model for the expression of that locus. That is, the AA genotype is the most likely to produce the asymptomatic phenotype and the least likely to produce the most severe phenotype. By contrast, the BB genotype is the most likely to produce the most severely symptomatic phenotype, less likely to produce the milder fidgety phenotype, and least likely to produce asymptomatic individuals.

Clearly, our findings are preliminary. The analytical approach needs much more exploration, and our sample needs to be much larger, including an oversampling of symptomatic individuals. A significant research goal in the near future has to be the attempt to further refine the behavioral assessment in order to make it possible to infer the latent genotype from the behavioral phenotype with even greater reliability. For this purpose, the CAPA includes a rich selection of indexes of onset, frequency, and impairment that we hope will help our pursuit of this goal.

References

Angold, A., Cox, A., Prendergast, M., Rutter, M., & Simonoff, E. (1989). *The Child and Adolescent Psychiatric Assessment.* Unpublished manuscript, MRC Child Psychiatry Unit, University of London, and Developmental Epidemiology Program, Duke University, Durham, NC.

Barkley, R. A. (1990). *Attention-deficit hyperactivity disorder: A handbook for diagnosis and treatment.* New York: Guilford Press.

Clogg, C. C. (1977). *Unrestricted and restricted maximum-likelihood latent structure analysis: A manual for users* (Tech. Rep. No. 1977-09). University Park: Pennsylvania State University, Population Issues Research Center.

Cloninger, C. R. (1986). A unified biosocial theory of personality and its role in the development of anxiety states. *Psychiatric Developments, 3,* 167–226.

Cloninger, C. R. (1987). Neurogenetic adaptive mechanisms in alcoholism. *Science, 236,* 410–416.

Eaves, L. J. (1982). The utility of twins. In V. E. Anderson, W. A. Hauser, J. K. Deny, & C. F. Sing (Eds.), *Genetic basis of the epilepsies* (pp. 249–276). New York: Raven Press.

Eaves, L. J. (1988). Dominance alone is not enough. *Behavior Genetics, 18,* 27–33.

Eaves, L. J., Silberg, J. L., Hewitt, J. K., Rutter, M., Meyer, J. M., Neale, M. C., & Pickles, A. (1993). Analyzing twin resemblance in multi-symptom data: Genetic applications of a latent class model for symptoms of conduct disorder in juvenile boys. *Behavior Genetics, 23,* 5–20.

Eysenck, H. J. (1952). *The scientific study of personality.* London: Routledge & Kegan Paul.

Goodman, L. A. (1974). The analysis of systems of qualitative variables when some of the variables are unobservable: Part 1. A modified latent structure approach. *American Journal of Sociology, 79*, 1179–1259.

Gray, J. A. (1970). The psycho-physiological basis of introversion–extraversion. *Behaviour Research and Therapy, 8*, 249–266.

Gray, J. A. (1981). A critique of Eysenck's theory of personality. In H. J. Eysenck (Ed.), *A model for personality*. New York: Springer-Verlag.

Haberman, S. (1979). *Analysis of qualitative data* (*Vol. 2*). San Diego, CA: Academic Press.

Lazarsfeld, P. F. (1960). Latent structure analysis and test theory. In H. Gulliksen & S. Messick (Eds.), *Psychological scaling: Theory and applications* (pp. 83–96). New York: Wiley.

Lazarsfeld, P. F., & Henry, N. W. (1968). *Latent structure analysis*. New York: Norton.

Rutter, M., & Gould, M. (1985). Classification. In M. Rutter & L. Hersov (Eds.), *Child and adolescent psychiatry* (pp. 304–319). London: Blackwell Scientific.

PART FIVE

Bridging the Nature–Nurture Gap

PART FIVE

Introduction

Theodore D. Wachs

The old heredity–environment controversy which, like Frankenstein's monster, often seems to be buried beneath tons of rubble but rises again for a sequel. (Hoffman, 1985, p. 127)

Although Hoffman's description is compelling, the parallel careers of the nature–nurture controversy and one of our great literary/cinematic characters are not totally isomorphic. Like Frankenstein's monster, the nature–nurture controversy has excited the popular imagination, at times even spilling into real-life experiments ("Whatever Happened to the Winton Children?"—*Yankee Magazine*, 1991). Also like Frankenstein's monster, nature–nurture has long aroused strong emotion. One of the earliest examples of the emotion-engendering aspects of nature–nurture is seen in the various chapters of the two-volume issue of the 1940 Yearbook of the National Society for the Study of Education, which presented provocative findings on the question of the modifiability of intelligence, as well as very strident critiques of these findings (e.g., "It appears characteristic of the Iowa group of workers that they so often find difficulty in reporting accurately either the data of others or their own"—Terman, 1940, p. 461).

At least on the surface, the initial appearances of the Frankenstein monster and the nature–nurture controversy appear to have been in the same century (Frankenstein was first published in 1818; the first formal appearance of the nature–nurture dichotomy is typically associated with the 1869 publication of Galton's *Hereditary Genius*; Plomin, DeFries, & McClearn, 1980). In fact, however, the nature–nurture controversy has been repeatedly emerging from the rubble in various guises since well before the 19th century (Fowler, 1983; Hunt, 1961). Furthermore, unlike the Frankenstein monster, whose ravages were confined to the printed page and later to the silver screen, the ravages of the nature–nurture controversy have all too often involved real people and real issues, such as the racial superiority of certain groups over other groups (Weizmann, Wiener, Wiesenthal, & Ziegler, 1990).

Unlike the Frankenstein monster, whose periodic reappearances can be traced back to a specific source (the sequels make money), the reasons for the durability of the nature–nurture controversy are unclear and are perhaps best left to historians of science. What has been understood, at least since the time of Aristotle, is that development must involve the contributions of both nature and nurture.

> The poets who divide ages by sevens are in the main right: but we should observe the divisions actually made by nature; for the deficiencies of nature are what art and education seek to fill up. (Aristotle, cited in McKeon, 1941, p. 1305)

Rather than asking why these continued reappearances, the critical question is, and always has been, what can be done to transform into something of scientific and social value the essentially useless dichotomy between nature and nurture?

Frankenstein was all too keenly aware that prior wisdom offered little in the way of potential solutions for the scientific questions he was dealing with.

> I have described myself as always having been imbued with a fervent longing to penetrate the secrets of nature. In spite of the intense labor and wonderful discoveries of modern philosophers, I always come from my studies discontented and unsatisfied. (Shelly, 1818/1985, p. 88)

Unlike Frankenstein, we currently have, and have had for some time, a potential solution for the transformation of the moribund nature-versus-nurture controversy into a viable scientific question. The solution lies in looking at genetic and environmental influences as a "how" process rather than as a set of static dichotomies.

> The heredity–environment problem is still very much alive. Its viability is assured by the gradual replacement of the questions which one and how much by the more basic and appropriate question how. (Anastasi, 1958, p. 206)
>
> Most developmentalists, if questioned would probably say that the classic nature–nurture issue is unresolvable as posed, and that the task is to determine exactly how genetic endowment interacts with experience in the course of development. (Hay, 1986, p. 152)

A refocusing of efforts toward answering the question how does not rule out studies that exclusively investigate environmental or genetic influences. Clearly, the how question holds as much for within-domains studies as for cross-domains studies. To understand the genetic contribution to gene–environment how studies, one must understand how genes contribute to variability in development; similarly, to understand environmental contributions to gene–environment how studies, one must understand how environments contribute to variability in development. Thus, there will always be unique questions that can best be answered only by purely genetic or environmental studies. For example, what are the changes in the nature of genetic influences over time?—a critical question that can best be answered by the use of standard behavioral genetics methodologies (Plomin, 1990). Similarly, the question of how higher order environmental influences mediate relations between microenvironment influences and development is one that can best be answered by the use of appropriate environmental methodologies (Wachs, 1992). Ultimately, if researchers are to progress in understanding the processes underlying developmental variability, it is clear that the genetic how and the environmental how will need to be integrated in ways that reflect both the unique and the combined influences of genes and environments. Integrating the genetic how and the environmental how is what is meant by bridging the nature–nurture gap.

The chapters contained in this section offer approaches to bridging this gap. Before writing their chapters, the authors were asked to consider three fundamental questions. First, what factors inhibit collaboration between behavioral geneticists and environmentally oriented researchers? Second, what are potential areas of collaboration between behavioral geneticists and environmentally oriented researchers? Third, what specific steps can be taken to facilitate such collaboration?

Chapter authors were chosen to reflect all sides of the nature–nurture spectrum. Thus, there are contributions from researchers whose primary identification is with the field of behavioral genetics (Goldsmith, chapter 17; Rowe and Waldman, chapter 19); there are also contributions from researchers whose primary identification is with the study of environmental influences (Bronfenbrenner and Ceci, chapter 16; Wachs, chapter 20), as well as a contribution by a researcher who has dealt with both the biological and the environmental spheres (Horowitz, chapter 18). Given the widely differing orientation of individual authors, it is not surprising that specific points made in individual chapters reflect a diverse set of arguments and solutions across all of these chapters. However, there also are a surprising number of areas of agreement. These include (a) the importance of directly measuring specific genetic and environmental contributions to developmental variability and (b) the need to shift from traditional main effect or additive models to more complex, systems-based models of genetic and environmental influences (possibly encompassing nonlinear or bidirectional components).

Readers expecting to find in these chapters a fully functioning bridge between nature and nurture will probably be disappointed. Rather than a bridge, what these chapters represent are individuals on both sides of the nature–nurture gap beginning to exchange plans for how this gap might be bridged. Unlike previous plans, which were grand in scope but vague in details, the present set of plans offers specific areas of focus that are both feasible and have the potential to provide answers that will advance our understanding of the how question. Whether these plans are actively implemented will ultimately determine whether researchers are able to realize Baron Frankenstein's dream of turning a monster into something of scientific value or whether the dichotomous nature–nurture monster will continue to ravage the scientific landscape for generations to come.

References

Anastasi, A. (1958). Heredity, environment, and the question "How?" *Psychological Review, 65*, 197–208.

Fowler, W. (1983). *Potentials of childhood* (Vol. 1). Lexington, MA: Lexington Books.

Galton, F. (1869). *Hereditary genius: An inquiry into its laws and consequences.* London: Macmillan.

Hay, B. (1986). Infancy. *Annual Review of Psychology, 37*, 135–162.

Hoffman, L. (1985). The changing genetics–socialization balance. *Journal of Social Issues, 41*, 127–148.

Hunt, J. McV. (1961). *Intelligence and experience.* New York: George Ronald.

McKeon, R. (1941). *The basic works of Aristotle.* New York: Random House.

Plomin, R. (1990). The role of inheritance in behavior. *Science, 248*, 183–188.

Plomin, R., DeFries, J., & McClearn, G. (1980). *Behavior genetics.* New York: Freeman.

Shelly, M. (1985). *Frankenstein.* New York: Penguin Books. (Original work published 1818)

Terman, L. (1940). Commentary. *Yearbook of the National Society for the Study of Education, 39*(Pt. 1), 460–461.

Wachs, T.D. (1992). *The nature of nurture.* Newbury Park, CA: Sage.

Whatever happened to the Winton children? (1991, February). *Yankee Magazine*, pp. 96–102.

Weizmann, F., Wiener, N., Wiesenthal, D., & Ziegler, M. (1990). Differential K theory and racial hierarchies. *Canadian Psychology, 31*, 1–13.

CHAPTER 16

Heredity, Environment, and the Question "How?"—A First Approximation

Urie Bronfenbrenner and Stephen J. Ceci

Three-and-a-half decades ago, Anne Anastasi (1958), the then-outgoing president of the American Psychological Association's Division of General Psychology, posed the just-mentioned question as a challenge to psychological science as a whole. Anastasi offered few answers. Instead, she urged her scientific colleagues to pursue what she saw as a more rewarding and necessary scientific goal. Rather than seeking

> to discover *how much* of the variance was attributable to heredity and how much to environment . . . a more fruitful approach is to be found in the question *"How?"* There is still much to be learned about the specific *modus operandi* of hereditary and environmental factors in the development of behavioral differences. (p. 197)

Today, 35 years later, Anastasi's challenge still stands despite the fact that recent developments both in science and society give it renewed importance. Thus, over the past decade, research in the fields of both behavioral genetics and human development has placed increased reliance on the traditional percentage-of-variance model (Plomin & Bergeman, 1991; Plomin, DeFries, & McClearn, 1990; *Psychological Science*, 1992; Scarr, 1992). The extensive body of research guided by this model—in particular, some of the general conclusions drawn from it—has evoked criticism, not only on scientific grounds but also on social and ethical grounds (*Child Development*, in press; *Behavioral and Brain Sciences*, 1991).

Social and ethical concerns notwithstanding, in our view, although the traditional model has made important contributions to the understanding of not only genetic but also environmental influences on human development (e.g., Plomin & Daniels, 1987), it nevertheless remains incomplete. In addition, some of its basic assumptions are subject to question. At the core of the problem lies precisely Anastasi's issue: the need to identify the mechanisms through which genotypes are transformed into phenotypes.

Overview

In this chapter, we take a first step in addressing that need by offering a possible conceptual framework for constructing a more systematic theoretical and operational model of genetic–environment interaction. Based on a bioecological perspective (Bronfenbrenner, 1989a, 1993; Ceci, 1990), the proposed framework replaces some of the key assumptions underlying the traditional paradigm of human behavioral genetics with formulations that we believe to be more consonant with contemporary theory and research in the field of human development. In addition to incorporating explicit measures of the environment conceptualized in systems terms and allowing for nonadditive synergistic effects in genetic–environment interaction, the model specifically posits empirically assessable mechanisms, called *proximal processes*, through which genotypes are transformed into phenotypes.

It is further argued, both on theoretical and empirical grounds, that heritability, defined by behavioral geneticists as "the proportion of the total phenotypic variance that is due to additive genetic variation" (Cavalli-Sforza & Bodmer, 1971), is in fact highly influenced by events and conditions in the environment. We propose that heritability (h^2) can specifically be shown to vary substantially as a direct function of the magnitude of proximal processes and the quality of the environments in which they occur, potentially yielding values of h^2 that, at their extremes, are both appreciably higher and lower than those heretofore reported in the research literature. Furthermore, what h^2 in fact measures is the proportion of variance attributable to observed individual differences in actualized genetic potential. It follows that the amount of unactualized potential remains unknown and cannot be inferred from the magnitude of h^2.

In formal expositions of the established behavioral genetics model, the point is usually made that the model is intended to apply only to individual differences in developmental outcome and not to differences between groups. Yet, to our knowledge, no systematic theoretical framework has been proposed by behavioral geneticists for conceptualizing and analyzing the role of heredity and environment in producing group differences in developmental outcomes. By contrast, a bioecological model explicitly conceptualizes both kinds of differences as interactive products of genetic–environment interaction and suggests research designs that permit the simultaneous investigation of both types of variation. (In the case of group differences, it is as yet possible to demonstrate environmental effects only; the assessment of the genetic contribution to group differences must wait on advances in molecular genetics and related fields; see the discussion of Hypothesis 5 later in this chapter.)

Finally, there is evidence that social changes taking place over the past 2 decades in developed societies as well as developing societies have undermined conditions necessary for the operation of proximal processes (Bronfenbrenner, 1989b, 1992). Hence, if it is valid, then the proposed model has importance for both science and society because it implies that humans have genetic potentials, in terms of both individual and group differences, that are appreciably greater than those that are presently realized and that progress toward such realization can be achieved through the provision of environments in which proximal processes can be enhanced, but which are always within the limits of human genetic potential.

The Bioecological Model

As previously noted, at the core of a bioecological model of human development is the concept of proximal process. At the outset, it is important to clarify how such processes differ from the classic physiopsychological processes of perception, cognition, emotion, and motivation. These processes are usually thought of as occurring primarily within the brain, which is also viewed as the "place" where development occurs. But, in our view, this is not the whole story because perception, cognition, emotion, and motivation involve psychological content: They are about *something*. And, from the beginning, much of that content is in the outside world. More specifically, in humans, the content turns out, early on, to be mainly about people, objects, and symbols. These entities exist initially only in the environment, that is, outside the organism. Hence, from its beginnings, development involves interaction between organism and environment. Moreover, interaction implies a two-way activity. The external becomes internal and becomes transformed in the process. But because, from its very beginnings, the organism begins to change its environment, the internal becomes external and becomes transformed in the process.

Thus far, we have been speaking in metaphors and deliberately so. We wish to convey to the reader a sense of the general schema in which our more systematic, substantive framework is cast. But the metaphor must also have some correspondence with reality, and for that purpose, it must take on more concrete forms. To make this transition from the abstract to the concrete, we return to the concept of interactive proximal processes and examine how they relate to the genetic endowment of the person on the one hand and to the environment on the other.

Genetic potentials for development that exist within humans are not merely passive possibilities but active dispositions expressed in selective patterns of attention, action, and response. However, these dynamic potentials do not spring forth full-blown like Athena out of Zeus's head from a single blow of Vulcan's hammer. The process of transforming genotypes into phenotypes is not so simple or so quick. The realization of human genetic potentials and predispositions for competence, character, and psychopathology requires intervening mechanisms that connect the inner with the outer in a two-way process that occurs not instantly but over time. This process is the focus of the first defining property of a bioecological model, which is formulated as follows.

Proposition 1

Especially in its early phases, and to a great extent throughout the life course, human development takes place through processes of progressively more complex reciprocal interaction between an active evolving biopsychological human organism and the persons, objects, and symbols in its immediate environment. To be effective, the interaction must occur on a fairly regular basis over extended periods of time. Such enduring forms of interaction in the immediate environment are referred to henceforth as proximal processes. Examples of enduring patterns of proximal process are found in parent–child and child–child activities, group or solitary play, reading, learning new skills, studying, athletic activities, and performing complex tasks.

Thus, to the extent that they occur in a given environment over time, proximal processes are postulated as the mechanisms through which human genetic potentials for effective psychological functioning are actualized.[1] In short, proximal processes are the primary engines of development. But, like all engines, they cannot produce their own fuel, nor are they capable of self-steering. A second defining property identifies the three-fold source of these dynamic forces.

Proposition 2

The form, power, content, and direction of the proximal processes that affect development vary systematically as a joint function of the characteristics of the developing person and the environment (both immediate and more remote) in which the processes are taking place and the nature of the developmental outcomes under consideration.

[1]Note that this formulation leaves unanswered the question of what mechanisms lead to the actualization of genetic potentials for functional incompetence. This issue is addressed in a more extended exposition of the bioecological model (Bronfenbrenner & Ceci, 1993).

Illustrative Research Designs and Hypotheses

Among the most consequential personal characteristics that affect the form, power, content, and direction of proximal processes is genetic inheritance. As yet, however, there are no concrete examples in which the bioecological model has been applied to samples composed of groups of contrasting consanguinity (e.g., identical vs. fraternal twins, biological vs. adopted children). Given the absence of such studies, we proceed as follows to illustrate the kinds of research designs that might be used for analyzing genetic–environment interaction in a bioecological model. First, we present a concrete example of findings obtained with what we call a *process–context model*, one in which the characteristics of the person (in this instance, children who differ in degree of consanguinity) have not yet been included in the design. We then present some examples of hypotheses derived from a bioecological model that could be tested once family members representing contrasting degrees of consanguinity have been incorporated into the design.

The results of the first step are shown in Figure 1. The data are drawn from a classic longitudinal study by Drillien (1964) of factors affecting the development of children of low birth weight compared with those of normal birth weight. For present purposes, only the data for the latter are shown. The figure depicts the impact of the quality of mother–infant interaction at 2 years of age on the number of observed problem behaviors at 4 years of age as a function of social class. As can be seen, in accord with Proposition 1, a proximal process (in this instance, mother–infant interaction across time) emerges as the most powerful predictor of developmental outcome. Furthermore, as stipulated in Proposition 2, the power of the process varies systematically as a function of the environmental context (in this instance, social class). Note also that proximal process has the general effect of reducing or buffering against environmental differences in developmental outcome.

Finally, the proximal process appears to have its greatest impact in the most unfavorable environment. From the perspective of a bioecological model, however, the greater effectiveness of proximal processes in poorer environments is to be expected only for indices of developmental dysfunction. For outcomes reflecting developmental competence,

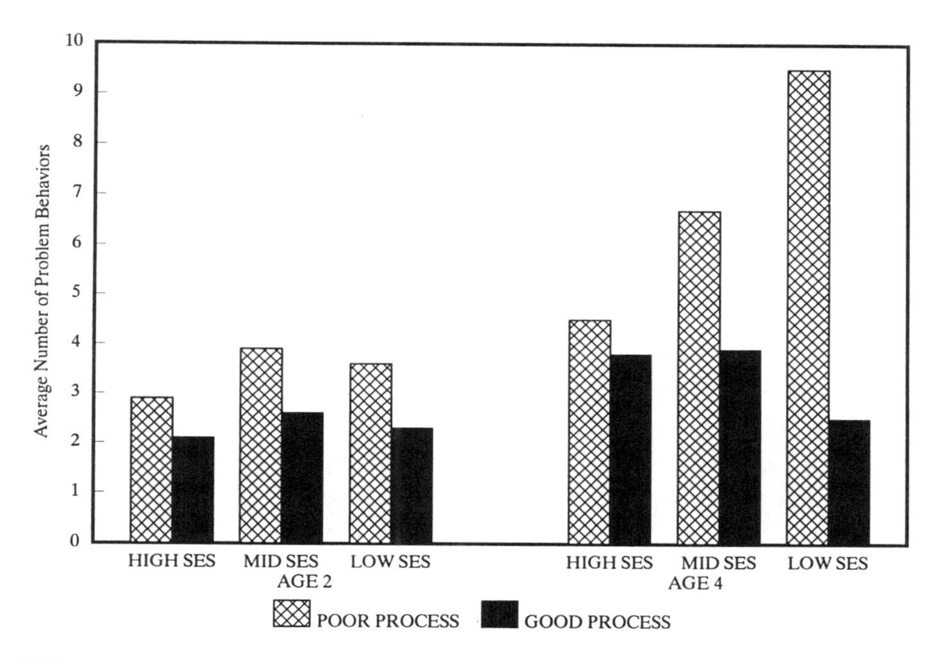

FIGURE 1. Effects of proximal process at 2 years of age on children's problem behaviors at 2 and 4 years of age by socioeconomic status (SES).

proximal processes are expected to have greater impact in more advantaged environments, primarily because the achievement of competence requires resources that exist in, and are drawn from, the broader external environment. For example, when the outcome is superior school achievement, mother–child interaction is most effective in families in which mothers have had some education beyond high school (Small & Luster, 1992).

In our second step, incorporating children of contrasting consanguinity into the research design requires the introduction of an additional dimension. Specifically, each of the six cells of the longitudinal design (two levels of care × three levels of social class) are now further stratified by the degree of genetic relationship (e.g., monozygotic vs. dizygotic twins, biological vs. unrelated children living in the same family). One would then proceed to calculate the value of h^2 for each of the original six cells.

What would be the anticipated results viewed from the perspective of a bioecological model? The expectations are based on two of the model's key assumptions: first, that proximal processes actualize genetic potentials for developmental competence and thereby reduce developmental dysfunction and, second, that h^2 is correctly interpreted as the proportion of variance attributable to actualized genetic potential. Given these assumptions, the bioecological model generates a series of empirically testable hypotheses. We cite five of them as examples.

Hypothesis 1

With respect to outcomes that reflect developmental competence, h^2 will be greater when levels of proximal process are high and smaller when such processes are weak. This prediction follows from the principle that proximal processes actualize genetic potentials for developmental competence, which thereby reduces variation attributable to the environment.

Hypothesis 2

The values of h^2 that are associated with high and low levels of proximal process will be more extreme (greater and smaller, respectively) than those previously reported in the literature (when proximal process was not taken into account). This hypothesis follows from Proposition 1, which stipulates that proximal processes have more powerful effects on development than do the characteristics of either the environment or the person.

Hypothesis 3

The power of proximal processes to actualize genetic potentials for developmental competence will be greater in advantaged and stable environments than in those that are impoverished and disorganized. For example, with respect to outcomes such as competence, we predict that the difference between values of h^2 that are associated with high versus low levels of proximal process will be even greater in middle-class environments than in lower class environments.

Hypothesis 4

Conversely, the power of proximal process to buffer genetic potentials for developmental dysfunction (such as Drillien's, 1964, index of children's problem behaviors) will be greater in disadvantaged and disorganized environments. Thus, for such outcomes, we predict that the difference between values of h^2 that are associated with high levels versus low levels of proximal process will be greater in lower class environments than in middle-class environments.

A final hypothesis, and its empirical investigation, are made possible by the specification of proximal processes as the mechanisms through which genetic potentials are actualized. Such processes not only exist in nature but can also be produced experimentally. For example, two randomly assigned groups, each including the same contrast in consanguinity but comparable in other respects, could be exposed to intervention strategies systematically differing in the degree to which they encourage the involvement of children in proximal processes in the home or other child-care settings. Although it would be difficult to assemble and sustain a sample of twins for this purpose, recent demographic changes in the United States are creating other research opportunities on a much larger scale. Thus, the growing number of families that contain both biological and stepchildren or adopted children can provide the contrasts in consanguinity necessary to assess the impact of proximal processes on the actualization of genetic potential (as assessed by the value of the corresponding h^2).

The applicability of an experimental strategy makes possible two important scientific gains. First, it provides a more rigorous test of the types of hypotheses already cited. The desirability of such a test arises from the following considerations. Although research has shown that environmental factors exert a substantial influence on proximal processes (Bronfenbrenner, 1986a, 1986b, 1989a, 1993), in accord with a bioecological model such processes—like all forms of human behavior—must necessarily also have a significant genetic component. Hence, stratification by levels of proximal process also results in some unknown level of genetic selection. By varying such levels experimentally, this source of bias is avoided. If, under these circumstances, groups that are randomly assigned to high versus low levels of proximal process show corresponding differences in levels of h^2, then this would constitute strong experimental evidence in support of the proposed conceptual framework.

Applying an experimental strategy can also shed light on the role of genetic–environment interaction in producing group differences in developmental outcomes. Thus, to the extent that experimentally induced increases in levels of proximal process can significantly reduce developmental differences that are associated with socioeconomic status, family structure, or other environmental contexts, this finding would indicate that such environmental differences are primarily a reflection of variation in proximal processes. These considerations lead to the following final hypothesis.

Hypothesis 5

Similarly high, experimentally induced levels of proximal process will substantially reduce differences between groups (e.g., social class effects) in the degree of actualized genetic potential. Such reductions will be manifested with respect to group differences in both developmental competence and developmental dysfunction. Conversely, similarly low levels of proximal process will substantially increase such group differences.

This last hypothesis has an unfortunate shortcoming. Unlike the others, it cannot be rigorously tested, at least as yet. It is surely possible to determine whether proximal processes reduce group differences in developmental outcome. Indeed, the results shown in Figure 1 illustrate precisely such an effect. Moreover, these results would be consistent with the last hypothesis. But research findings that are merely "consistent" with a particular hypothesis are of course not sufficient to establish its validity. Thus, in the present instance, true validation of the hypothesis requires an assessment of the extent of actualized genetic potential in groups that do not differ systematically in degree of consanguinity. In this case, no such assessment is possible at present (i.e., there is no way to calculate an estimate corresponding to an h^2). The issue cannot be resolved solely on the basis of phenotypic data and must wait on further scientific advances in methods for analyzing human genotypes.

It is appropriate that we end this chapter on a note of uncertainty, for it sounds the underlying theme of the chapter as a whole. We would of course be gratified if our theoretical constructs and hypotheses turn out to have some validity. But that is not the main purpose of the undertaking. Indeed, our aim, and that of developmental science as well, might be better served if the concepts and hypotheses were to be found wanting. For our principal intent is not to claim answers but to provide a theoretical framework that might enable our colleagues in the field and ourselves to make some further progress in discovering the processes and conditions that define the scope and limits of human development, and to develop a corresponding operational model that permits our position to be falsified. We hope that our colleagues will be sufficiently intrigued, or perhaps provoked, to join in that effort.

References

Anastasi, A. (1958). Heredity, environment, and the question "How?" *Psychological Review, 65*, 197–208.

Behavioral and Brain Sciences. (1991). Special section of commentary on Plomin and Bergeman (1991). *14*, 386–414.

Bronfenbrenner, U. (1986a). Ecology of the family as a context for human development. *Developmental Psychology, 22*, 723–742.

Bronfenbrenner, U. (1986b). Recent advances in research on the ecology of human development. In R. K. Silbereisen, K. Eyferth, & G. Rudinger (Eds.), *Development as action in context* (pp. 287–309). Berlin: Springer-Verlag.

Bronfenbrenner, U. (1989a). Ecological systems theory. In R. Vasta (Ed.), *Six theories of child development: Revised formulations and current issues* (pp. 185–246). Greenwich, CT: JAI Press.

Bronfenbrenner, U. (1989b). Who cares for children [Invited address to UNESCO] (Bilingual publication No. 188). Paris: UNESCO.

Bronfenbrenner, U. (1992). Child care in the Anglo-Saxon mode. In M. Lamb, K. J. Sternberg, C. P. Hwang, & A. G. Broberg (Eds), *Child care in context: Cross-cultural perspectives* (pp. 281–291). Hillsdale, NJ: Erlbaum.

Bronfenbrenner, U. (1993). The ecology of cognitive development: Research models and fugitive findings. In R. H. Wozniak & K. Fischer (Eds.), *Thinking in context* (pp. 3–24). Hillsdale, NJ: Erlbaum.

Bronfenbrenner, U., & Ceci, S. J. (1993). *Nature–nurture reconceptualized: Toward a new theoretical and operational model.* Manuscript submitted for publication.

Cavalli-Sforza, L. L., & Bodmer, W. F. (1971). *The genetics of human populations.* San Francisco: Freeman.

Ceci, S. J. (1990). *On intelligence . . . more or less: A bioecological treatise on intellectual development.* Englewood Cliffs, NJ: Prentice-Hall.

Child Development. (in press). Special section of commentary on Scarr.

Drillien, C. M. (1964). *The growth and development of the prematurely born infant*. Edinburgh, UK: Livingston.

Plomin, R., & Bergeman, C. S. (1991). The nature of nurture: Genetic influence on "environmental" measures. *Behavioral and Brain Sciences, 14*, 373–427.

Plomin, R., & Daniels, D. (1987). Why are children in the same family so different from one another? *Behavioral and Brain Sciences, 10*, 1–16.

Plomin, R., DeFries, J. C., & McClearn, G. E. (1990). *Behavior genetics*. New York: Freeman.

Psychological Science. (1992). Special section: Ability testing. *3*, 266–278.

Scarr, S. (1992). Developmental theories for the 1990s: Development and individual differences. *Child Development, 63*, 1–19.

Small, S., & Luster, T. (1992). *Effects of parental monitoring on adolescent development: The process–person–context model applied*. Unpublished manuscript.

CHAPTER 17

Nature–Nurture Issues in the Behavioral Genetics Context: Overcoming Barriers to Communication

H. H. Goldsmith

As behavioral genetics has gained more exposure in the field of psychology, most psychologists have become acquainted with the rudiments of twin, family, and adoption studies. They know that modest-to-moderate heritability (often ranging from about 30% to about 60%) has been documented for a variety of traits in the cognitive and personality domains. More recently, most researchers have probably become aware that the findings from behavioral genetics studies also hold implications for how the environment operates. Thus, it has become difficult for environmentally oriented psychologists simply to acknowledge the existence of genetic effects in a vague manner and carry on as before. The struggle to understand behavioral genetics beyond the surface level that is accessible to anyone with psychometric training is not trivial. Such understanding requires knowledge not only of quantitative/population genetics but also of evolution and molecular genetics. The behaviorist tradition placed some realms of American psychology apart from the other life sciences, and thus a generation or more of researchers has been unfamiliar with these concepts and methods from a sister science.

On the other hand, as behavioral geneticists have begun to draw conclusions about the way the environment operates, they have struggled to understand in detail current conceptualizations of environmental influence. Unfortunately, these conceptualizations are typically difficult to reduce to the structural biometric equations that are behavioral geneticists' customary tools.

This chapter offers some understandings and opinions that might facilitate further integration of nature- and nurture-based approaches to human behavior. It is addressed to psychologists rather than behavioral geneticists, and it omits technical discussion that can be found in developmentally oriented texts (e.g., Plomin, 1986) and review chapters (e.g., Goldsmith, 1988).

Goals of Behavioral Genetics

There is widespread agreement that past implementations of behavioral genetics techniques need to be supplemented to address the functioning of the environmental forces more meaningfully. As a prelude to considering the treatment of environment in current behavioral genetics, I specify the goals of the field. Of course, the overarching goal of behavioral genetics is to understand genetic influences on behavior, both human and animal. This cannot be done without considering the complementary influence of the environment. Current methods (twin, family, and adoption studies) explicate individual differences rather than species-general behavioral patterns. The overarching goal of understanding genetic influences on behavior can be divided into more specific goals, some of which involve the environment more than others. These goals include the following.

1. *Explicating the biometric architecture of traits*. Behavioral geneticists seek to infer the relative influences of various classes of genetic and environmental influences from patterns of covariation among individuals of varying degrees of genetic and environmental relationship.
2. *Providing clues to the selection history of the trait during evolution*. The

amount of additive genetic variance and the amount of dominance genetic variance both have implications for the degree and type of selection that might have operated on the trait (details are beyond the scope of this chapter).

3. *Predicting the response to artificial selection in agriculturally valuable species*. More heritable traits respond more rapidly to selective breeding.
4. *Identifying likely traits for molecular genetics and neurochemical analyses*. Quantitative genetic analysis serves its "signpost" function for biological investigation most clearly when a Mendelian pattern of single-gene inheritance is discovered. However, it also encourages the search for specific genes for highly heritable forms of psychopathology as more promising than that for less heritable disorders. Thus, it seems that a search for biological mechanisms in highly heritable bipolar affective disorder might have a greater chance of early success than a similar search for biological mechanisms for less heritable unipolar depression. Linkage and association methods that are now available should be able to detect a gene or quantitative trait locus that accounts for an appreciable portion—but by no means all—of the variance in traits. (A quantitative trait locus is a segment of DNA that contains a gene affecting a continuous trait but that also may contain other DNA such as closely linked genes or noncoding regions of DNA; see Paterson et al., 1988.)
5. *Providing relative risk figures and other data for genetic counseling*. A related goal is providing data useful for designing treatment programs.
6. *Yielding taxonomies based on genetic relationship rather than simply on observed covariation*. Behavioral genetics analysis can help determine, for example, whether schizoaffective disorder is related to schizophrenia or the affective disorders or whether certain personality characteristics belong in a "spectrum" that shades into the abnormal. In the normal realm, questions about the genetic distinctiveness of, say, the Big Five personality dimensions or of cognitive skills can be addressed.
7. *Helping to understand how environmental influences operate in families* (see later discussion for more detail).

8. *Studying the extent of gene–environment interaction and covariation and identifying specific instances of such interaction and covariation.*
9. *Understanding more about the nature of developmental transitions in longitudinal analyses by documenting changes in heritability and the ways that different traits are interrelated during development.* This goal will probably be realized as growth curve and survival analyses are implemented more often in a behavioral genetics context.

It is noteworthy that these nine goals are all addressed by current methodology in behavioral genetics. As methodology becomes more sophisticated, the list should increase. Of course, behavioral genetics methodology does not fully address each of these issues. As in most areas of science, fuller understanding requires the joint perspectives of several methodologies. Included among the methodologies are epidemiology, physiology, and various paradigms for investigating the effects of experience.

Overcoming Barriers to Joint Nature–Nurture Investigation

Goals 7, 8, and 9, in particular, require the joint efforts of researchers who specialize in the nature and nurture of behavior. Unfortunately, there are remaining barriers to communication between these groups of researchers, although the barriers are dissipating rapidly as young researchers are trained. One fundamental barrier to communication is overinterpretation of the basic biometric model. Although the behavioral genetics framework is valuable for addressing at least the nine just-listed goals, it is not a comprehensive model for behavior. When we overinterpret the behavioral genetics model to try to answer questions beyond its scope, misunderstandings occur. The most frequent overinterpretation is the attempt to make process-oriented interpretations from nonsupplemented, traditional behavioral genetics data. Other misunderstandings involve failures to appreciate subtle distinctions between differences among individuals and the development of an individual.

Consider the following generic (hypothetical) criticism of behavioral genetics by developmental psychologists:

> Genes exert their effects on behavior via complex pathways. The biological systems involved in gene expression are highly interactive. Feedback loops, threshold effects, and various other highly contingent processes make these systems inherently nonlinear. Also, the nature of development is constructive, with behavior as an emergent property of systems that are context bound. Given this view of development, the basic assumptions of behavioral genetics are greatly oversimplified. It is unrealistic to assume that genetic and environmental effects can be disentangled. It is also unrealistic and misleading to assume that genetic effects are substantially linear, as behavioral geneticists typically do.

So, what is wrong with this hypothetical criticism? It thoroughly confuses individual development and individual differences. The statements about biological systems are probably true of the nature of individual development, but they are not oriented toward individual differences. Individual differences are the stuff of behavioral genetics, and classic behavioral genetics inferences are confined to genetic and environmental effects on phenotypic variance, not genes and environments per se. There is no contradiction in analyzing individual differences by linear regression of outcome on sources of variation, even when the individual differences result from highly contingent developmental processes operating in the life of individuals. In fact, psychologists frequently do analogous exercises. For example, both earlier IQ and quality of schooling might predict later academic achievement of children in a linear fashion. Computing the relevant regression and interpreting the partial regression coefficients is a legitimate and potentially useful exercise even though the actual learning experiences of the children were highly contingent, interpersonal, and context bound. Analyzing the nature of the contingencies and contextual influences is simply a different task.

Another barrier to communication and cooperation is captured by the common statement that "behavioral genetics methods slight the environment because they estimate environmental effects as what remains after genetic effects have been estimated." This statement does contain an element of truth, but it is overly broad and wrong in some contexts. The element of truth is that environmental influences specific to the individual (i.e., not shared by a cotwin, sibling, stepsibling, or whatever other kin is included in the design) are estimated as a residual term, and sometimes—but not always—the estimation is confounded with measurement error. Note that being a residual term does not mean that the magnitude of an effect is diminished artifactually. Indeed, the residual term that contains unshared environmental variation frequently accounts for the greatest portion of variance in behavioral traits.

Contrary to the implication of the statement that environment is "what remains" after genetic effects are estimated, typical behavioral genetics methods treat genetic and shared environmental effects in an evenhanded manner, given the assumptions about how these effects can be partitioned. The more cogent criticism is that genetic partitioning is based on sound theories of Mendelian inheritance, whereas environmental partitioning is based on familial units that might not be the most important markers of environmental influence.

In classic studies, neither genes nor environmental factors are measured directly. However, there is no necessary barrier to incorporating direct measures of both environments and genotypes into models. In fact, behavioral genetics models that were published in the mid-1970s allowed direct measures of the environment (e.g., Morton, 1974). On the near horizon are models that will incorporate direct measures of genes detected by molecular techniques as well as genetic factors inferred by resemblance among relatives.

This mention of molecular genetics brings up another barrier to understanding. One attitude seems to be that quantitative genetics, the standard methodology of current-day behavioral genetics, is only a handmaiden that will soon give way to the new molecular techniques. Again, this attitude contains an element of truth but obscures the larger perspective. It was once believed that biometrics (viewed as the study of continuous variation) and the principles of Mendel (dealing with the inheritance of discrete, single-gene traits) were incompatible. However, in the 1930s, a synthesis proved possible and led to much of the current understanding of genetics and evolution. A similar synthesis is likely molecular and quantitative methodologies. Despite the enormous importance that mapping the human genome will have for society in general and behavioral science in particular (Kevles & Hood, 1992; Shapiro, 1991), I believe it unlikely that molecular genetics will replace quantitative genetics as an empirical basis for understanding inherited effects on behavior. Application of the molecular techniques already requires much of the framework of quantitative genetics for inference about human behavior. Also, behavioral geneticists are hard at work developing new models to incorporate the effects of single genes understood at the molecular level (Vogler, 1992).

Environments: Shared and Nonshared, Familial and Extrafamilial

It is reasonably well documented that the modest similarity of siblings (as well as of parents and offspring) is mostly accounted for by shared genes rather than shared environmental factors for many aspects of personality and intellectual skills, especially when assessment is via paper and pencil. Although conclusions would be premature, emerging evidence from studies using objective behavioral assessment has not invalidated this notion. Despite extensive explication of the issue (Plomin & Daniels, 1987), the meaning of shared and nonshared environments continues to create confusion. Behavioral geneticists have unwittingly contributed to this confusion by sometimes referring to the shared environment as the "between-families," or simply the *familial* environment. However, all experience that is shared by a pair of relatives (say, cotwins) does not occur in the family context, and experiences within the family often affect family members differently.

Furthermore, the variance component referred to as shared environment differs from one kinship design to the next. Thus, the environment shared by cotwins has a somewhat different quality than the environment shared by ordinary siblings or adopted siblings. Of course, these differences in the quality of the shared environment might well be irrelevant to the behavior under study. Fortunately, this "armchair" criticism can be subjected to empirical test, and such tests are common in the more technical behavioral genetics literature (see Loehlin, 1992, for an accessible illustration in the personality domain).

Another statistic that is viewed as relevant to environmental effects is *vertical environmental transmission variance.* Along with genetic variance, vertical environmental transmission variance accounts for parent–offspring similarity. Conceptually, all factors (mechanisms operating inside or outside the family) that contribute to parent–offspring similarity are included in vertical environmental transmission variance. The processes underlying vertical environmental transmission variance and shared environmental variance of siblings could be similar (e.g., social class) or very different. Moreover, the estimate of vertical environmental transmission variance might well be zero at the same time that parents exert strong effects on the offspring behavior in question. For example, hypothetically, aggressiveness in fathers might induce inhibited behavior in children of a certain age. Such a hypothetical effect would not emerge in a univariate analysis of parent–offspring of either aggressiveness or inhibition. Of course, part of the solution is multivariate analysis, which has emerged with statistical and computing advances (e.g., McArdle & Goldsmith, 1990; Neale & Cardon, 1992). Still largely remaining is the task of integrating compelling theories of how the environment works into these multivariate designs.

In summary, the distinctions between shared and nonshared environmental effects and transmitted environmental effects in behavioral genetics designs do not map well onto some issues concerning the nature and effects of interaction among family members (Hoffman, 1991). However, the empirical findings highlighting the importance of nonshared over shared effects must now be accommodated by socialization researchers.

Objective and Effective Environments and Genes

The estimate of shared environment that emerges from a behavioral genetics analysis does not refer to common, overt experience but to the effect of that experience in creating similarity between relatives. For instance, suppose that we studied shyness in twins. A particular pair of identical twins might jointly experience, say, their family's move into a new neighborhood. However, if the effect of that move—for whatever reasons—is that one twin becomes more shy and the other less shy, then the seemingly common family event (moving to a new neighborhood) contributes to the nonshared environmental variance component in a twin analysis. Thus, classical behavioral genetics studies yield estimates of the effects of environments.

The same is true of the estimates of genetic variance from behavioral genetics studies. We know that identical twins, for instance, have identical structural genotypes, barring somatic mutations. However, due to genetic regulatory mechanisms, identical twins do not necessarily have identical effective (or functioning) genotypes when a behavior is measured, and it is the effect of the genotype that contributes to the genetic variance estimate in twin analyses. Of course, these principles also apply to family and adoption designs.

It should be emphasized that the recognitions about the environment in the previous paragraphs are universally appreciated among behavioral geneticists. However, it is evident that they have not always been adequately communicated to other psychologists.

Genetic Mediation of Environmental Effects and Vice Versa

Behavioral geneticists have long realized that gene expression is always environmentally mediated. One classic "textbook" illustration of this idea is that heritability increases in expressive environments (environments that allow full expression of the genotype) and decreases in restrictive environments.

Another illustration of the ubiquity of environmental mediation of genetic effects is the debate among behavioral geneticists about the usefulness of the idea of *active genotype–environment correlation* (Plomin, DeFries, & Loehlin, 1977). Although versions of this idea have captured the attention of many developmental psychologists (e.g., Scarr & McCartney, 1983), some behavioral geneticists believe that active genotype–environment correlation cannot be meaningfully distinguished from "direct" genetic effects. That is, even direct genetic effects are always instances of genes correlated with environments, although the environments might occasionally be entirely biological in nature. In the case of social environments, suppose that genotypic differences are correlated with, say, antisocial behavioral tendencies. These tendencies might be manifested, in part, by seeking peers who are experienced in antisocial behaviors themselves. The association with peers might be the most proximal influence on antisocial acts. This scenario would usually be characterized as active gene–environment correlation of the type in which the individual selects an environment on the basis of genetically influenced behavioral predispositions. But don't all genetic effects on behavior involve selection of relevant environments, in the sense that genes and their proximal and distal products must be expressed in a supportive context, where the "context" may range from the physiological to the social? Perhaps so, but it may still be useful to retain the concept of active gene–environment correlation for scenarios in which the environment is a measurable experience.

It is less clear that investigators who study the role of experience in individual differences have long appreciated that (a) the effects of experience may differ depending on genotype of the person undergoing the experience (Plomin & Bergeman, 1991) or (b) genetic relationships between family members may moderate the effects of their joint experience.

Political Barriers to the Acceptance of Behavioral Genetics

In this final section, I consider issues at the interface of science and politics—issues considered often by behavioral geneticists (Plomin, DeFries, & McClearn, 1993; Scarr & Kidd, 1983). The reason for including these considerations is that some controversy about behavioral genetics research is essentially political rather than scientific. The key issue is genetic determinism. This discussion does not deal with determinism at a philosophical level but will simply point out some features of the current debate. Practically all scientists who write about the issue agree that genetic influence should not be equated with lack of modifiability. Behavioral geneticists have used concepts that counter overly deterministic thinking (e.g., reaction range; see Turkheimer & Gottesman, 1991, for a recent explication of the concept). Even when deterministic thinking is replaced by probabilistic thinking, some social scientists object to the idea that genetic factors can be used to predict behavioral differences. These social scientists seem to say, in effect,

> We realize that genetic effects should not be equated with determinism, and we realize that behavioral geneticists view the effects of genes in a probabilistic manner. Unfortunately, however, documentation of genetic effects on behavior will be misunderstood by others, including persons with political power who might use the information to the detriment of disadvantaged persons.

This line of objection is heard frequently enough that it should be taken seriously.

At one level, the critics argue that scientific understanding of certain issues should not be pursued, and thus, some investigators tend to dismiss the criticism as antiscientific. Another perspective is that, regardless of traditional behavioral genetics research, the human genome mapping project and associated research will eventually force researchers to deal with documented genetic differences among humans, and avoidance of the issue is futile.

At the other level, a more sophisticated anti–behavioral genetics argument claims that the permissible inferences from human behavioral genetics designs are too abstract to be practically useful. Note that this line of criticism implicitly recognizes that the inferences from quantitative genetic designs are valid and can be immensely valuable when directed toward increasing the economic or practical value of agricultural species (Goal 3, at the beginning of this chapter). In considering this criticism, the reader must judge the importance of the nine goals enumerated at the beginning of this chapter.

Returning to the issue of genetic determinism, several issues and questions are sometimes overlooked. It would seem crucial to know what the general public, as well as political leaders, currently believe about the relative influence of inheritance and experience in molding behavior. It is not so clear that the public embraces experience over inheritance. Some of my experience suggests that an accurate description of current behavioral genetics findings to public groups outside academia often moves them toward a less hereditarian position. Of course, an accurate description of current behavioral genetics findings emphasizes that genetic effects on individual differences in most behaviors are quite moderate and that genes and experience are intimately intertwined during an individual's development.

Whatever the public's intuitions about genetics, one can also question the critics' contention that widespread appreciation of genetic influences on behavior would have negative consequences. There seems to be little objection to identifying the specific genes responsible for such diseases as cystic fibrosis, Duchenne muscular dystrophy, retinoblastoma, and some cases of early onset familial Alzheimer's disease. Indeed, identification of the genes that are responsible for these disorders is considered a step in designing effective treatment. Slightly more controversy attends genetic investigations of behavioral disorders such as schizophrenia and affective disorders. Nevertheless, few doubt the evidence that genes play a role in various forms of psychopathology. For example, in schizophrenia research, knowledge about genetics can aid in developing treatment approaches and evaluating familial risk (Gottesman, 1991). Another benefit of documenting genetic contributions to liability to psychopathology such as schizophrenia and childhood autism is relief of unfounded guilt in families of persons whose mental illness was previously blamed on family dynamics, usually centered on maternal deficiencies.

At present, there is extensive publicity about the origins of homosexuality, particularly male homosexuality. The scientific issue is unresolved. Interestingly, the media juxtaposes the issue of origins of homosexuality with the protection of civil liberties. Whether or not their position is well reasoned, various activist gay political groups are reportedly strongly invested in portraying homosexuality as biological or genetic—the distinction is often not respected. The notion is that biological roots legitimize homosexual orientation as "natural" and thus buttress legal arguments to extend the same civil protection to sexual orientation as currently afforded race, religion, national origin, and physical disability.

Of course, many would argue that the source of human difference is irrelevant to whether civil liberties should be protected. However, the relevant point here is that heritability (and genetic influence more generally) is viewed as a positive and desirable feature of homosexuality by these activist groups. Members of these groups might well view attempts to suppress genetic research on homosexuality as an implicit attack on their threatened civil liberties (Gelman, 1992).

I believe that these arguments suggest that there are instances in which knowledge or belief that genes affect behavioral differences relieve personal suffering and support a progressive political agenda.

One might object that most of the illustrations I have used pertain to behavioral problems rather than variation in adaptive behavior. What are the political consequences of further documenting the moderate effects of genes on differences in personality and cognitive abilities? This is a difficult question because, as mentioned earlier, one might suspect that the public as well as policymakers already believe that genes play a role in observed behavioral variation. Also, researchers have little experience to guide them. However, one can look to the physical domain for analogies. There is widespread consensus that American society overemphasizes physical appearance. What are the political ramifications of the widespread belief that physical appearance is largely hereditary in origin? Does believing that unattractive people are unattractive chiefly through the luck of their parentage rather than through experience affect the way society regards these persons?

Perhaps the question is better stated as a fear voiced by environmentalists: Belief that genes affect behavior will undermine intervention and educational efforts. It seems, however, that the rationale for intervention should be demonstrated effectiveness. Once effectiveness is documented, then ill-founded suppositions that are based on deterministic thinking (of either the genetic or environmental kind) should be put to rest.

One might ask what implication genetic differences hold for the type of intervention that might be effective. Perhaps the answer is that simple knowledge of heritability holds no such implication. A trait might be highly heritable (e.g., height) and still susceptible to a uniform intervention (e.g., improved nutrition). Although genotype–environment interaction is notoriously difficult to demonstrate, this may be partly because there is little information about which features of the genotype and environment might interact. As genetic processes are better understood, clues about relevant environments should emerge. Thus far, successes are confined to single-gene traits and the biochemical environment, as illustrated by the well-known case of phenylketonuria, in which the behavioral effects of the mutant gene are ameliorated by deleting phenylalanine from the diet.

A final question is, what should be the political implications of learning that persons differ in personality or skill partly because of inheritance? If inheritance is properly understood as involving active gene–environment correlation, and if one views society as obliged to facilitate adaptive development, then it would seemingly follow that society should support efforts to provide a menu of opportunities to individuals. Provision of such menus would probably require more rather than less support.

Conclusion

Enlightened genetic and environmental approaches to understanding behavior complement rather than conflict. Our vast ignorance of behavior cautions modesty on all sides.

References

Gelman, D. (with D. Foote, T. Barrett, & M. Talbot). (1992, February). Born or bred? *Newsweek*, pp. 46–53.

Goldsmith, H. H. (1988). Human developmental behavioral genetics: Mapping the effects of genes and environments. *Annals of Child Development, 5*, 187–227.

Gottesman, I. I. (1991). *Schizophrenia genesis: The origins of madness*. New York: Freeman.

Hoffman, L. W. (1991). The influence of the family environment on personality: Accounting for sibling differences. *Psychological Bulletin, 110*, 187–203.

Kevles, D. J., & Hood, L. (1992). *The code of codes*. Cambridge, MA: Harvard University Press.

Loehlin, J. C. (1992). *Genes and environment in personality development*. Newbury Park, CA: Sage.

McArdle, J. J., & Goldsmith, H. H. (1990). Alternative common factor models for multivariate biometric analyses. *Behavior Genetics, 20*, 569–608.

Morton, N. E. (1974). Analysis of family resemblance: I. Introduction. *American Journal of Human Genetics, 26*, 318–330.

Neale, M. C., & Cardon, L. R. (1992). *Methodology for genetic studies of twins and families.* Norwell, MA: Kluwer Academic.

Paterson, A. H., Lander, E. S., Hewitt, J. D., Peterson, S., Lincoln, S. E., & Tanksley, S. D. (1988). Resolution of quantitative traits into Mendelian factors by using a complete linkage map of restriction fragment length polymorphisms. *Nature, 335*, 721–726.

Plomin, R. (1986). *Development, genetics, and psychology.* Hillsdale, NJ: Erlbaum.

Plomin, R., & Bergeman, C. S. (1991). The nature of nurture: Genetic influences on environmental influences. *Behavioral and Brain Sciences, 14*, 373–427.

Plomin, R., & Daniels, D. (1987). Why are children in the same family so different from each other? *Behavioral and Brain Sciences, 10*, 1–16.

Plomin, R., DeFries, J. C., & Loehlin, J. C. (1977). Genotype–environment interaction and correlation in the analysis of human behavior. *Psychological Bulletin, 84*, 309–322.

Plomin, R., DeFries, J. C., & McClearn, G. E. (1989). *Behavioral genetics: A primer* (2nd ed.). New York: Freeman.

Scarr, S., & Kidd, K. K. (1983). Developmental behavioral genetics. In M. M. Haith & J. J. Campos (Eds.), *Handbook of child psychology: Vol. 2. Infancy and developmental psychobiology* (4th ed., pp. 345–433). New York: Wiley.

Scarr, S., & McCartney, K. (1983). How people make their own environments: A theory of genotype → environment effects. *Child Development, 54*, 424–435.

Shapiro, R. (1991). *The human blueprint: The race to unlock the secrets of our genetic script.* New York: St. Martin's Press.

Turkheimer, E., & Gottesman, I. I. (1991). Individual differences and the canalization of human behavior. *Developmental Psychology*, 27, 18–22.

Vogler, G. P. (1992). Partitioning genetic variance of quantitative traits using marker loci in twins [Abstract]. *Behavior Genetics, 22*, 760.

CHAPTER 18

The Need for a Comprehensive New Environmentalism

Frances Degen Horowitz

The idea that there is a "gap" to be bridged between nature and nurture in approaching research on behavioral development is conceptually flawed. It reflects a relatively simplistic understanding of the dynamics of human behavioral development. This is all the more surprising during an era when considerably more sophisticated conceptualizations involving the relationships of genetic and environmental variables in behavioral development are readily available. An understanding of the historical basis and the conceptual and methodological issues related to the persistence of simplistic analyses is important. A consequence of this understanding will be the development of conceptualizations of a new environmentalism that recognizes the complex interactions and transactions involving genetic, biological, and environmental variables as they contribute to behavioral development.

Historical Perspective and the Task at Hand

In the 1920s and 1930s, when modern genetics was but a glimmer on the academic horizon, the nature–nurture issue in psychology had already been boldly drawn. Gesell (1933) had made the case for nature, and Watson (1924) had made the case for nurture. By the 1940s, Watson's behaviorism had triumphed, and the case for nurture dominated. Twenty years later, however, the tables turned. Piaget's organismically (if not genetically) driven developmental view ascended (Flavell, 1963; Piaget, 1952). Chomsky (1965) made claims for a "nature-driven" language acquisition. The case for an organismic world view, if not a nature world view in the typical nature–nurture controversy, was well made. Overton and Reese (1973) declared that the two world views of organismic psychology and behavioristic psychology were inherently incompatible. A branch of behaviorists, who derived their sustenance from the research of B. F. Skinner, went off to apply principles informed by operant conditioning to a wide variety of behaviors and in a wide variety of settings and became the remaining, if lonely, champions of the nurture point of view (Martin & Pear, 1978). Their collective efforts provided strong evidence for the ability to manipulate and control behavior by the application of operant principles. The successes were especially dramatic with respect to individuals with mental retardation, developmental delay, and developmental disabilities, for whom there were demonstrations of significantly improved behavioral competence, developmental progress, and a capacity for independent behavior (Baer, 1973; Bricker, 1982; Hall & Broden, 1967). The use of the principles in the classroom was also shown to result in significant improvement in academic achievement (Hall, Lund, & Jackson, 1968; Ullman & Krasner, 1965).

At the same time, but along a quite separate intellectual trajectory, there was growing evidence that the broad outlines of normal development were deeply rooted in organismically controlled characteristics. Although Gesell's (1928) evidence on this point had been decisive much earlier with respect to motor development, results from research influenced by Piaget and Chomsky were suggestive of a strong universal patterning of development in cognition and language (Dale, 1976; Gelman & Baillargeon, 1983).

Yet Hebb (1949) had suggested that early experience affected developmental outcome. The findings on the influence of early experience supported Hebb's basic position. This evidence, as well as studies of enrichment at later ages on brain organization, learning, and resistance to stress, has provided for a considerable extension of Hebb's basic position (Diamond, 1988). An evaluation of the adult status of children who had tested as below normal in the infancy and preschool years, but who were subjected to an environmental intervention, showed that those who had experienced the intervention were functioning at a much higher level as adults than those who had not (Skeels, 1966). The efforts to translate earlier reports of such evidence into social policy fueled the establishment of Project Head Start as a program aimed at affecting developmental outcome (Zigler & Valentine, 1979) that was measured largely in terms of achievement in school.

It would appear that despite the very strong case being made for nature as being in charge of development, various sources of evidence and a belief in environmental influence on development were strong enough to affect social policy. Two reviews of the evidence from the earliest of these social policy–driven efforts came to opposite conclusions. Jensen (1969) declared that early intervention programs had failed to raise intelligence. Horowitz and Paden (1973) claimed that the effects of the intervention programs was proportional to their intensity.

Both conclusions continue to have some validity. Few of the subsequent studies of the effects of early intervention have shown massive, long-term impact on IQ scores, although in some studies there is some evidence for modest-to-moderate gains (Garber, 1988). On the other hand, environmental interventions during the preschool and early school years have been shown to make a significant difference on functional outcomes such as less need for special education, diminished incidence of delinquency, and higher probabilities of avoiding welfare in the early adult years (Consortium for Longitudinal Studies, 1983).

During this same period, the notion of development as the result of a "transaction" between the organism and the environment became a widely accepted metaphor. The metaphor was suggested by Sameroff and Chandler (1975) as a result of their review of the literature on postnatal development of infants born at risk. They concluded that environmental factors, more than the factors of physical compromise, were responsible for the ultimate level of developmental outcome among infants who were born with compromising medical and physical problems. Their analysis led them to suggest that the organism and the environment served as mutual influences, and they advanced the notion that developmental progress was the result of a transaction between the organism and the environment, with the environment mediated through care giving. In a curious way, the transactional model sidestepped the nature–nurture question. On the one hand, the notion of a transaction appears to be clearly one that seems to fall toward the nurture side of the explanation for developmental outcome. But it is often espoused by those who combine a transactional model with a systems theory analysis in such a manner as to seem to be making the case for nature (Sameroff, 1983).

At the same time as these developments were taking place, methodological advances in population statistics contributed to a growth of research in the field of behavioral genetics (Plomin, 1983). Evidence of greater similarities in a variety of behavioral areas among monozygotic twins, compared with dizygotic twins and nontwin siblings, and studies that compared adopted and nonadopted populations had strengthened the case for the influence of nature over nurture in development. The most responsible of the investigators in this research area have taken great pains to note the limitations of the claims that can be made from the data on behalf of genetic control of behavior (Plomin, 1989). Nevertheless, some less responsible investigators and the popular hunger for statements of certainty have fed the resurgence of tipping the scales more toward the nature point of view, with great potential for that point of view to influence social and educational policy.

The interpretation of scientific evidence and the effect of the interpretation never occurs in a social vacuum. At any period in history, evidence from scientific investigations is used to support a variety of social and political agendas (Horowitz, 1991; Kamin, 1974). This is now especially the case at a time when communities across the United States are facing so many pressing and seemingly intractable social and economic problems. It is therefore incumbent upon behavioral scientists to approach the nature–nurture issues critically, carefully, and with an eye to preventing oversimplified interpretations from becoming the basis of ill-conceived social policies.

In light of this, it is both troubling and consequential that the nature–nurture issue has been drawn too simply and too facilely given the inherent complexity of development—especially human development. The fullest recognition of the complexity of the nature–nurture controversy requires clarification of both conceptual and methodological issues.

Conceptual Issues in the Nature–Nurture Controversy

Some of the confusion in the nature–nurture controversy revolves around the use of the term *development* to refer to both individual development and generic characteristics of behavioral development. For this reason, I have chosen to make reference to behavioral development as involving *universal* and *nonuniversal* behaviors (Horowitz, 1987). The choice of terminology here purposely avoids the use of the terms *phylogenetic* and *ontogenetic*, which might simplistically be seen as comparable. The notion of universals and nonuniversals in behavioral development is not analogous to phylogenetic and ontogenetic even though the claim for their being nonanalogous might for some appear to rest on minor distinctions.

In the structural/behavioral model (Horowitz, 1987), the universal behaviors are those that are "acquired" by all normal humans in all minimally *normal environments*. (The use of italics here is meant to signify that in addition to this being a conceptual issue, the terms involved also signify a methodological issue as discussed later.) Although opportunities to learn are involved in the acquisition of these behaviors, the probability of their being acquired by normal humans in normal environments is close to 100%. The "genetic" contribution to the universal behaviors is of two kinds: that which is species specific and that which is based on the individual's biological parents. The species-specific contribution is, itself, the result of the genetic characteristics of the species as well as prenatal developmental processes that occur in an environmental context (Gottlieb, 1992; Kuo, 1967; Oyama, 1985). Human speech and human motor characteristics (with the underlying human morphology typical of the species) exemplify the universals that typify all normal humans. The genetic contributions from biological parents determine some of the individual differences in the universal behaviors.

The nonuniversal behaviors will only be acquired as a result of opportunities to learn. There is no 100% probability of acquisition unless the environment provides the proper conditions. These conditions involve the presence of the universal behaviors necessary for the acquisition of the nonuniversal behavior; they also involve the individual differences that are the result of both genetically influenced characteristics and environmentally determined individual histories as well as appropriate, environmentally provided, opportunities to learn.

There is increasing evidence that genes influence only the expression of characteristics in an environmental context and that the environmental context contributes to individual differences even among the behaviors in the universal behavioral repertoire (Gottlieb, 1992; Oyama, 1985). Thus, one may ask whether there is any sound conceptual basis for assigning behavioral outcomes to genetic as opposed to environmental contributions except in those instances in which clear genetic aberrations and abnormalities render the environmental context ineffective or severely limit its influence.

Gene expression in an environmental context requires an understanding of how the transaction occurs and is regulated. In some instances, the role of the environment may be largely an enabling one whereby as long as a normal environmental context is present the genetic contribution will be relatively unvarying. For example, eye color is an expressly genetic characteristic inherited from biological parents. The color in the proper environmental context will be expressed. However, in behavioral areas involving such things as temperament or even alcoholism, genetic factors, biological factors (which is not synonymous with genetic), environmental factors, and the multiple levels of interaction of these factors with each other all need to be understood to account for these characteristics.

From this point of view, any discussion of "bridging the nature–nurture gap" is conceptually flawed and is scientifically obsolete.

More than 40 year ago, Anastasi talked of "nature and nurture" (Anastasi, 1958; Anastasi & Foley, 1948). The modern version, in light of recent advances in biological genetics, recommends that researchers view the effects of nature (genes) factors in the context of nurture (environment) factors and view the effects of nurture factors in the context of nature factors. In other words, development of the universal and nonuniversal behavioral repertoires is the result of multivariate processes, many of which are nonlinear in their functional relationships. Recent advances in genetics support this conceptual approach (see Gottlieb, 1992).

Methodological Issues in the Nature–Nurture Controversy

There are several methodological issues that follow from a discussion involving an attempt at conceptual clarification. They involve questions of how one measures the environment and questions of how one measures genetic influence.

Measuring the Environment and Its Influence

There are many more standard and accepted measures of individual behavior than there are of environments. One has only to attempt a study of environmental influences on human infant behavior, even when the environment is defined in terms of parental care-giving behavior and environmental provisions made by parents for their children, to realize that there is no consensus about standard measures of the environment. Except for the Home Observation for Measurement of the Environment Scale (Bradley, Caldwell, & Elardo, 1979), there is no standard measure of home environments. Most investigators devise their own measurement instrument to do observations of child/environment interactions.

There is also no agreement about how best to parse or define the environment in which a person is growing up. There are no standard units of environmental variables, and the definitions of developmentally facilitative environments find no consensus. At best, the efforts to manipulate environmental variables focus on gross dimensions; at worst, the efforts to define and manipulate the environment are laden with value judgements about good (positive/effective) and bad (negative/ineffective) that have minimal empirical support.

For obvious reasons, investigators using animals have made considerably more progress on defining and manipulating environmental variables (Gottlieb, 1978; Kuo, 1967; Marler, Zoloth, & Dooling, 1981). It is this kind of research that leads one to appreciate how much of behavior and behavioral development, which was heretofore considered "innate" and unalterably genetically influenced, is subject to environmental influence in its development and function. These results illuminate the critical role of environmental variables in developmental processes that heretofore appeared to be under automatic control of organismic variables.

The task of trying to understand exactly how environmental variables function in human behavioral development will be a difficult and complex one. Central to the effort will be better and more standardized measurements of the environment. Two schemes for parsing levels of the environment have been proposed (Bronfenbrenner & Crouter, 1983; Horowitz, 1987). They each segment the environment in terms of levels of complexity but not much progress has been made toward the development of standard measuring instruments for any of the levels. The scheme proposed by Horowitz (1987) involves thinking about the environment as ranging in complexity from single stimuli to the cultural patterning of environmental experience. It is at the cultural level that researchers will encounter the greatest challenges for useful measurements, but it is probably at the level of cultural patterning that the most powerful and subtle environmental variables may well operate.

Measuring Genes and Their Influence

The methodological strategies that have been used in behavioral genetics studies of behavior and developmental outcome have gained wide acceptance. Yet at best, they are distal measures that require a great deal of population selection and extensive statistical analysis before any interpretation of the results can be made. They require the use of unsupported assumptions about the constancy of shared environments in families (Hoffman, 1991). The utility of these studies is extremely limited for researchers who are interested in illuminating the contributions that genetic factors make to individual behavior and developmental outcome.

Researchers need to question whether there is much more to be gained from further studies involving the current techniques used in human behavioral genetics research that are aimed at evaluating the role of genes in individual behavioral development. It is unlikely that continuing with the present line of human behavioral genetics research will allow for a better understanding of the role of genes in behavioral development. Progress in understanding the role of genes in developmental processes is now more likely to be made in animal research for which direct experimental control and manipulation of the relevant variables is possible.

Reconceptualizing Environmentally and Genetically Oriented Investigations to Common Ground and Common Goals

Methodological issues cannot always be separated from conceptual issues. This is particularly the case with respect to questions involving the contribution of the environment and genes to behavioral development. If, in fact, genetic material is expressed only in the context of environment, and if genetic expression differs as a function of environmental context, then attempts to evaluate the influence of genes on behavioral development independent of the environment are conceptually flawed. It may be much more fruitful to abandon questions that seek answers about the sources of influence and focus instead on the processes that are involved in development and on questions of how genetic and environmental variables function to account for development and developmental outcome. Anastasi called for a similar research program in 1958! In other words, the goal of ascribing portions or percentages of outcome variance to environmental and genetic sources should be relinquished. Rather, researchers should be asking how genetic and environmental variables "coact" and how manipulations of genetic and environmental coactional systems alter outcome. It is unlikely that the current techniques used in behavioral genetics will result in answering these kinds of questions.

In seeking a common ground for environmentally and genetically oriented investigations, genetically and environmentally oriented research need not cease. However, the conceptual and methodological considerations reviewed here do suggest that there is little more to be gained from the research that is now described as behavioral genetics as it is conducted with current methodologies, measures, and assumptions. As already noted, the real progress in understanding how genetic factors influence development and how environmental factors condition genetic contributions will not be found in behavioral genetics research but in gene–environment studies that are focused on the coacting nature of variables. The distal nature of current behavioral genetics research does not really reveal anything about the contribution of genes to behavior because the efforts have been aimed at estimating the portion of the outcome variance that can be attributed to genetic influence. These efforts are based on a conceptually flawed model.

The suggestions being made here are not meant to negate the contributions of behavioral genetics to date. They have provided a useful antidote to unbridled environmentalism. They have been suggestive of sources of influence. But a conceptually more adequate model would use the notion of *constitutional* influences rather than genetic influences. *Constitutional* and *genetic* are not synonymous descriptors; nor are *organismic* and *genetic*. The terms constitutional and organismic do not exclude environmental influences. In fact, if behavioral genetics investigators would discuss their findings as suggestive of constitutional rather than genetic contributions, then important conceptual progress would result. This would be particularly helpful in studies for which the dependent variables have been relatively uncontroversial, such as in the instances of alcoholism and depression.

Reconceptualizing behavioral genetics research as research involving constitutional contributions to behavior and development would greatly enhance the search for an understanding of the multiple contributions of organismic and environmental variables to the processes that result in particular behaviors and particular developmental outcomes. The advances in biological genetics make it increasingly inappropriate to believe that there is a single, particular gene for any given behavior or developmental outcome except in cases of clear genetically driven syndromes such as Down's syndrome. The current understandings make it meaningless to believe that there exists a gene for alcoholism, a gene for depression, a gene for schizophrenia, and most certainly, a gene for intelligence. Yet, it is undeniable that there are constitutional contributions in all of these areas and characteristics, provided that one defines constitutional as reflecting effects of environmental variables on the organism. The strategy I suggest is one that can be considered a *comprehensive new environmentalism*. In using this conceptual approach, the understanding, manipulation, and control of developmental outcome must be sought by considering relationships in multiple realms.

How to Proceed

The basic task at hand is to reframe the nature–nurture question in terms of a comprehensive environmentalism that recognizes that development is the result of complex processes involving organisms and environments. Genetic material is involved in those processes. Environmental contributions are involved in those processes. Genetic material does not get expressed nor does it influence the processes in the absence of an environment. The question is not one of bridging the gap between nature and nurture because there is no gap to be bridged. The very notion of a gap calls forth images of getting from one side to the other. But nature and nurture are not end points separated by the chasm of our ignorance. Nor are they points on a continuum. Rather, they are terms that now require redefinition and that must be reconceptualized with respect to the processes of development and with respect to methodological strategies for studying behavioral development. Scientific progress in understanding behavioral development and the determinants of developmental outcome requires the adoption of new and conceptually different kinds of questions. Until there is some consensus about how to frame such questions, there will be no end to a debate that has been fruitless at best and scientifically misguided and socially destructive at worst.

References

Anastasi, A. (1958). Heredity, environment, and the question "How?" *Psychological Review*, *65*, 197–208.

Anastasi, A., & Foley, J. P. (1948). A proposed reorientation in the heredity–environment controversy. *Psychological Review*, *55*, 239–249.

Baer, D. M. (1973). The control of developmental process: Why wait. In J. Nesselroade & H. W. Reese (Eds.), *Life-span developmental psychology: Methodological issues* (pp. 185–193). New York: Academic Press.

Bradley, R. H., Caldwell, B. M., & Elardo, R. (1979). Home environment and cognitive development in the first 2 years: A cross-lagged panel analysis. *Developmental Psychology, 15*, 246–250.

Bricker, D. (1982). *Intervention with at-risk and handicapped infants.* Baltimore: University Park Press.

Bronfenbrenner, U., & Crouter, A. C. (1983). The evolution of environmental models in developmental research. In W. Kessen (Ed.), *Handbook of child psychology: Vol 1. History, theory and methods* (4th ed., pp. 357–414). New York: Wiley.

Chomsky, N. (1965). *Aspects of a theory of syntax.* Cambridge, MA: MIT Press.

Consortium for Longitudinal Studies. (1983). *As the twig is bent: Lasting effects of preschool programs.* Hillsdale, NJ: Erlbaum.

Dale, P. (1976). *Language development* (2nd ed.). New York: Holt, Rinehart & Winston.

Diamond, M. C. (1988). *Enriching heredity.* New York: Free Press.

Flavell, J. (1963). *The developmental psychology of Jean Piaget.* Princeton, NJ: Van Nostrand.

Garber, H. L. (1988). *The Milwaukee Project.* Washington, DC: American Association on Mental Retardation.

Gelman, R., & Baillargeon, R. (1983). A review of some Piagetian concepts. In J. H. Flavell & E. M. Markman (Eds.), *Handbook of child psychology: Vol. 3. Cognitive development* (4th ed., pp. 167–230). New York: Wiley.

Gesell, A. (1928). *Infancy and human growth.* New York: Macmillan.

Gesell, A. (1933). Maturation and the patterning of behavior. In C. Murchinson (Ed.), *A handbook of child psychology.* Worcester, MA: Clark University Press.

Gottlieb, G. (1978). Development of species identification in ducklings: Change in species-specific perception caused by auditory deprivation. *Journal of Comparative and Physiological Psychology, 92*, 375–387.

Gottlieb, G. (1992). *Individual development and evolution: The genesis of novel behavior.* New York: Oxford University Press.

Hall, R. V., & Broden, M. (1967). Behavior changes in brain-injured children through social reinforcement. *Journal of Experimental Child Psychology, 5*, 463–479.

Hall, R. V., Lund, D., & Jackson, D. (1968). Effects of teacher attention on study behavior. *Journal of Applied Behavior Analysis, 1*, 1–12.

Hebb, D. O. (1949). *The organization of behavior.* New York: Wiley.

Hoffman, L. W. (1991). The influence of the family environment on personality: Accounting for sibling differences. *Psychological Bulletin, 110*, 187–203.

Horowitz, F. D. (1987). *Exploring developmental theories: Toward a structural/behavioral model of development.* Hillsdale, NJ: Erlbaum.

Horowitz, F. D. (1991, May). *Combining behavioral and organismic approaches: Implications for research and clinical practice* [Invited address]. Presented at the annual meeting of the Association for Behavior Analyses, San Francisco, CA.

Horowitz, F. D., & Paden, L. Y. (1973). The effectiveness of environmental intervention programs. In B. M. Caldwell & H. N. Ricciuti (Eds.), *Review of child development research* (Vol. 3, pp. 331–402). Chicago: University of Chicago Press.

Jensen, A. R. (1969). How much can we boost I.Q. and scholastic achievement? *Harvard Educational Review, 39*, 1–123.

Kamin, L. J. (1974). *The science and politics of IQ*. New York: Wiley.

Kuo, Z. Y. (1967). *The dynamics of behavior development.* New York: Random House.

Marler, P., Zoloth, S., & Dooling, R. (1981). Innate programs for perceptual development: An ethological view. In E. Gollin (Ed.), *Developmental plasticity* (pp. 135–172). New York: Academic Press.

Martin, G., & Pear, J. (1978). *Behavior modification.* Englewood Cliffs, NJ: Prentice-Hall.

Overton, W. F., & Reese, H. W. (1973). Models of development: Methodological implications. In J. R. Nesselroade & H. W. Reese (Eds.), *Life-span developmental psychology: Methodological issues* (pp. 65–86). New York: Academic Press.

Oyama, S. (1985). *The ontogeny of information.* Cambridge, MA: Cambridge University Press.

Piaget, J. (1952). *The origins of intelligence in children.* New York: International Universities Press.

Plomin, R. (1983). Developmental behavioral genetics. *Child Development, 54*, 253–259.

Plomin, R. (1989). Environment and genes: Determinants of behavior. *American Psychologist, 44*, 105–111.

Sameroff, A. (1983). Developmental systems: Contexts and evolution. In W. Kessen (Ed.), *Handbook of child psychology: Vol. 1. History, theory and methods* (4th ed., pp. 237–294). New York: Wiley.

Sameroff, A., & Chandler, M. (1975). Reproductive risk and the continuum of caretaking casualty. In F. D. Horowitz (Ed.), *Review of child development research* (Vol. 4., pp. 187–244). Chicago: University of Chicago Press.

Skeels, H. M. (1966). Adult status of children with contrasting early life experiences: A follow-up study. *Monographs of the Society for Research in Child Development, 31*(Serial No. 105).

Ullman, L. P., & Krasner, L. (Eds.). (1965). *Case studies in behavior modification.* New York: Holt, Rinehart & Winston.

Watson, J. B. (1924). *Behaviorism.* New York: Norton.

Zigler, E. F., & Valentine, J. (Eds.). (1979). *Project Head Start: A legacy of the war on poverty.* New York: Free Press.

CHAPTER 19

The Question "How?" Reconsidered

David C. Rowe and Irwin D. Waldman

In her 1986 presidential address to the Behavior Genetics Association, Sandra Scarr (1987b) celebrated how successfully behavioral genetics had become part of the mainstream of the social sciences. She wrote, "The mainstream of psychology has joined our tributary, and we are in danger of being swallowed up in a flood of acceptance" (p. 228). Behavioral genetics studies were being published in developmental psychology journals rather than only in specialized publications (Rowe, 1987), and social scientists in general were paying greater attention to behavioral genetics research findings. Thus, the fields of nature and nurture were moving together.

But despite the sense of progress trumpeted in Scarr's (1987b) speech, environmentally oriented[1] researchers and behavioral geneticists

[1]Following Wachs (1983), we use the term *environmentally oriented* to refer to researchers who are primarily interested in the effects of specific environmental aspects and mechanisms on specific developmental outcomes.

remain apart. The separation can be seen, for instance, in most introductory child development textbooks. Behavioral genetics theory, data, and methods are usually treated in an isolated chapter, and once introduced, they are promptly forgotten in the coverage of other topics. Various behavioral genetics research designs (e.g., such as twin and adoption studies) are used by a relatively small cadre of behavioral geneticists; they are not used by the majority of psychologists. A recent statement of a behavioral genetics view of child development (Scarr, 1992) elicited strong critiques from environmentally oriented scholars (Baumrind, in press; Bronfenbrenner, Lenzenweger, & Ceci, 1993).

Hindered Communication Between Research Orientations

There are several reasons for a continuing estrangement between behavioral geneticists and environmentally oriented researchers. Political reasons are external to the practice of science; methodological and conceptual reasons are internal to it. Although in this chapter we focus on the latter sources of estrangement, we acknowledge that genetic explanations of individual differences have been a flash point for dispute over the potential social malleability of behavioral traits (Degler, 1991). Moreover, since Galton (1869) first advocated eugenic measures to improve the human species genetically, behavioral genetics has been popularly associated with conservative social policies and with the direct manipulation of human reproduction (Kevles, 1985). For these reasons, some social scientists prefer to avoid behavioral genetics methods because they believe, rightly or wrongly, that behavioral genetics approaches to understanding development are too ideologically unpalatable. Ideological and political disputes continue to separate behavioral geneticists from other behavioral scientists, but these disputes may be less important now than they were previously because heritability is no longer equated with a lack of trait malleability. Nevertheless, political debates may reignite over the genetic bases of racial and ethnic differences or over the potential benefits of changing child-rearing styles in influencing developmental outcomes.

Ideological views are unlikely to be swayed by bodies of data, but other sources of methodological and conceptual separation may be more amenable to efforts at increasing communication and collaboration between environmentally oriented researchers and behavioral geneticists. With this in mind, we have identified several areas of conceptual misunderstanding in which researchers can strive to bridge the nature–nurture gap. They stem from Anastasi's (1958) complaint that behavioral genetics is too concerned with "how much" variation in a trait is environmental or genetic and not enough concerned with the fruitful question of "how" genetic and environmental influences on a trait interact in development.

The first area is the emphasis in behavioral genetics on abstract components of variance, with its associated heavily mathematical treatment of data. Heritability is an index of the proportion of a trait's total variation that is due to genetic variation among the members of some population. This variation results from the substitution of one allele for another at each of a large number of genetic loci relevant to the trait. But behavioral geneticists are as yet unable to identify the specific loci affecting quantitative traits in humans. Although new molecular genetics methods may permit their identification in the future, for the present, they are unknown.

In biometrical genetic models, environmental variation is divided into that due to shared environmental influences, symbolized as c^2, and that due to nonshared environmental influences, symbolized as e^2. Shared environmental influences are experienced in common by family members and increase their similarity on a trait, whereas nonshared environmental influences are experienced uniquely by each family member and cause them to differ on a trait. Measurement errors are contained in the nonshared environmental component of variance. In biometrical models, the difference between the monozygotic (MZ) twin correlation and a trait's statistical reliability estimates the nonshared variance component not due to measurement error. This component is about 10–20% of the variance in childhood IQ and 20–30% of that in nonintellectual personality traits.

As is true for genetic influences, a behavioral genetics study may identify shared and/or nonshared environmental influences without specifying the mechanisms underlying these influences. This gives environmentally oriented researchers the impression that behavioral geneticists are not seriously studying the environment. It also reinforces a widespread belief that behavioral geneticists are interested only in how much variation is environmental or genetic and not in how environmental or genetic processes operate in development.

The use of components of variance raises another block to collaboration: their population dependence. As is well known, heritability estimates vary from one study to another, trivially from sampling variation, but more critically because true genetic or environmental variation can differ among populations. This population specificity is, of course, equally true of parameter estimates of shared and nonshared environmental influences. Nevertheless, the abandonment of these estimates is sometimes advocated because they lack universal applicability. In this vein, Baumrind (1991) argued: "At best the linear model that estimates heritability is a *local* analysis that pertains to the actual distribution of genotypes and environments in the particular population of twins or adoptees sampled" (p. 387, italics added). Critics of heritability seem to make the inferences that (a) such estimates are necessarily nongeneralizable and (b) "local" results are ones also useless and without practical value.

The second criticism follows from the common lack of specific environmental measures in behavioral genetics studies. Environmentally oriented researchers developed both observational and self-report measures of family environment, and they continue to explore methods of improving environmental measures. Many times, behavioral geneticists seem content to calculate estimates of environmental influence in a study that lacks a single direct measure of the environment. Thus, the behavioral genetics results do not appear to address the concerns of the environmentally oriented researcher, and vice versa. Indeed, with only measures of abstract variance components, how can behavioral geneticists consider developmental processes?

A third complaint is not about method but about results. Behavioral geneticists have found few cases in which shared environment (c^2) substantially influences behavioral variation (Rowe, 1990). Notable exceptions include shared environmental influences on childhood IQ, antisocial behavior, and psychiatric disorders such as depression and alcoholism. If environmentally oriented researchers dislike behavioral genetics approaches, then it may be because their findings challenge the notion that favorite environmental variables create variation between families rather than within them.

Consider a simplified example: *Number of books in the home* is a shared environmental influence that should make children in one family more intelligent than those in another family. Yet for childhood intelligence, estimates of shared environmental influences (approximately 25%) are small in relation to estimates of heritable influences (approximately 50%); and for middle-class families, the shared environmental influences appear to be almost negligible during adolescence and adulthood (Plomin, 1986; cf. Scarr, Weinberg, & Waldman, in press). Hence, family differences in number of books is likely not to be a substantial source of differences in children's IQs because its influence is necessarily limited by the environmental variance.

Environmentally oriented researchers are not the only ones to greet these results with some consternation. Richard Rose, a respected and widely published behavioral geneticist, expressed his disbelief over the lack of shared environmental influences and argued that its lack defies common sense (Rose, Kaprio, Williams, Viken, & Obremski, 1990). In their large Scandinavian twin studies, Rose and colleagues found evidence for a particular kind of shared influence—one due to twin siblings' mutual contact. But sibling mutual effects are not "traditional" familial environmental influences, as are parental rearing styles that are thought to strongly direct children's behavioral outcomes.

To draw these lines of argument together, environmentally oriented researchers contend that behavioral geneticists (a) ignore the contextual dependence of estimates of genetic and environmental influences, (b) neglect the effects of specific environmental mechanisms on developmental processes, and (c) produce results threatening to mainstream developmental theory.

Actual or Potential Areas of Common Ground for Environmentally Oriented Researchers and Behavioral Geneticists

We believe that the best solution to these sources of separation is not for behavioral geneticists to deny the concerns of environmentally oriented researchers but rather for them to propose research designs that can address these concerns with relevant empirical data. Waldman and Rowe (1993) described in detail a variety of research strategies for bridging the nature–nurture gap. Our designs are illustrated using data from twins and full siblings– and half siblings–groups that are relatively more accessible and less expensive to study than are rare adoptees and that are relatively more informative genetically than are distant biological relatives.

In our examples, we propose several extensions of a multiple regression approach to analyzing twin data proposed by DeFries and Fulker (1985). Although the multiple regression approach does not yield different information from maximum-likelihood model fitting (Cherny, DeFries, & Fulker, 1992), it is simple to use, flexible, and provides results that are easily communicated to a wider community of social science researchers. The DeFries and Fulker equation takes this form:

$$P_1 = b_1P_2 + b_2R + b_3P_2R + b_0, \qquad (1)$$

where P_1 is the phenotype of one sibling, P_2 is the phenotype of the other sibling, R is the coefficient of relation corresponding to the level of genetic relatedness (i.e., 1.0 for MZ twins, 0.5 for dizygotic [DZ] twins and full siblings, and 0.25 for half siblings), and P_2R is the product of the siblings' phenotype and the coefficient of relationship.[2] The equation's intercept is b_0. The unstandardized regression coefficient, b_3, for the interaction

[2]In its original formulation, this regression model was proposed for selected samples of proband and cotwin (i.e., samples in which a twin pair was selected because at least one twin was extreme on the trait being studied). In our extensions, we have proposed regression models for nonselected sibling pairs. Nonselected data are "double entered" so that each sibling is once the dependent variable (P_1) and once the independent variable (P_2). The sample size is then twice the number of sibling pairs. Standard errors given in computer runs must be corrected to the true sample size (i.e., the number of sibling pairs).

term (P_2R) directly estimates heritability because it indicates the degree to which sibling similarity on the trait differs by level of genetic relatedness. Hence, it is similar to other methods of estimating heritability that are based on comparisons of correlations between classes of relatives (e.g., MZ and DZ twins) that differ in genetic relatedness (Falconer, 1989). The unstandardized regression coefficient, b_1, for siblings' phenotype (P_2) directly estimates the shared environmental influences because it indicates the part of sibling similarity not due to genetic influences. The expectation of the b_2 regression weight is more complex and is not used in most applications of this method.

Environmental Context and Estimates of Genetic and Environmental Influences

Estimates of both genetic and environmental components of variation may depend on environmental context. Of course, not every new environmental context can be anticipated. As shown by the internal collapse of Soviet communism, a change unanticipated by most departments of political science, human affairs always contain a large unpredictable element. Nonetheless, the focus of most social science research is on environmental contexts as they exist now, on what Robert Plomin has called "what is" versus "what can be." Behavioral geneticists are most interested in the extent to which changes in genetic and environmental influences are a function of differing environmental contexts in existing societies with their current social traditions, technologies, and economic practices. The critics of behavioral genetics studies suggest that heritability estimates may be extremely limited and that they may apply only to middle-class individuals. Although this may be true, the extent of variation in environmental and genetic parameters by context is an empirical issue. If shared environmental influences on achievement or antisocial behavior emerge in working-class communities or among the urban poor, then theories of development should understand why. Public policies, in particular, depend on good information about genetic and environmental influences in specifiable social contexts.

To discover the range of applicability of genetic and environmental parameters, studies need to sample large numbers of subjects from diverse population groups; this goal is realistically attainable only if efforts are shared among research groups. Simple analytic techniques, such as the DeFries and Fulker (1985) regression approach, can be modified to include variables representing social contexts, such as communities' social class, urbanization, and ethnic composition. Using *EC* to represent some specific measure of environmental context, several terms can be used to augment the basic DeFries and Fulker regression equation:

$$P_1 = b_1P_2 + b_2R + b_3P_2 \bullet R + b_4EC + b_5P_2 \bullet EC + b_6R \bullet EC + b_7P_2 \bullet R \bullet EC + b_0, \quad (2)$$

where $P_1, P_2, R, P_2 \bullet R, b_1, b_2, b_3$, and b_0 are as in Equation 1. In this augmented regression equation, *EC* is a specific measure of some environmental context, $P_2 \bullet EC$ is the two-way interaction of siblings' phenotype and the environmental context measure, $R \bullet EC$ is the two-way interaction of the coefficient of relationship and the environmental context measure, and $P_2 \bullet R \bullet EC$ is the three-way interaction of siblings' phenotype, the coefficient of relationship, and the environmental context measure. The unstandardized regression coefficient, b_7, for the three-way interaction indicates the degree to which heritability varies by environmental context, whereas the coefficient b_5 indicates the degree to which shared environmental influences vary by environmental context. Thus, both terms (b_7 and b_5) provide evidence on which environmental conditions may modify genetic or environmental influences.

One difficulty with detecting moderating effects of environmental context is the low statistical power associated with interaction terms in regression equations (e.g., Wahlsten, 1991). McClelland and Judd (1993) provided a useful insight into conditions influencing our ability to detect statistical interactions. They showed mathematically that, when testing an interaction term for significance, statistical power depends on the joint distribution of the two variables forming the product. In the example just given of shared environmental influences, this interaction is the product of environmental context and sibling phenotype. When most cases fall into the four corners of this joint distribution, statistical power is high. As in a well-designed experiment, the ideal design occurs when each person falls into one of four extremes on the environmental context and phenotype: low–low, high–low, low–high, and high–high. Figure 1 illustrates this situation. The x axis represents environmental context; the y axis represents the trait phenotype. The bar height is proportional to the number of cases. In the upper panel of Figure 1, the statistical power is excellent, and the number shown (1.0) represents the efficiency of this design in relation to an optimal design. In the bottom panel of Figure 1, the statistical power is poor (relative efficiency = 0.06) because the joint distribution is approximately multivariate normal, with the majority of cases near the center of the bivariate plot. McClelland and Judd's research carries an important implication: Field studies must be designed to oversample extremes and undersample distributional centers if interaction effects are to be found. Moreover, a good sampling procedure must guide the study; statistical legerdemain will not compensate for data in which the joint distribution on the variables entering into the interaction is already poor.

There is a lack of data on whether environmental contexts substantially modify heritability (h^2) or shared and nonshared environmental influences (e.g., c^2 and e^2, respectively), but the methodological tools to find the answer do exist. Because distributional extremes tend to have great social implications (regardless of whether they are the contributions to society of productive geniuses or the harm to society of rapacious criminals), understanding genetic and environmental components at these extremes, using measured qualities of environmental contexts, is an important future goal for collaboration between environmentally oriented researchers and behavioral geneticists.

Specific Environmental Measures as Mediators of Genetic and Environmental Influences

Latent variables are not much use to developmental science unless they are tied to developmental theories and measured variables. As a latent variable, heritability represents the effects of allelic substitutions at a multitude of genetic loci; but the particular genes themselves are unidentified. To understand the relevant neurobiology underlying a trait, the genes must be identified; and around the country, behavioral geneticists

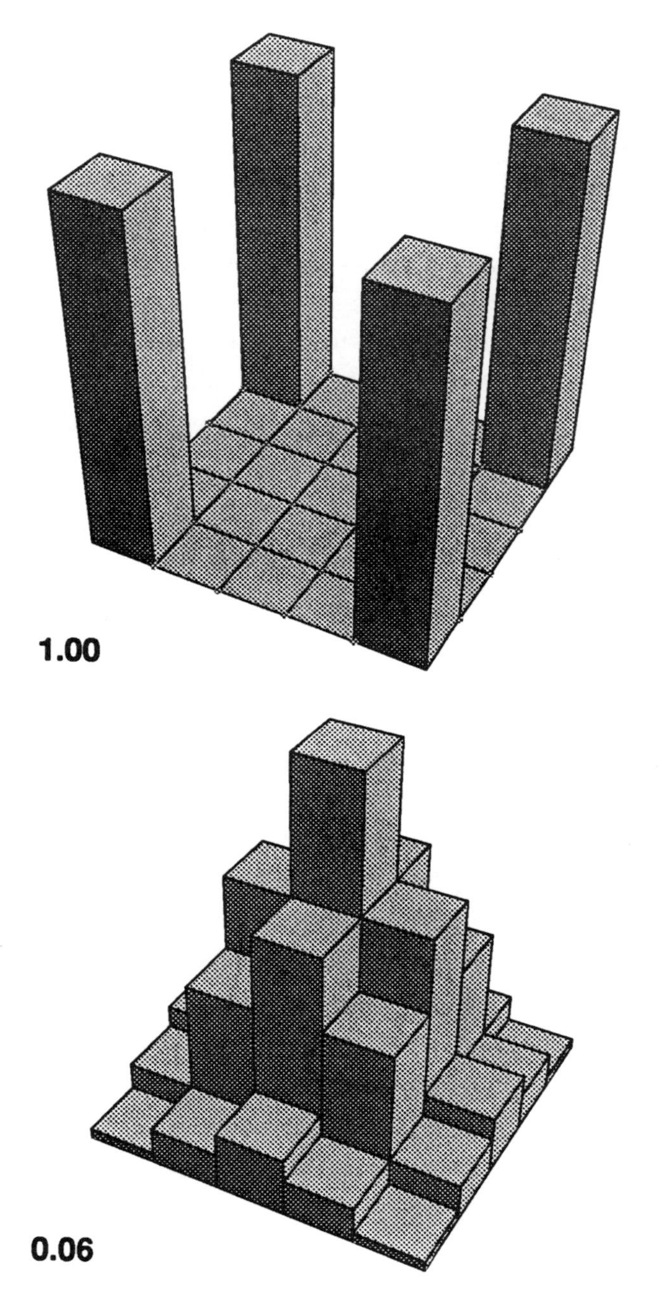

FIGURE 1. Power analysis for joint distributions of phenotype and environmental context. (Top panel, relative efficiency of design = 1.00; bottom panel, relative efficiency of design = 0.06.)

and molecular geneticists have begun to collaborate in their "hunt" to locate genes involved in behavioral traits.

Environmental influences become more interesting and useful when they, too, are elaborated from abstract components of variance to specific, measured environmental mechanisms and processes. The detection of composite shared (c^2) or nonshared (e^2) environmental influences (viz., variance components) should be the first step toward identifying specific environmental mechanisms and processes. Fleshing out abstract genetic and environmental variance components into specific mechanisms and processes can be considered a form of statistical mediation (Baron & Kenny, 1986). Mediation occurs when at least part of the effects of one variable on another are expressed through a third variable.

A general strategy for detecting specific environmental mechanisms that mediate environmental influences is to add a measured environmental variable to the DeFries and Fulker (1985) basic regression equation (Equation 1). This measured variable should be added only after the abstract environmental variance component has been found to be statistically significant. If a measured variable is to mediate shared environmental influences, then the term for c^2 should be statistically significant in the DeFries and Fulker basic regression equation. If a measured variable is to mediate nonshared environmental influence, then the variance unexplained in the basic regression equation (i.e., $1 - R^2$) should exceed estimates of unreliability of measurement. This logic shows the essential value of abstract variance components; their existence must be verified before specific environmental influences are identified, just as heritability analysis must precede research in molecular genetics.

The measured environmental variable will differ computationally depending on whether one is investigating mediation of shared or nonshared environmental variation. For mediation of shared environmental variation, it is sufficient to add to the DeFries and Fulker (1985) equation a measure representing a specific environmental influence that similarly affects both siblings' phenotypes. This augmented equation would be

$$P_1 = b_1P_2 + b_2R + b_3p_2 \bullet R + b_4M + b_0. \qquad (3)$$

For example, the Home Observation for Measurement of the Environment (HOME) Scale (Bradley, Caldwell, & Elardo, 1979) assesses aspects of the familial environment that are relevant to IQ. Assuming a statistically significant shared environmental variance component (b_1 in Equation 1), adding each sibling pair's HOME score to the DeFries and Fulker (1985) basic regression equation permits a test for the mediational influence of these aspects of family environment on IQ. With this variable (M) in the equation, the c^2 parameter estimate (b_1) should be reduced because this measured variable "takes up" some of the explanatory variance associated with abstract shared environmental influences (as indicated by the significance test of b_4, the regression coefficient for the mediator variable).

For mediation of nonshared environmental influences, the added variable (M) would be a signed difference score for sibling A's score minus sibling B's score on the specific environmental measure (Rodgers, Rowe, & Li, 1993). For the HOME Scale, a difference score can be constructed from those environmental assessments on which each child receives a separate rating (e.g., parental affection, but not the number of books the family owns). The use of difference scores can remove most of the within-families association of the environmental measure with family environmental deviations from the population mean. But the within-families environmental differences could predict sibling differences in developmental outcomes. In the DeFries and Fulker (1985) model, this difference is contained in the variance that is unexplained by the basic regression equation (Equation 1). That is, sibling differences are part of the residual of nonshared environmental influences. Adding the difference score, therefore, should leave the estimates of shared environmental influence and heritability relatively unchanged. The total variance accounted for by the regression equation would increase, however, as the measured variable "takes up" some of the explanatory variance associated with nonshared environmental influences (as indicated by the significance of b_4).[3] A more extensive discussion of these and other techniques for exploring measured environmental variables in the context of behavioral genetics research designs is provided by Waldman and Rowe (1993).

[3]Nonshared genetic influences also appear in this residual. One way to evaluate them is to add the interaction of the specific environmental measure that mediates nonshared environmental influences with the coefficient of genetic relatedness to the augmented DeFries and Fulker regression model (i.e., to Equation 3).

An Example of Exploring Shared Environmental Influences

To illustrate the possibility of exploring variance components with measured variables, we reanalyzed data from Rowe's (1983) twin study of delinquent behavior. Table 1 shows estimates of heritability and shared environmental influences for male twins obtained from the DeFries and Fulker (1985) basic regression equation. Shared environmental influences accounted for 28% of variance in delinquent behavior. This number suggests that some kind of shared environmental influence on delinquency exists but does little to illuminate the specific psychosocial processes behind it. Thus, in themselves, variance components may be intellectually dissatisfying to environmentally oriented researchers.

If one looks for specific environmental measures that either moderate or mediate this shared environmental component, then one can begin to build more sophisticated theories about it. To test potential moderators of shared environmental influences, we used the augmented DeFries and Fulker equation (Equation 2). For our measure of environmental context, we first tried the average of mother's and father's *years of education.* This variable had no statistically significant moderating

TABLE 1

Shared Activities Moderate the Environmentability Parameter Estimate

Model/ Parameter	Regression coefficient	Term	Estimate
Basic			
h^2	b_3	$R \bullet P_2$	37%
c^2	b_1	P_2	28%
Augmented			
h^2	b_3	$R \bullet P_2$	19%
	b_1	P_2	−0.032
	b_5	$P_2 \bullet EC$	0.025

Note. Interaction terms in the basic DeFries and Fulker (1985) equation (Equation 1): delinquency, genetic relatedness, Genetic Relatedness × Delinquency. Interaction terms in the augmented DeFries and Fulker equation: delinquency, shared activities, genetic relatedness, Genetic Relatedness × Delinquency, Delinquency × Shared Activities. The three-way interaction was not statistically significant and was dropped from the augmented equation. The interaction terms were not statistically significant at conventional levels ($p < .05$). The b_3 and b_5 terms were at about the .25 and .1 levels, respectively.

effect. Next, we tried *shared activities*, a scale that asks twins how often they engage together in ordinary teenage activities such as movie going and attending sporting events. This variable had a moderating effect. Table 1 shows the unstandardized regression coefficient (b_5) that is associated with the two-way interaction term based on shared activities and delinquency ($P_2 \cdot EC$). The positive unstandardized regression coefficient (0.025) indicates that the shared environmental influence increased when twins had more mutual contact.

Table 2 gives new estimates of c^2 that are based on whether twins shared few, an average number, or many activities (number of shared activities [SA] = 12, 18, and 24, respectively). Separate c^2 parameters can be estimated for these levels of shared activities using the equation

$$c^2 = (b_1 + b_5 EC), \tag{4}$$

where b_1 is the beta weight on P_2 in the augmented equation, b_5 is the weight on the $P_2 \cdot EC$ interaction term, and EC is the shared activities variable. The shared environmental effect for twins with an average level of contact ($SA = 18$) was 42% (i.e., $-0.032 + 0.025 \cdot 18$). Shared activities dramatically moderated the degree of shared environmental component, which ranged from 27% to 57% of the total variation of self-reported delinquency.

In a separate study of nontwin siblings, Rowe and Gulley (1992) further explored the effects of sibling influence on delinquency. They also found "sibling effects" for both sisters and brothers but not for mixed-sex siblings. Sibling mutual warmth, as well as overlapping peer groups,

TABLE 2

Shared Environment Estimates for Male Twins From Augmented DeFries and Fulker (1985) Equation Shared Activities Parameter Estimate

Number of shared activities	Parameter estimate (c^2)
Few	27%
Average	42%
Many	57%

Note. Number of shared activities = 18 at mean (*Average*), 12 at 1.5 standard deviations below the mean (*Few*), and 24 at 1.5 standard deviations above the mean (*Many*).

appeared to create the conditions for sibling mutual influence. Our point, however, is not to work out all of the details in the process of sibling influence on antisocial behavior. Rather, we emphasize, as many environmentally oriented researchers would agree, that finding shared environmental influences as an abstract component of variance (viz., c^2) is the beginning, not the end, of one's search for causal process. By tying this abstract variance component to measured variables, we were able to eliminate some hypothesized environmental influences (in this instance, social class) while building support for other hypothesized environmental influences (siblings' shared activities).

Specific Steps to Foster Collaboration Between Environmentally Oriented Researchers and Behavioral Geneticists

Thus far, we have presented several research designs that should aid behavioral geneticists and environmentally oriented researchers in collaborative efforts by incorporating specific environmental contexts and mechanisms into behavioral genetics analyses. Although these research designs should be helpful, we think that there are several preconditions to a successful collaboration. These thoughts are not new (e.g., Wachs, 1983; Waldman & Weinberg, 1991), but they bear repeated mention because they are so crucial to behavioral geneticists and environmentally oriented researchers sharing a common perspective on developmental research.

The first precondition to successful collaboration is for environmentally oriented researchers to fully acknowledge the necessity of controlling genetic influences in assessing environmental effects on development. Although this point has been made consistently by behavioral geneticists, environmentally oriented developmental researchers continue to use correlations between parental characteristics and children's development. This research strategy may be profitably carried out within the context of behavioral genetics designs (e.g., studies of adoptive and nonadoptive family members or studies of families with twins as parents or twins as children), but when nuclear families are sampled—as in most studies—one cannot uniquely infer environmental influences. It is not sufficient for environmentally oriented researchers to acknowledge the role of genetic influences on developmental outcomes and to proceed by studying nuclear families, with the rationale that they are interested specifically in examining environmental influences. Rather, environmentally oriented investigators must make explicit efforts to disentangle genetic and environmental influences, efforts that are best accomplished through the use of behavioral genetics designs.

A second precondition for successful collaboration is for behavioral geneticists to acknowledge the very limited appeal that abstract variance components have for developmentalists. Developmentalists are interested in the roles that specific mechanisms and processes play in producing continuity and change in developmental outcomes. Many environmentally oriented researchers appear skeptical of the composite indices of environmental influences that emerge from behavioral genetics studies and appear to feel that many of the findings from behavioral genetics studies (e.g., negligible shared environmental influences on various developmental outcomes) may be primarily a function of the failure to include specific measures of the environment in such analyses. Even when environmentally oriented researchers accept behavioral genetics methods for the estimation and characterization of composite indices of genetic and environmental influences, they are often left dissatisfied by the inability to make statements regarding the effects of specific environmental mechanisms and processes on specific developmental outcomes.

The obvious implication is that behavioral geneticists should include specific environmental measures in their research to a greater degree. It also has been suggested (Baumrind, 1991; Scarr, 1987a) that behavioral geneticists should sample subjects from a wider range of environments to increase the likelihood that familial environmental characteristics will emerge as important influences on developmental outcomes. We have tried to facilitate these advances by presenting several research designs and analytic models that are especially useful for examining the effects of specific environmental contexts and measures on developmental outcomes. There is reason for optimism because a number of important developmental behavioral genetics studies (e.g., the Colorado Adoption

Project [Plomin, DeFries, & Fulker, 1988] and the Minnesota Transracial Adoption Study [Weinberg, Scarr, & Waldman, 1992]) have included specific measures of the familial environment, and recent advances in behavioral genetics models allow them to take better advantage of specific environmental measures in biometrical models (e.g., Loehlin, Horn, & Willerman, 1989). Furthermore, behavioral genetics designs permit the investigation of genetic and environmental influences on environmental measures and correlations of these measures with developmental outcomes (e.g., Plomin & Bergeman, 1991; Plomin, Loehlin, & DeFries, 1985) and the examination of how the relation of specific environmental measures and developmental outcomes changes over time (e.g., Loehlin et al., 1989).

The distance is not so great, then, between environmentally oriented researchers and behavioral geneticists that the two camps cannot share a common perspective on development, communicate in a common language of statistical techniques, and apply common research designs to solving the unknowns of behavioral development. A new unity of social science researchers could lead to a secure knowledge base in developmental psychology and to theories that reveal underlying principles of human development.

References

Anastasi, A. (1958). Heredity, environment, and the question "How?" *Psychological Review, 65*, 197–208.

Baron, R., & Kenny, D. (1986). The moderator–mediator variable distinction in social psychological research. *Journal of Personality and Social Psychology, 51*, 1173–1182.

Baumrind, D. (1991). To nurture nature. *Behavioral and Brain Sciences, 14*, 386–387.

Baumrind, M. (in press). The average expectable environment is not good enough: A response to Scarr (1992). *Child Development.*

Bradley, R. H., Caldwell, B. M., & Elardo, R. (1979). Home environment and cognitive development in the first 2 years: A cross-lagged panel analysis. *Developmental Psychology, 15*, 246–250.

Bronfenbrenner, U., & Ceci, S. J. (1993). *Nature–nurture reconceptualized: Toward a new theoretical and operational model.* Manuscript submitted for publication.

Cherny, S. S., DeFries, J. C., & Fulker, D. W. (1992). Multiple regression analysis of twin data: A model-fitting approach. *Behavior Genetics, 22*, 489–497.

DeFries, J. C., & Fulker, D. W. (1985). Multiple regression analysis of twin data. *Behavior Genetics, 15*, 467–473.

Degler, C. N. (1991). *In search of human nature: The decline and revival of Darwinism in American social thought*. New York: Oxford University Press.

Falconer, D. S. (1989). *Introduction to quantitative genetics* (3rd ed.). New York: Wiley.

Galton, F. (1869). *Hereditary genius: An inquiry into its laws and consequences*. London: Macmillan.

Kevles, D. J. (1985). *In the name of eugenics: Genetics and the uses of human heredity*. New York: Knopf.

Loehlin, J. C., Horn, H. M., & Willerman, L. (1989). Modeling IQ change: Evidence from the Texas Adoption Project. *Child Development, 60*, 993–1004.

McClelland, G. H., & Judd, C. M. (1993). *The statistical difficulties of detecting moderator variables*. Manuscript submitted for publication.

Plomin, R. (1986). *Development, genetics, and psychology*. Hillsdale, NJ: Erlbaum.

Plomin, R., & Bergeman, C. S. (1991). The nature of nurture: Genetic influence on "environmental" measures. *Behavioral and Brain Sciences, 14*, 373–427.

Plomin, R., DeFries, J. C., & Fulker, D. W. (1988). *Nature and nurture during infancy and early childhood*. New York: Cambridge University Press.

Plomin, R., Loehlin, J. C., DeFries, J. C. (1985). Genetic and environmental components of "environmental" influences. *Developmental Psychology, 21*, 391–402.

Rodgers, J. L., Rowe, D. C., & Li, C. (1993). *Beyond nature versus nurture: DF analysis of nonshared influences on problem behaviors*. Manuscript submitted for publication.

Rose, R. J., Kaprio, J., Williams, C. J., Viken, R., & Obremski, K. (1990). Social contact and sibling similarity: Facts, issues, and red herrings. *Behavior Genetics, 20*, 763–778.

Rowe, D. C. (1983). Biometrical genetic models of self-reported delinquent behavior: A twin study. *Behavior Genetics, 13*, 473–489.

Rowe, D. C. (1987). Resolving the person–situation debate: Invitation to an interdisciplinary dialogue. *American Psychologist, 42*, 218–227.

Rowe, D. C. (1990). As the twig is bent?: The myth of child rearing influences on personality development. *Journal of Counseling and Development, 68*, 606–661.

Rowe, D. C., & Gulley, B. L. (1992). Sibling effects on substance use and delinquency. *Criminology, 30*, 217–233.

Scarr, S. (1987a). Distinctive environments depend on genotypes. *Behavioral and Brain Sciences, 10*, 38–39.

Scarr, S. (1987b). Three cheers for behavioral genetics: Winning the war and losing our identity. *Behavior Genetics, 17*, 219–228.

Scarr, S. (1992). Developmental theories for the 1990s: Development and individual differences. *Child Development, 63*, 1–19.

Scarr, S., Weinberg, R. A., Waldman, I. D. (in press). IQ correlations in transracial adoptive families. *Intelligence.*

Wachs, T. D. (1983). The use and abuse of environment in behavior–genetic research. *Child Development, 54*, 396–407.

Wahlsten, D. (1991). Sample size to detect a planned contrast and a one degree-of-freedom interaction effect. *Psychological Bulletin, 110*, 587–596.

Waldman, I. D., & Rowe, D. C. (1993). *Behavioral genetics designs for examining environmental mechanisms in childhood conduct disorder.* Unpublished manuscript.

Waldman, I. D., & Weinberg, R. A. (1991). The need for collaboration between behavioral geneticists and environmentally oriented investigators in developmental research. *Behavioral and Brain Sciences, 14*, 412–413.

Weinberg, R. A., Scarr, S., & Waldman, I. D. (1992). The Minnesota Transracial Adoption Study: A follow-up of IQ test performance at adolescence. *Intelligence, 16*, 117–135.

CHAPTER 20

The Nature–Nurture Gap: What We Have Here Is a Failure to Collaborate

Theodore D. Wachs

There is a curious paradox in the operation of what is called the nature–nurture debate. The debate has historically been framed in terms of nature versus nurture (see introduction to Part Five). This dichotomy continues to be debated (see Turkheimer, 1991) in spite of nearly half a century of increasing agreement that this dichotomy is meaningless and that development is a function of the combined influence of nature and nurture (Anastasi, 1958; Hebb, 1949; Plomin, 1990; Wachs, 1983). Why this continued debate on what is essentially a meaningless dichotomy? What I argue in this chapter is that, although there is general agreement on the ideal of development as a joint function of nature and nurture, in reality this ideal is honored more in the abstract than in practice. A small example will illustrate what I mean.

There recently appeared in my mail an announcement for a postdoctoral position: "To investigate genetic and environmental factors in the development of psychopathology." Applicant qualifications were a doctoral degree in psychology "plus advanced training and strong research skills in any two of these following areas: psychopathology, psychophysiology, quantitative methods, and behavioral genetics." Although those running this project accepted the ideal that phenotype (psychopathology) is a joint function of genetic and environmental factors, none of the required training or research skills involved an understanding or expertise in the study of environmental influences.

What this example illustrates for me is that the nature–nurture gap represents not so much a disagreement about conceptual issues but rather the fact that there is virtually no cross-disciplinary research or theoretical collaboration between genetically and environmentally oriented researchers. The futility of the nature–nurture question is a concept that researchers can all agree on; but like world peace and a clean environment, few, if any, researchers are willing to do anything about it. If development is influenced by the joint action of nature and nurture, then it would seem important that the research processes mirror the nature of this reality, namely joint collaboration between genetically and environmentally oriented researchers (Wachs, 1993).

What I hope to illustrate in this chapter are some of the reasons why such collaboration has rarely occurred, what common areas of focus are ripe for collaboration, and what specific steps can be taken to facilitate such collaboration. Although most of my comments will be from the viewpoint of an environmentally oriented researcher (caveat emptor), they are written from the perspective of an environmentalist who has some knowledge of behavioral genetics and who has argued that an understanding of environmental influences requires an integration of biological and experiential factors (Wachs, 1992).

Factors Inhibiting Collaboration

Perhaps the major limiting factor to cross-disciplinary collaboration is the fact that most environmentally and genetically oriented researchers have only a sketchy knowledge of the models, methods, and issues that are central for researchers in the other domain (Goldsmith, 1988). This knowledge gap has at least two implications. First, there may be a natural reluctance to venture outside of the safe confines of one's own area of expertise. Second, for those few brave genetic or environmental researchers who attempt to integrate concepts or methods from the other domain into their own designs, all too often their lack of expertise leads to inappropriate choices. These inappropriate choices, in turn, may result in conclusions about genetic or environmental influences that are not accepted by researchers in the other domain because they are based on concepts and methods that are viewed as outdated or incorrect. Because genetic and environmental researchers rarely communicate directly, what has emerged are two parallel literatures, with widely discordant conclusions, that only further increase miscommunication. What I illustrate in the following section are examples of how knowledge gaps about current methods and concepts in the study of environmental influences lead behavioral genetics researchers to conclusions that are discordant with the conclusions drawn by environmentally oriented researchers. My aim is not to criticize the field of behavioral genetics but rather to show how knowledge gaps lead to suspect conclusions, which in turn inhibit the chances of finding common grounds for collaboration. The points that I raise are drawn from an environmental perspective, but I must stress that barriers to collaboration exist on both sides of the nature–nurture equation. Concerns about potentially suspect conclusions drawn from environmental studies can be, and have been, raised by behavioral genetics researchers, but these concerns are not explicated here because of space limitations. Detailed discussion of concerns that have been raised by behavioral genetics researchers can be found in a number of sources (e.g., Goldsmith, 1988; Plomin & Bergeman, 1991).

Methodological Gaps

Over the past 10 years, environmentalists have pointed out that behavioral genetics studies that discuss "environmental" influences rarely, if ever, directly measure the environment (Hoffman, 1991; Wachs, 1983). Rather, environment is estimated, either as residual variance or from parental phenotype (e.g., adoptive parent IQ). In spite of repeated demonstrations of the inappropriateness of this approach, sweeping statements continue to be made about the nature of environmental influences, even when environment is never assessed directly (e.g., Braungart, Plomin, DeFries, & Fulker, 1992; Cardon, Fulker, DeFries, & Plomin, 1992). In those few behavioral genetics studies in which environment actually is measured, all too often the measurement is either inadequate (e.g., socioeconomic status) or does not encompass current theoretical formulations on the multilevel structure of the environment (Bronfenbrenner, 1993; Wachs, 1992).

What kinds of conclusions are drawn from these types of studies, and how do these conclusions differ from those found in environmental research, wherein the environment is directly measured? Several examples will suffice. Behavioral genetics researchers have pointed out the high IQ correlations of identical twins who were reared in different homes as evidence for a limited role of the environment on cognitive development (Loehlin, Willerman, & Horn, 1988). However, Bronfenbrenner (1986) has provided evidence that illustrates how these correlations are dramatically reduced when one considers the mediating influence of higher order (macrosystem) components of the environment rather than assuming that being reared in different homes means encountering totally different environments. Rose and colleagues have similarly demonstrated how radically different conclusions about the role of environmental influences on personality can result when the environment is measured rather than estimated (Rose, Kaprio, Williams, Viken, & Obrenski, 1990; Rose, Koskenvuo, Kaprio, Sarna, & Langinvainio, 1988).

Perhaps the most critical example is seen in the recent assertion by a number of behavioral genetics researchers that environment has an influence on developmental variability only at the extremes; from this viewpoint, within most normal family situations, environment is seen as having little to do with children's development (Scarr, 1992; Turkheimer & Gottesman, 1991). I explore this assertion in some depth as an illustration of how misinformation leads to incorrect conclusions, which in turn hamper collaboration.

Environmental researchers have no difficulty accepting the point that extreme environments can influence developmental variability. However, from an environmental perspective, there are two reasons to strongly disagree with the assertion that environments operate only at the extremes. First, the statement typically is based on studies in which environment is estimated rather than measured directly. When environment is measured directly, one can find relations between specific components of the environment and specific aspects of development, even in normal, nonextreme family situations (e.g., Bornstein & Tamis-LeMonda, 1990; Gottfried & Gottfried, 1984; Wachs, 1987). Even more striking is the fact that the same environmental parameters that relate to children's development in normal family situations in the United States and western Europe also relate to development in the same way for children living in nonwestern, less developed countries (Wachs et al., 1993).

Not only is the assertion that environmental influences are relevant only at the extremes incorrect from an empirical standpoint, but the assertion is also incorrect when viewed within the framework of current environmental theory. I refer here to the conceptual distinction made by environmentally oriented psychobiologists between experience-expectant versus experience-dependent development (Greenough, Black, & Wallace, 1987). Experience-expectant development is based on organisms using environmental information that has been commonly available to their species throughout much of the species' evolutionary history. In contrast, experience-dependent development involves the storage and usage of information that is unique to the individual's own developmental history. If one wishes to talk only about experience-expectant development, then a statement that environment has little relevance in normal family situations makes some sense. At this level, the kinds of experiences that are involved in influencing experience-expectant development would be common in most families (e.g., language), and the individual would be "tuned" to be responsive to these types of stimulation. In contrast, for experience-dependent development, the assertion that experience is irrelevant in normal family situations makes little sense because this type of development is directly dependent on the unique experiences that each individual encounters within his or her family and nonfamily environments.

Environmentalists can easily accept the argument that some aspects of development are influenced by aspects of the environment that are commonly encountered by most children. For experience-expectant aspects of development, unique environmental contributions may occur only under extreme conditions, with outcome variability primarily being a function of nonenvironmental factors (Horowitz, 1987). However, based on existing theory and research, environmentalists see little validity in a general statement that environment relates to variability in development only under extreme situations. Not only is such a statement incorrect, but there is little point in developing collaborative relations with researchers who accept this misinformation because there is little room for environment, as environmentalists understand it, in this type of system. What is shown here is a clear example of how misinformation, derived from a lack of cross-domain knowledge, inhibits the chances of collaboration between environmental and behavioral genetics researchers.

Conceptual Gaps

As researchers have attempted to conceptualize the relation of nature to development, they have over the centuries shifted their viewpoint from preformationism to predeterminism to a probabilistic multidetermined system of influences (Hunt, 1961). Within this probabilistic systems framework neither genetic nor environmental determinism is acceptable because both genetic and environmental influences on development are intermingled (Gottlieb, 1991; Oyama, 1985). Unfortunately, the knowledge that genetic (and environmental) determinism has died has apparently not reached all quarters. In a recent article, Scarr (1992) argued for the viability of what I call a "masked" model of genetic determinism. What Scarr proposed was a model in which environment is included but only insofar as environment is "driven" by genes: "The theory of genotype → environment effects holds that genotypes drive experiences" (p. 9). Within this framework of masked genetic determinism, the influence of the child's environment is merely a reflection of the parent's and child's individual, genetically driven, phenotypic characteristics.

Scarr's (1992) model is clearly incorrect in a number of areas. As Scarr herself pointed out in an earlier publication, genes do not directly code for behavioral phenotype (or environments); the road from gene to behavior is extremely complex and can be influenced by the action of multiple biological and environmental factors (Scarr & Kidd, 1983). Also contrary to Scarr's argument, nonextreme environments are related to developmental variability (see earlier discussion). Furthermore, the relation between child characteristics and environment is not nearly as consistent or straightforward as Scarr implies (Crockenberg, 1986), nor can all domains of the microenvironment be reduced to individual constructions (Wohlwill & Heft, 1987). Perhaps most critical is the fact that, from a historical point of view, Scarr's model is clearly regressive, in the sense of attempting to return to an outdated paradigm of genetic determinism. Environmentalists can easily work within a multidimensional systems framework because such a framework allows for both unique and combined biological–environmental contributions to development. Environmentalists cannot work within a framework of genetic determinism, however well masked, because there is little room for unique or even combined environmental contributions within such a framework.

Most environmentalists do not reject the importance of biological (genetic) influences when these are viewed within the framework of a multidimensional system of influences on individual development (Horowitz, 1987; Wachs, 1992). However, accepting biological–genetic contributions as a necessary part of a multidimensional probabilistic system is not the same thing as accepting genetic determinism. An insistence on genetic determinism is not only outdated and incorrect but also hinders all attempts at cross-disciplinary collaboration. Obviously, the same conclusion holds in the case of environmentalists who insist on pure environmental determinism.

Potential Areas of Common Focus

Within the framework of a probabilistic multidetermined system, there appear to be two domains with great potential for collaboration between genetically and environmentally oriented developmental researchers. The first is gene–environment covariance; the second is gene–environment interaction.

Gene–Environment Covariance

Behavioral genetics researchers (Plomin, DeFries, & Loehlin, 1977) have been among the leaders in identifying specific covariances between genotypes and environments that may exist in nature. Perhaps the most familiar is passive gene–environment covariance. In this situation, offspring receive both specific genes and specific environments from their parents (e.g., children who receive both "smart genes" and "enriched environments" from their parents). A second type is reactive covariance, wherein children with specific gene-related characteristics are more likely to elicit specific reactions from their environment (e.g., hostile children are more likely to elicit hostility from their peers). Least studied is active covariance, wherein children with certain gene-related individual characteristics are more likely to seek out certain situations than children without these characteristics (e.g., children with high verbal abilities are more likely to spend time reading).

Whatever the type, developmental variability that is associated with gene–environment covariance cannot be ascribed to either genetic or environmental factors. Rather, gene–environment covariance is best viewed as a unique source of influence, wherein the extent and nature of the correlation between genetic and environmental factors is the cause of developmental variability (Wachs, 1992).

Although reviewers have noted the importance for development of gene–environment correlation (McCall, 1991), for the most part there has been remarkably little collaborative study on this question. Most research in this area involves the study of organism–environment covariance, and even here collaborative research is the exception rather than the rule (for a review of this area see Wachs, 1992). Both the potentials and pitfalls for collaborative research on gene–environment correlation can be seen by focusing on specific domains such as child temperament or psychopathology. There is agreement that both temperament (Bates, 1989; Buss & Plomin, 1984) and some domains of child psychopathology have a genetic basis (Folstein & Rutter, 1988; Rutter et al., 1990). Conceptually, both temperament and child psychopathology should be domains that are highly sensitive to certain types of gene–environment covariance. For example, children whose parents have major psychoses should be at greater risk for developing some form of psychopathology because these children have a higher probability of receiving both risk genes and abnormal environments from their parents (passive covariance; Asarnow, 1988). Similarly, infants with difficult temperaments should have a greater likelihood of eliciting negative interactions from their caregivers than do children with easy temperaments (reactive covariance; Crockenberg, 1986).

Although individual studies can be found that support the just-mentioned hypotheses, reviews have also noted major inconsistencies in available research findings (Crockenberg, 1986; Rutter, 1990; Wachs, 1992). What is suggested by these reviews is that the process of gene–environment covariance may be more complex than would be indicated by the relatively simple concepts of passive, reactive, and active covariance. For example, available evidence suggests that not all children with disturbed parents have a genetic liability for the disorder in question (Wachs & Weizmann, 1992). Other evidence indicates that relations between child temperament and subsequent caregiver behaviors are multidetermined, being influenced not only by the child but also by individual parental characteristics and by the general environmental context within which caregivers function (Wachs, 1992).

If researchers are interested in understanding the processes underlying developmental variability, and if these underlying process variables covary with each other, then it may be a mistake to separate statistically what nature has joined together (Scarr, 1985). Furthermore, as just suggested, if the covarying processes are themselves part of a larger system of influences, then it may be just as much a mistake to study covarying genetic and environmental processes in isolation. As a result, I argue that domains such as child temperament and psychopathology are ripe for collaborative research between genetic and environmental researchers, with the research focus centered around questions on the nature and influence of gene–environment covariance.

In developing collaborative research strategies that center around the concept of gene–environment covariance, it becomes important to build on the unique contributions of each domain. From the behavioral genetics side, one can obtain delineation of which specific child or caregiver characteristics have the clearest linkages to individual genotypes and whether there are age periods when these linkages are particularly salient. From the environmental side, one needs detailed measures of specific characteristics of the child's environment that are likely to directly covary with individual caregiver or child characteristics, as well as information on higher level aspects of the environmental context that are likely to moderate existing covariation. Joint collaboration is essential in developing a conceptual framework that allows researchers to specify, in advance, for a particular individual characteristic, the nature of the covariance process that is being dealt with, and whether the covariance process will be mediated by higher order biological or environmental processes. Similarly, in understanding the causes and consequences of child temperament or psychopathology, it would be very important to assess what biological or environmental factors influence whether specific gene-related child characteristics fit positively (match) or negatively (mismatch) specific characteristics of the child's environment.

Gene–Environment Interaction

Main-effects research looks at the effects of single variables considered in isolation. Coactive research looks at the additive contribution of two or more variables (Rutter, 1983). Research on gene–environment interaction refers to the combined nonlinear impact on development of genetic and environmental influences (Rutter, 1983). Interactions between genes and environments may be synergistically negative, such that the effect of two risk factors taken together is greater than the impact of either risk factor taken in isolation or additively; alternatively, gene–environment interactions may be positive, with the impact of a protective factor buffering the individual against the impact of biological or environmental risks.

For the most part, neither behavioral geneticists nor environmentally oriented researchers have been comfortable with the concept of gene–environment interaction (Wachs, 1983). In part, this is because the existence of gene–environment interaction makes it extremely difficult to argue for either main-effects genetic or environmental theories. For example, if there is differential individual reactivity to objectively similar environmental stimulation, then one really cannot talk about main effects being associated with these environments without considering the characteristics of the individual upon whom the environment impinges. In spite of decades of attempting to ignore the existence of interactions

(Cronbach, 1957), there is ample evidence for the operation of organism–environment interaction in development (Rutter & Pickles, 1991; Wachs, 1992). Evidence for interactions appears in spite of the fact that both conceptual and methodological factors continually operate to diminish the chances of finding existing interactions. Factors that have been identified as masking existing interactions include the following:

1. The use of inappropriate or nonprecise measures of the environment or of individual characteristics.
2. The lack of statistical power in designs testing for interactions.
3. The possibility that interactions may be higher order in nature (encompassing multiple environmental and organismic factors).
4. The atheoretical nature of most studies investigating organism–environment interaction.

A detailed review of these and other factors appears in a recent edited volume on this topic (Wachs & Plomin, 1991).

As with covariance, most research on interaction has involved organism–environment interaction rather than gene–environment interaction. Also, like covariance, gene–environment interaction is a unique influence on development that cannot be assigned to the genetic or environmental side. It is the nonlinear interaction between genes and environments that is critical, not the separate genetic and environmental influences taken in isolation. Finally, like gene–environment covariance, the study of gene–environment interaction is a research area in which the nature of the research process should mirror the nature of the phenomena, namely collaborative interactions between genetically and environmentally oriented researchers. This is particularly true for process-oriented research. As an example, I suggest developmental psychopathology. Available reviews have suggested that the occurrence, nature, and degree of child psychopathology is a function of the combined interactive operation of genetic and environmental factors (Asarnow, 1988; Rutter, 1990; Wachs & Weizmann, 1992). What is not clear is the nature of this interaction. A number of questions assessing the nature of this interaction could best be handled by collaborative research. For example, what are the developmental outcomes associated with specific combinations of genetically based biological liabilities and buffers interacting with specific environmental liabilities and buffers (Rutter, 1990)? Alternatively, are children with genetically based biological vulnerabilities at higher risk as a function of greater vulnerability to environmental stress, less ability to use existing environmental supports, or some combination of these two processes? Answers to these types of questions are not likely to be generated by genetic and environmental researchers working in isolation, if only because isolation decreases the chances of adequately measuring appropriate genetic and environmental characteristics and the chances of developing appropriate multidomain systems theories to frame the research questions. Answers to these types of questions are more likely to emerge from direct collaboration between genetically and environmentally oriented researchers.

The Process of Developing Collaborative Research

Although calls for the necessity of collaboration between behavioral geneticists and environmentally oriented researchers are not new, for the most part such collaboration has not occurred, at least not to the degree needed. A behavioral geneticist calling up an environmental colleague to ask for an easy way to measure a child's environment or an environmental researcher visiting his or her neighborhood geneticist to ask whether a subset of twins in his or her sample is monozygotic or dizygotic is not what is being called for in this chapter. What is needed is active collaboration, involving the formulation of models to delineate the processes whereby combinations of genetic and environmental factors influence development, as well as the development of appropriate methodologies to test these models. This type of collaboration cannot occur if researchers from different domains are busy defending their turf against what they see as uninformed intrusions by researchers from other domains. The trick is to find questions of common interest and ways of educating colleagues from other domains on appropriate models and methods in one's own domain.

Clearly, the study of gene–environment covariance and gene–environment interaction offers questions of common interest. Furthermore, the nature of covariance and interaction processes precludes statements about genetic or environmental primacy because both genes and environments are inseparably commingled in these processes.

As far as concerns behavioral geneticists learning about environmental models and methods, and vice versa, the most appropriate format might be one that involves problem solving. I suggest that a critical problem that needs to be dealt with before collaboration can actually occur involves differences in the types of research designs favored by genetically and environmentally oriented researchers. For the most part, behavioral genetics research is based on large-scale twin or adoption studies, often involving measurement of a variety of phenotypes (Goldsmith, 1988). In contrast, the majority of environmental studies involve detailed measurement of natural family environments (Wachs, 1983), which by necessity involves relatively small sample sizes (relatively small compared with the sample sizes used in behavioral genetics studies). Statistical power in behavioral genetics studies comes from the large sample sizes that are used. Statistical power in environmental studies comes from the relative precision found with the use of detailed repeated measurement of an individual's environment. Unfortunately, there is a clear trade-off between sample size and the ability to use detailed repeated measurement of the environment. Unless researchers have at their disposal infinite resources, they cannot use repeated detailed environmental measurements with large samples, primarily because of cost considerations (both time and money). The sample size–precision tradeoff, although serious, is not insurmountable. At least for the area of organism–environment interaction, a number of possibilities have been suggested that might well allow collaborative research without sacrificing the unique design requirements of both behavioral geneticists and environmentally oriented researchers (Wachs & Plomin, 1991). These include the following:

1. Using smaller sample sizes but having the sample consist of children at the extremes of the genetically based individual characteristics that are being studied to maximize the probability that researchers are getting relatively "noise-free" measures of the child's contribution.
2. Avoiding statistical models that focus on percentage of variance ac-

counted for or that depend on reaching a precise level of statistical significance in favor of models based on confidence intervals.

3. Using specific environmental and individual characteristic marker variables across different studies, which will allow researchers to aggregate individual precise small sample studies into larger "meta" studies.

Given that the problem of finding a common set of design procedures is not insurmountable, I argue that solving this problem would be best handled by a conference involving behavioral geneticists and environmentally oriented researchers. Coming up with a collaborative design cannot be accomplished without geneticists learning something about environmental models and methods or without environmentalists learning something about genetic models and methods. As I argued earlier, this type of knowledge is essential for successful collaboration. I suspect that researchers could make great strides toward developing collaborative research ties by locking a select group of behavioral geneticists and environmentally oriented researchers in a room and then letting them emerge only when they have agreed on a set of common research questions and a set of common models and methods to analyze these questions. (Such a procedure is successfully used by the College of Cardinals when selecting a new pope.)

Most cross-disciplinary dialogue is currently conducted through exchanges in the scientific press or through intermittent meetings such as an American Psychological Association round table on nature–nurture. These strategies have not been particularly useful in producing collaborative research, if only because attempts at building bridges have been hypothetical–conceptual in nature. Researchers need to stop constructing bridges in the air and begin to construct actual empirical bridges between nature and nurture if they are to progress in this area.

References

Anastasi, A. (1958). Heredity, environment, and the question "How?" *Psychological Review, 65*, 197–208.

Asarnow, J. (1988). Children at risk for schizophrenia—converging lines of evidence. *Schizophrenia Bulletin, 14*, 613–631.

Bates, J. (1989). Concepts and measures of temperament. In G. Kohnstamm, J. Bates, & M. Rothbart (Eds.), *Handbook of temperament in childhood* (pp. 3–36). New York: Wiley.

Bornstein, M., & Tamis-LeMonda, C. (1990). Activities and interactions of mothers and their first born infants in the first six months of life. *Child Development, 61*, 1206–1217.

Braungart, J., Plomin, R., DeFries, J., & Fulker, D. (1992). Genetic influence on tester rated infant temperament as assessed by Bayley Infant Behavior Record. *Developmental Psychology, 28*, 40–47.

Bronfenbrenner, U. (1986). Ecology of the family as a context for human development. *Developmental Psychology, 22*, 723–742.

Bronfenbrenner, U. (1993). The ecology of cognitive development: Research models and fugitive findings. In H. R. Woznaik & K. Fisher (Eds.), *Specific environments: Thinking in contexts* (pp. 3–24). Hillsdale, NJ: Erlbaum.

Buss, A., & Plomin, R. (1984). *Temperament: Early developing personality traits*. Hillsdale, NJ: Erlbaum.

Cardon, L., Fulker, D., DeFries, J., & Plomin, R. (1992). Continuity and change in cognitive ability from 1 to 7 years of age. *Developmental Psychology, 28*, 64–73.

Crockenberg, S. (1986). Are temperamental differences in babies associated with predictable differences in caregivers. In J. Lerner & R. Lerner (Eds.), *Temperament and psychosocial interaction in children* (pp. 53–73). San Francisco: Jossey-Bass.

Cronbach, L. (1957). The two disciplines of scientific psychology. *American Psychologist, 12*, 671–684.

Folstein, S., & Rutter, M. (1988). Autism–family aggregation and genetic implications. *Journal of Autism and Developmental Disabilities, 18*, 3–30.

Goldsmith, H. (1988). Human developmental behavior genetics. *Annals of Child Development, 5*, 187–227.

Gottfried, A., & Gottfried, A. (1984). Home environment and cognitive development in young children of middle socioeconomic status families. In A. Gottfried (Ed.), *Home environment and early cognitive development* (pp. 57–116). New York: Academic Press.

Gottlieb, G. (1991). Experiential canalization of behavioral development. *Developmental Psychology, 27*, 4–13.

Greenough, W., Black, J., & Wallace, C. (1987). Experience and brain development. *Child Development, 58*, 539–559.

Hebb, D. (1949). *The organization of behavior*. New York: Wiley.

Hoffman, L. (1991). The influence of the family environment on personality. *Psychological Bulletin, 110*, 187–203.

Horowitz, F. (1987). *Exploring developmental theories*. Hillsdale, NJ: Erlbaum.

Hunt, J. McV. (1961). *Intelligence and experience*. New York: Ronald.

Loehlin, J., Willerman, L., & Horn, J. (1988). Human behavior genetics. *Annual Review of Psychology, 39*, 101–134.

McCall, R. (1991). So many interactions, so little evidence: Why? In T. D. Wachs & R. Plomin (Eds.), *Conceptualization and measurement of organism–environment interaction* (pp. 142–161). Washington, DC: American Psychological Association.

Oyama, S. (1985). *The ontogeny of information*. Cambridge, UK: Cambridge University Press.

Plomin, R. (1990). The role of inheritance in behavior. *Science, 248*, 183–188.

Plomin, R., & Bergeman, C. (1991). The nature of nurture: Genetic influences on "environmental" measures. *Behavioral and Brain Sciences, 14*, 373–427.

Plomin, R., DeFries, J., & Loehlin, J. (1977). Genotype–environment interaction and correlation in the analysis of human development. *Psychological Bulletin, 84*, 309–322.

Rose, R., Kaprio, J., Williams, C., Viken, R., & Obrenski, K. (1990). Social context and sibling similarity. *Behavior Genetics, 20*, 763–778.

Rose, R., Koskenvuo, M., Kaprio, J., Sarna, S., & Langinvainio, H. (1988). Shared genes, shared experiences and similarities of personality. *Journal of Personality and Social Psychology, 54*, 161–171.

Rutter, M. (1983). Statistical and personal interactions. In D. Magnusson & V. Allen (Eds.), *Human development: An interactional perspective* (pp. 295–320). New York: Academic Press.

Rutter, M. (1990). Commentary: Some focus and processes considerations regarding effects of parental depression on children. *Developmental Psychology, 26*, 60–67.

Rutter, M., MacDonald, H., Le Couteur, A., Harrington, R., Boulten, T., & Badey, A. (1990). Genetic factors in child psychiatric disorders. *Journal of Child Psychology and Psychiatry, 31*, 39–84.

Rutter, M., & Pickles, A. (1991). Person–environment interactions: Concepts, mechanisms and implications for data analysis. In T. D. Wachs & R. Plomin (Eds.), *Conceptualization and measurement of organism–environment interaction* (pp. 105–141). Washington, DC: American Psychological Association.

Scarr, S. (1985). Constructing psychology: Making facts and fables for our times. *American Psychologist, 40*, 499–512.

Scarr, S. (1992). Developmental theories for the 1990s. *Child Development, 63*, 1–19.

Scarr, S., & Kidd, K. (1983). Developmental behavior genetics. In M. Haith & J. Campos (Eds.), *Handbook of child psychology: II. Infancy and developmental psychobiology* (pp. 345–434). New York: Wiley.

Turkheimer, E. (1991). Individual and group differences in adoption studies of IQ. *Psychological Bulletin, 110*, 392–405.

Turkheimer, E., & Gottesman, I. (1991). Individual differences and the canalization of human behavior. *Developmental Psychology, 27*, 18–22.

Wachs, T. D. (1983). The use and abuse of environment in behavior genetic research. *Child Development, 54*, 396–407.

Wachs, T. D. (1987). Specificity of environmental action as manifested in environmental correlates of infant's mastery motivation. *Developmental Psychology, 23*, 782–790.

Wachs, T. D. (1992). *The nature of nurture.* Newbury Park, CA: Sage.

Wachs, T. D. (1993). Determinants of intellectual development: Single determinant research in a multideterminant universe. *Intelligence, 17*, 1–10.

Wachs, T. D., Bishry, Z., Sobhy, A., McCabe, G., Galal, O., & Shaheen, F. (1993). Relation of rearing environment to adaptive behavior of Egyptian toddlers. *Child Development, 64*, 586–604.

Wachs, T. D., & Plomin, R. (1991). *Conceptualization and measurement of organism–environment interaction.* Washington, DC: American Psychological Association.

Wachs, T. D., & Weizmann, F. (1992). Prenatal and genetic influences upon behavior and development. In C. Walker & M. Roberts (Eds.), *Handbook of clinical child psychology* (2nd ed., pp. 183–198). New York: Wiley.

Wohlwill, J., & Heft, H. (1987). The physical environment and the development of the child. In I. Altman & J. Stokols (Eds.), *Handbook of environmental psychology* (pp. 281–328). New York: Wiley.

PART SIX

The Interplay of Nature and Nurture: Redirecting the Inquiry

PART SIX

Introduction

Michael Rutter

The psychological literature of the past is replete with books and articles reflecting an ongoing battle between geneticists and environmentalists over the relative importance of nature and nurture in the determination of individual differences in psychological characteristics and in different forms of psychopathology. From the vantage point of today, the literature has a distinctly dated and old-fashioned feel to it. No present-day serious student of human behavior doubts that both nature and nurture are strongly influential, and it is accepted that much of the interest lies in studying the interplay between the two. Because of that recognition, the last decade has seen a major increase in collaborative research between geneticists and others. These collaborative ventures are already beginning to pay off in casting much-needed light on a variety of key psychological issues. However, because the rejection of the false "nature-versus-nurture" dichotomy has led to an opening up of new avenues of investigation, it has also raised new questions that need to be considered. The charting of the large known territory where there is a high level of agreement among clinicians and researchers from quite diverse backgrounds has served as a reminder that there remains a substantial area of new ground where pioneers have opened up some important paths to be followed but which has yet to be systematically mapped. Many chapters in this book outline those new paths, but those in this section represent a taking-stock of where researchers are now and where they should go from here. The authors of the chapters in this section were asked to consider four general groups of questions, which were deliberately expressed in provocative form.

First, in what ways can behavioral genetics aid the understanding of developmental, risk, or protective processes? Because all human behavior is to some extent genetically influenced, what value is there in providing quantitative estimates that heritability is 30% rather than, say, 50%? Indeed, because heritability estimates are confined to the population providing the data, and because estimates will vary if there is any major change in environmental risk factors or their distribution, is there any point at all to assessing heritability? Because researchers cannot do anything about humans' genetic endowment (apart from the "Star Wars" technology of gene therapy), why should behavioral genetics research be of any interest to policymakers, clinicians, or environmental researchers? Even for genetics researchers, is not behavioral genetics a bit passé and old-fashioned now that there is the power of molecular genetics to identify individual genes?

Second, most psychologists now consider themselves interactionists of one kind or another; but how can the interplay between nature and nurture actually be studied? What are the ways in which there may be such an interplay, and how may the mechanisms be identified? Given that it has been so hard to find gene–environment interactions in behavioral genetics research up to now, is there really any future in this line of inquiry? Are the interaction concepts wrong, or have researchers tackled the issues the wrong way? Is there any way in which the study of nature–nurture interplay may aid the understanding of environmental risk or protective mechanisms in relation to psychopathology or developmental processes more generally?

Third, in what ways, if any, can genetics research aid the understanding of the organization of behavior or the delineation of psychopathological categories? Because all genetics research has to start with some definition of the phenotypical trait or disorder to be investigated, is not genetics research hopelessly tied to the concepts of the day, which most researchers agree are inadequate in many key respects? Psychologists are divided on the extent to which there are continuities or discontinuities between normality and psychopathology; is there any way in which genetics research can help with respect to this important issue? Epidemiological studies in both childhood and adult life have shown the major extent to which there is substantial comorbidity between supposedly separate psychiatric disorders; does that not hopelessly hamper genetics research, or are there ways in which genetic findings might actually aid the understanding of the meaning of comorbidity patterns? Despite years of endeavor, researchers do not seem to be any nearer to discovering the mode of transmission of disorders such as schizophrenia or bipolar affective disorder, for which there are strong genetic contributions of some kind. Is there any expectation that researchers can do any better in the future? Molecular genetics research with both disorders has led to premature claims that have then had to be withdrawn; how will things get any better for the future?

Fourth, how may genetics research aid the understanding of developmental change? It is clear that development must, and does, involve a complicated mix of continuities and discontinuities and that genetic factors play a part in both. However, will it be possible for genetics research to take researchers beyond that general statement? How can genetics research be used to test hypotheses about developmental mechanisms or processes? Some psychopathologists have recently become interested in the notion of "heterotypic continuity," and developmental psychologists have also come to appreciate the ways in which the surface manifestations of an underlying trait may change over the course of development (e.g., attentional and habituation processes in infancy are related to problem-solving intelligence in later childhood). However, it seems very likely that the general concept of heterotypic continuity involves a variety of different mechanisms; is there any way in which genetics research may help to disentangle these and test competing hypotheses about their operation? What are the implications for genetics research strategies if developmental questions are to be investigated?

The Part Six chapter authors represent a range of backgrounds and perspectives. John K. Hewitt (chapter 21) writes as an experienced behavioral geneticist; the others, in a variety of different ways, have been collaborators in projects involving some genetic perspective or have been consumers of genetic ideas. Hewitt outlines what is involved in a psychometric approach to behavior and indicates how behavioral genetics analyses can shed new light on both the organization of behavior and the developmental processes as they apply to both normal behavior and psychopathology. Genetic analyses are by no means constrained by the limitations of prevailing psychiatric concepts; to the contrary, they can do much to refine and develop them. Attention is drawn to some important methodological advances and to the ways in which these are likely to provide information that is relevant to the work of clinicians.

David Reiss (chapter 22) responds to the questions from the perspective of a psychiatrist with a special expertise in the study of family systems and as someone who is now codirecting a large-scale behavioral genetics study on nonshared environmental influences. He notes the issues that must be addressed in using both twin and adoptee designs to tackle genetic questions, and he indicates the ways in which the restrictions of each design can be overcome. He goes on to emphasize two major recent findings that are particularly influential: first, the demonstration that, for many behaviors, environmental influences that are common to siblings in the same family are less important than those that impinge differentially on each child and, second, the finding that many supposedly environmental variables are influenced significantly by genetic factors. Some important implications are discussed in relation to research that is still ongoing. He suggests that over the next few years there is likely to be a rapid gain in knowledge about environmental mechanisms involved in the expression of genetic influences as well as the achievement of a better understanding of environmental effects on development that do not reflect genetic processes. More specifically, he argues that the dynamics of sibling relationships may well become the fulcrum of new psychosocial models of development.

Michael Rutter (chapter 23, with Judy Silberg and Emily Simonoff) discusses the issues from the viewpoint of a child psychiatrist who has used epidemiological and longitudinal research strategies to investigate the effects of psychosocial adversities but whose collaborative research in recent years has made increasing use of genetic research strategies. Silberg and Simonoff are both behavior geneticists involved in those collaborations. These authors approach the topic by considering the critiques of behavioral genetics that have been expressed from the perspectives of molecular genetics, environmental research, psychopathology, and medicine. They go on to consider some of the key priorities for future research, discussing how genetic designs may be used to test for environmental effects. They further consider how these designs can lead to a better understanding of diagnostic issues, cast crucial light on the very important issue of the diverse mechanisms that may be involved in the overlap between supposedly different forms of psychopathology, and provide information on the mechanisms involved in developmental change and heterotypic continuity and on the continuities and discontinuities between normality and disorder. They argue that the main strength of genetic research strategies lies in their power to provide leverage on a wide range of questions in developmental psychopathology that, at first sight, do not seem to have anything very much to do with genetics as such. As a consequence, progress in the future is likely to be aided by effective collaborations between clinical researchers, developmental investigators, and behavioral geneticists.

The importance of both nature and nurture is now generally accepted, and attention is being paid to the specific mechanisms involved in their interplay. These provide the agenda for future research, as discussed in the chapters in this section and in other parts of this book.

CHAPTER 21

The New Quantitative Genetic Epidemiology of Behavior

John K. Hewitt

In proposing a structure for the chapters in this section, Rutter posed some general and specific questions about the role of behavioral genetics research in things such as the understanding of the organization of behavior and the delineation of psychopathology, behavioral development, and the interplay between nature and nurture. More provocatively, he asked why anyone should be interested in the (nonmolecular) genetics of human behavioral traits at all. In this chapter, I examine these questions and discuss some fruitful avenues for further research.

The organization of behavior and its assessment, the development of behavior, and the delineation of psychopathology are important and recurring themes in behavioral genetics that go beyond simple considerations of heritability. Each provides an opportunity to illustrate some of the exciting conceptual advances that have occurred recently. The continuing development of these conceptual advances, together with their practical application, is defining a new quantitative genetic epidemiology of behavior that will complement any gains in understanding of the biology of behavior that arise from the exploitation of molecular genetics technologies.

The Organization of Behavior

The use of genetically informative designs in epidemiological research or, more generally, adopting a genetically informed perspective from which to view behavior leads to important new insights about its organization and development. Specifically, I consider what is gained from multivariate genetic analyses over traditional "nongenetic" or phenotypic analyses.

By way of introduction, consider the traditional psychometric approach to understanding the organization of individual differences in behavior through factor analyses of the phenotypic intercorrelations between variables. In Figure 1, I illustrate the essence of the definition of a psychometric factor in the form of a path diagram. (Readers not familiar with this kind of schematic presentation should see, for example, Loehlin, 1987, for an introductory account of the use and interpretation of path diagrams representing factor and other nongenetic models or Neale & Cardon, 1992, for a more detailed exposition in the context of genetic studies.) A goal of psychometric analysis is to determine how many such factors are needed to represent the communality or correlations among the variables and, given this determination, the extent to which the variance of each measured variable may be determined by the common underlying factors. This leads naturally to the definition of traits or syndromes and broad behavioral summaries such as those derived from the Child Behavioral Checklist (CBCL) in the area of child and adolescent psychopathology (Achenbach & Edelbrock, 1983). Here, children's behavioral problems are summarized in terms of two broad factors of externalizing and internalizing problems with contributing syndromes such as aggressive, depressed, or hyperactive behavior. Similarly, in the cognitive domain, factor analyses result in an understanding of the organization of behavior in terms of a broad underlying general factor, *g*, and group factors describing verbal abilities, spatial abilities, and so on. Again,

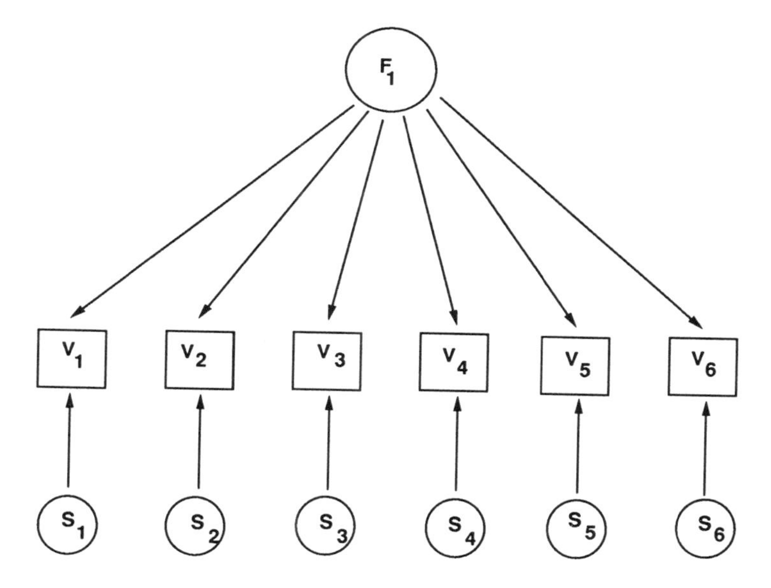

FIGURE 1. The "psychometric" or "phenotypic" or "latent phenotype" model. By convention, the rectangular boxes, V_1, V_2, and so on, represent measured variables, and the circles represent latent or unmeasured explanatory variables. F_1 is a common factor and S_1, S_2, and so on, represent sources of variation specific to each variable. The arrows or paths (which, for simplicity, are not labeled) from the latent variables to the observed variables would be quantified as the partial correlation coefficients, or factor loadings.

in the domain of psychometrically assessed adult personality, there are the "Big Five" factors, for which one set of labels would be Extraversion, Agreeableness, Conscientiousness, Neuroticism, and Openness (see Loehlin, 1992, for a review of behavioral genetics analyses of these personality factors).

However, implicit in this kind of approach is an untested assumption that the genetic and environmental factor structures—or principles of organization—are congruent or similar at least. The constraints imposed by this assumption can be seen if one compares the top and bottom panels of Figure 2. The top panel of Figure 2 shows genetic and environmental factors having a similar proportional impact on each variable. By contrast, in the bottom panel of Figure 2 the proportional impact of the

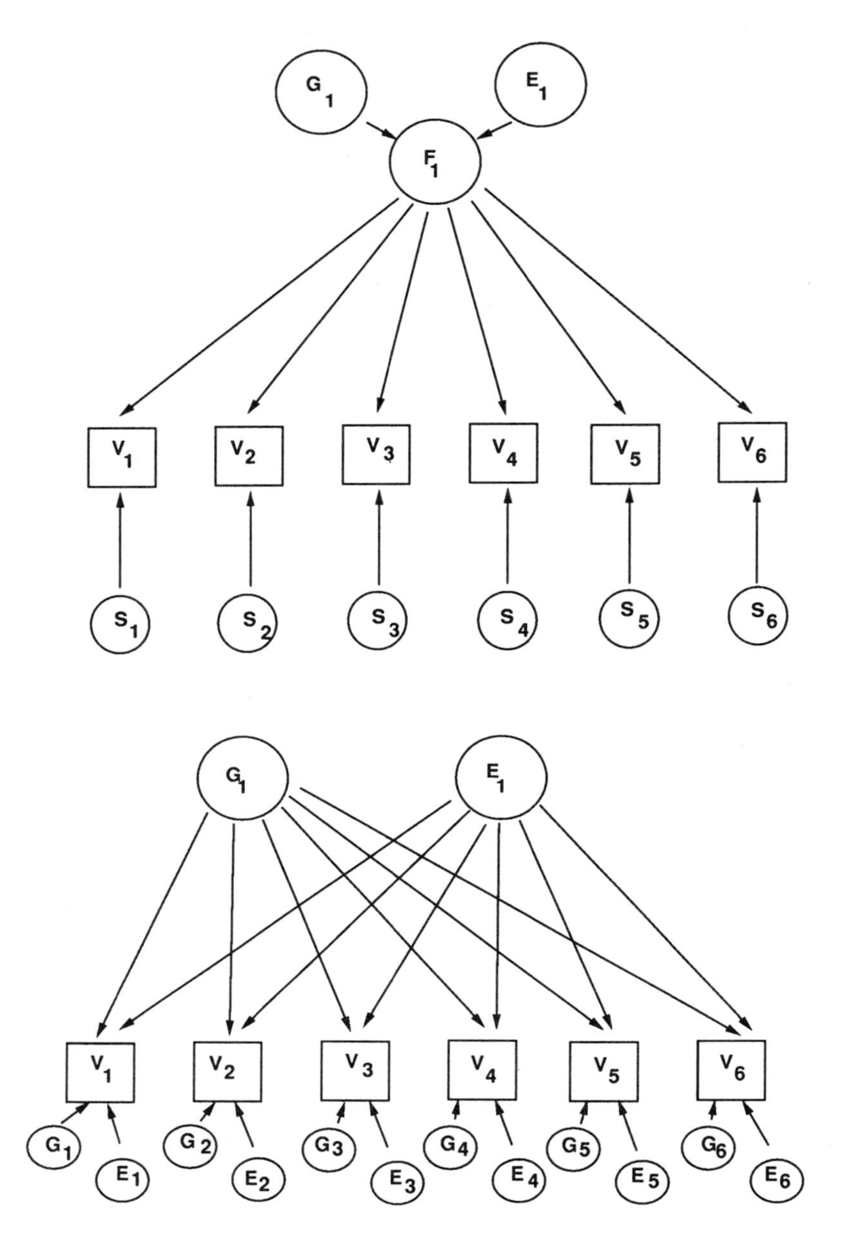

FIGURE 2. Top panel: The psychometric factor model, which illustrates the implicit assumption that, relatively, the influence of genetic (G) variation and environmental (E) variation on the factorial or syndromic (F) organization of behavior is the same. (V = variable; S = variation specific to each variable.) Bottom panel: The independent pathway or biometric factor model, which illustrates the possibility that the pleiotropic action of genes may cause a different syndrome pattern from that resulting from the manifold effects of environmental influences. (V = variable.)

genetic and environmental factors may differ from variable to variable. Psychometric factor models implicitly assume that the ratios of genetic factor loadings, as one goes from one variable to the next, are equal to those of the environmental factors. Thus, for example, one could not adequately characterize the circumstance that some correlations among variables are largely determined by the pleiotropic action of genes, whereas others largely reflect manifold effects of environmental insults. Equally, phenotypic factor analyses cannot account for whether the complexity of genetic influences (i.e., the number of genetic factors) is the same as the complexity of the environmental influences.

An example of the interesting insights that genetically informative data can provide is found in the study of symptoms of anxiety and depression reported by Kendler, Heath, Martin, and Eaves (1987). Real data from the real world are seldom as elegant and systematic as one's conceptual apparatus would like them to be, and the Kendler et al. full report is no exception. However, at the risk of oversimplifying, I summarize the conclusions of their multivariate genetic analysis of self-reported symptoms of anxiety and depression from a population-based Australian National Twin Registry sample.

First, traditional phenotypic factor analysis suggested that depression and anxiety tend to form separate symptom clusters or syndromes. However, taking into account the additional information that the Kendler et al. (1987) genetically informative twin design provided (i.e., the correlations of item responses across individuals with different degrees of genetic relatedness), they were able to reject the psychometric factor model in favor of the independent pathways model. The reason for this was largely because a single genetic factor appeared to predispose individuals toward endorsement of both anxiety *and* depression items, whereas two distinct environmental factors contributed either mainly to depression or mainly to anxiety, respectively. Hence, the authors' overall rhetorical characterization of their results was "same genes, different environments."

Although I must emphasize caution about the extent to which the detailed empirical results are simplified in this summary and that they have been only partially replicated (Kendler, Neale, Kessler, Heath, &

Eaves, 1992), the implications of results such as these could be profound. In the first place, the understanding of the mechanism of genetic influence is different because, genetically, anxiety and depression are seen as largely pleiotropic expressions of the same genes (leaving aside the possibility of genetic linkage disequilibrium). At the least, research strategies for identifying major genes or quantitative trait loci (QTL) influencing these symptoms would be different if one approached them as genetically distinct traits. The other side of this coin is that in seeking to identify the etiologically significant components in the environment, one must expect that risk factors for anxiety symptoms will tend to be independent of those for depression. Such conclusions will not necessarily surprise clinicians, but it is important to note that the traditional nongenetic analysis falls short of being able to arrive at this practically and conceptually important distinction.

The Assessment of Behavior

A related issue, which has a special relevance to the study of children, is that of how best to understand disagreements among different informants about the behavior of given individuals. This is perhaps most obvious in the study of developmental psychopathology, wherein primary sources of information are parent, teacher, or professional descriptions of a child's behavior. Clinicians know that agreement across different categories of informant is often low (i.e., with correlations around 0.3; Achenbach, McConaughy, & Howell, 1987) and that even correlations for similar informants (e.g., mothers with fathers) are modest (i.e., around 0.5–0.6).

There are, of course, good reasons why the levels of agreement should be low (Cox & Rutter, 1985): Different informants, such as the child's parents, teachers, or peers, have different situational exposure, different degrees of insight, and different perceptions, evaluations, and normative standards that may create differences in assessment (Hewitt, Silberg, Neale, Eaves, & Erickson, 1992). The point here is that behavior genetic approaches, making use of multiple informants and genetically informative designs such as the twin study, can test the assumption that raters are assessing the same essential characteristic of the child, perhaps unreliably and perhaps with bias (Hewitt et al., 1992). Alternatively, this same approach can determine the degree of genetic and environmental specificity for the assessments from different informants that, again, as with the standard multivariate case already discussed, can provide important information about the sources of the disagreements among informants. Figure 3 illustrates two extreme views of parental and teacher ratings of children's behavior. The first, analogous to the psychometric or latent phenotype factor model, assumes that each rater is rating the same phenotype of child, P_1 (which could even be the child's actual behavior), and that there would also be sources (not shown in the figure) of unreliability or bias in each individual's rating that would reduce the correlation among the raters. The second (bottom panel of Figure 3) illustrates the most general or agnostic model that allows for the possibility of any degree of genetic or environmental similarity or specificity of the behaviors rated.

In a practical example of this kind of analysis applied to parental ratings of children's behavior, Hewitt et al. (1992) showed that (a) to a good approximation, mothers and fathers can be assumed to be rating the same underlying phenotype for a global assessment of internalizing behavior problems using the CBCL (cf. top panel of Figure 3); (b) the contributions of rater bias and unreliability could be separated from the shared and nonshared environmental components of variation in a behavior genetic analysis; and (c) when this was done, the heritability of the consistently rated behavior in young boys may be as high as 70% compared with the 47% estimate that would have been obtained from maternal assessments alone. Although the corresponding analyses were not reported in detail, there was evidence of genuine informant specificity, over and above biases and unreliability, in the rating of externalizing behavior problems (cf. bottom panel of Figure 3). Again, if these results hold up under replication, they will provide important new information to guide clinicians' interpretations of parental disagreements about different kinds of behavior problems.

As with other multivariate genetic analyses, extending and developing those kinds of methods, in conjunction with the collection of high-quality multiple informant twin, family, and adoption data, will advance

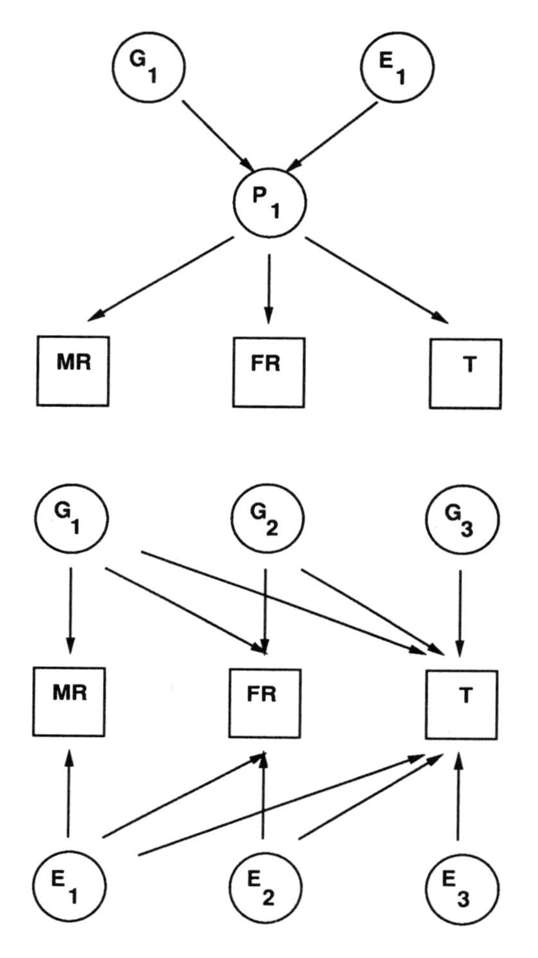

FIGURE 3. Top panel: Maternal (MR), paternal (FR), and teacher (T) ratings assumed to be assessing equivalent behaviors (cf. the psychometric model in the top panel of Figure 2). (P = phenotype.) Bottom panel: An agnostic biometric model, known as a Cholesky or triangular decomposition, that allows for any degree of genetic (G) or environmental (E) specificity in the different ratings.

the understanding of the extent to which psychological traits should be viewed not so much as fixed objective phenotypes (e.g., height or weight) but as properties of individuals that may be expressed or perceived differently in relation to different individuals or informants. These are empirical questions, and behavioral geneticists are developing the methods to address them.

Understanding Development

Just as behavioral genetics analysis can shed new light on the organization of behavior and on the proper interpretation of multiple sources of information about behavior, it has recently been appreciated how powerful quantitative genetic epidemiology can be in aiding the understanding of developmental processes. As one example, the theoretical work of Eaves, Long, and Heath (1986) and its subsequent explication (Hewitt, Eaves, Neale, & Meyer, 1988), application (Boomsma, Martin, & Molenaar, 1989), and extension (Eaves, Hewitt, & Neale, 1988) have put researchers in a strong position to specify some key developmental issues.

In particular, the distinction between *persistent* effects, whether genetic or environmental in origin, and *transient* effects is proving to have significant power in explaining developmental changes in phenotypes ranging from those of traditional concern for psychologists, such as cognitive ability (Eaves et al., 1986) and delinquency (Rowe & Britt, 1991), to those of interest to behavioral medicine, such as blood pressure (Hewitt, Carroll, Sims, & Eaves, 1987) and obesity (Fabsitz, Carmelli, & Hewitt, 1992). In addition to traditional models that ascribe changes in parent–offspring resemblance to specific genes being switched on at key periods of biological change such as adolescence, there are two different mechanisms that a priori predict different patterns of age-to-age correlations and parent–offspring resemblance. They may operate on genetically and environmentally caused variation. First, correlations between different ages may arise because the effects of expressed phenotypes (e.g., drug use) are persistent and thus influence behavior at later ages. This developmentally transmitted effect combines with newly arising variation at later ages (e.g., leading to drug-abusing "careers"). Intervention to prevent expression of the behavior at one age would have directly transmitted benefits at later ages. Alternatively, age-to-age correlations may be a consequence of the same genes and environments directly influencing the behavior in the same way at each age, without the persistent effects from earlier behavior. In this case, intervening to prevent the expression of behavior (e.g., hyperactivity) at one age would not confer subsequent benefits in the absence of continued intervention. Although the two mechanisms may act together, the pattern of correlations during development differs for the two mechanisms, and developmental transmission can be detected even against a background of continuously acting genetic (or environmental) influences (Eaves et al., 1986; Hewitt et al., 1988). In psychometric terms, the essential difference between the patterns generated by the two mechanisms is that between a simplex pattern for the developmentally transmitted, persistent effects of early behavior on later behavior and a common factor pattern for the nondevelopmental uniform impact of genetic or environmental influences. These alternatives are illustrated as Models II and III in Figure 4.

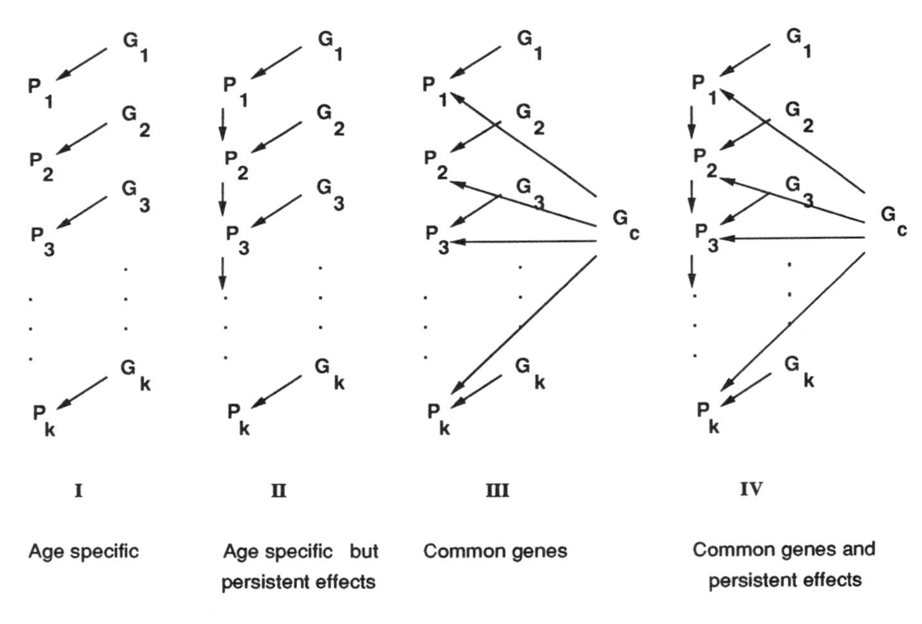

FIGURE 4. Four component path models for the genetics of a developing phenotype. In this figure, P_1, P_2, and so forth, are the genetically determined aspects of the phenotype on Occasions 1, 2, and so on; G_1, G_2, and so on, are the occasion-specific genetic effects; and G_c represents genetic effects common to all occasions. Corresponding to these genetic mechanisms, there may similarly be different kinds of environmental contributions during development. The first model shows effects that are entirely specific to a given age or occasion. Model II shows age-specific genetic effects whose effects persist over time. Model III shows common genes influencing the phenotype in the same way at different ages. Model IV includes all of these effects.

These developmental models also predict particular patterns of change in parent–offspring resemblance that are not accounted for by traditional factor models (e.g., Martin & Eaves, 1977) extended to measurements at different ages. Moreover, the pattern of monozygotic and dizygotic twin correlations need not be in accord with the parent–offspring correlation expected by traditional nondevelopmental genetic models. The developmental perspective provides a theoretical framework for understanding, and in fact predicting, these apparent discrepancies.

The multivariate extension of the developmental model (Eaves et al., 1988) provides researchers with a method for testing alternative causal hypotheses about the interrelationships among dispositions, environmental indexes, and outcome variables, taking into account genetic and developmental effects. Longitudinal twin, family, and adoption data will provide data sets designed for this purpose (an example of these analyses is provided in this book in chapter 5).

Latent Class Models and the Delineation of Psychopathology

Perhaps one of the most exciting new developments in the quantitative genetic epidemiology of behavior is the application by Lindon Eaves (Eaves et al., 1993) of latent class models to genetically informative data from twins and, potentially, other family relationships. The motivation behind this is an attempt to accommodate categorical approaches to clinical diagnosis within a quantitative genetic framework without having to impose a series of dimensional assumptions, as would be done in the standard polygenic threshold model (e.g., Falconer, 1989).

Temporarily setting aside the underlying dimensional assumptions on which most multivariate and developmental analyses are predicated, Eaves et al. (1993) delineated a procedure for analyzing multiple-symptom data on the assumption that individuals fall into one of a set of classes, each of which is free to present its own particular symptom profile (which need not necessarily be simply a more extreme set of symptoms across the board of the kind a traditional dimensional threshold model would predict).

An example of the application of this analysis to symptoms of hyperactivity is given in chapter 15 of this book, and full details of the methodology are reported by Eaves et al. (1993). Suffice it here to quote from that article to provide a sense of the potential for the delineation of psychopathology that this approach may offer.

> Freed from the assumptions of normality and additivity characteristic of the threshold model, we can consider a tantalizing array of specific mechanistic hypotheses which may be tested by appropriate application of the latent class model to kinship data. Of course, there is no guarantee that the alternatives are any better in summarizing complex sets of behavioral ratings and, indeed, they may be worse. But at least they should be considered.
>
> Although twins are not necessarily the first approach that most researchers would consider for testing discontinuous genetic models for complex phenotypes, the approach we have outlined seems to have great potential for resolving important nosological and etiological questions in clinical areas requiring the integration of data from multiple items and covariates some of which may only be gathered on a contingent basis. We find especially appealing the possibility of teasing apart the interaction of genes and environment underlying certain forms of etiological heterogeneity. (Eaves et al., 1993, p. 17)

In other words, (a) are conduct-disordered children, for example, simply showing more extreme, more pervasive, and more frequent conduct problems than nondisordered children or, by contrast, are disordered children showing distinct patterns of problem behavior different from that of nondisordered children; (b) are there etiologically distinct categories of disordered children within the overall diagnostic group; and (c) how do genetic and environmental risks contribute to these categories and etiological distinctions? These questions become even more important when one considers longitudinal observations for which genetic or environmental etiological heterogeneity may be related to different developmental trajectories, patterns of comorbidity, familial risk, and eventual outcome.

Where Do We Go From Here?

I have described some methodological advances in multivariate behavioral genetics analyses related to the organization of behavior, the interpretation of multiple sources of information about behavior, behavioral development, and categorical approaches to understanding psychopathology. In the space available to me, I have not discussed other, equally noteworthy, advances in behavioral genetics analysis such as the regression techniques of DeFries and Fulker (1985, 1988) for estimating heritabilities from selected samples, analysis of genotype–environmental interactions, detecting differences in heritability in extreme groups compared with the normal range of individuals, and, more recently, detecting linkage in a quantitative trait (Fulker et al., 1991). Likewise, there has been a plethora of research activity aimed at improving the understanding of issues such as sex-limited expression of behavior (e.g., depression or alcoholism), testing hypotheses about the direction of causation between (putative) risks and (putative) outcomes (Heath et al., 1993), the buffering or amplifying effects of protective or risky environments on the expression of genotype predispositions (Heath, Jardine, & Martin, 1989), and so on. Interested readers may turn to Neale and Cardon's (1992) compilation of this fast-moving field of methodological research.

All of these conceptual advances are just now beginning to be applied to data sets that are adequate in size and whose collection and measurement has been designed with their application in mind. However, all of these new analytical strategies beg the fundamental question of the heritability of the behaviors under study. It is, of course, axiomatic that the presence of statistically significant heritability is required for the success of genetic analyses. It is also the case that the more substantial the heritability, the more powerful the genetic analysis will be, regardless of whether it is quantitative or categorical, or even if one is concerned with finding genetic linkages with single major genes or QTL. Currently in the field of developmental psychopathology, for example, knowledge of the basic properties of genetic and environmental control is woefully inadequate. Is hyperactivity in children largely genetic? What about depression? Is it different before and after puberty? Or different in boys than girls? Are there aspects of behavior (e.g., conduct problems or delinquency in adolescence) that are considerably influenced by environments shared by siblings? How does this change with the transition to adulthood? All of these kinds of questions are, in essence, heritability-type questions and, as such, they should be an integral part of the understanding of behavioral development.

Clinicians desperately need more of the new kinds of basic quantitative genetic epidemiological information I have discussed. This information will come from longitudinal multivariate (and, if necessary, multiple informant) twin, family, and adoption studies. These data are needed to map out the general characteristics of the genetic and environmental influences on individual behaviors, their comorbidity, and their development.

References

Achenbach, T. M., & Edelbrock, C. S. (1983). *Manual for the Child Behavior Checklist and Revised Child Behavior Profile*. Burlington: University of Vermont, Department of Psychiatry.

Achenbach, T. M., McConaughy, S. H., & Howell, C. T. (1987). Child/adolescent behavioral and emotional problems: Implications of cross-informant correlations for situational specificity. *Psychological Bulletin, 101*, 213–232.

Boomsma, D. I., Martin, N. G., & Molenaar, P. C. M. (1989). Factor and simplex models for repeated measures: Application to two psychomotor measures of alcohol sensitivity in twins. *Behavior Genetics, 19*, 79–96.

Cox, A., & Rutter, M. (1985). Diagnostic appraisal and interviewing. In M. Rutter & L. Hersov (Eds.), *Child and adolescent psychiatry* (pp. 233–248). Oxford, UK: Basil Blackwell.

DeFries, J. C., & Fulker, D. W. (1985). Multiple regression of twin data. *Behavior Genetics, 15*, 467–473.

DeFries, J. C., & Fulker, D. W. (1988). Multiple regression analysis of twin data: Etiology of deviant scores versus individual differences. *Acta Geneticae Medicae et Gemellologiae, 37*, 205–216.

Eaves, L. J., Hewitt, J. K., & Heath, A. C. (1988). The quantitative genetic study of human developmental change. In B. S. Weir, E. J. Eisen, M. M. Goodman, & G. Namkoong (Eds.), *Proceedings of the Second International Conference on Quantitative Genetics* (pp. 297–311). Sunderland, MA: Sinauer.

Eaves, L. J., Long, J., & Heath, A. C. (1986). A theory of developmental change in quantitative phenotypes applied to cognitive development. *Behavior Genetics, 16*, 143–162.

Eaves, L. J., Silberg, J. L., Hewitt, J. K., Rutter, M., Meyer, J. M., Neale, M. C., & Pickles, A. (1993). Analyzing twin resemblance in multi-symptom data: Genetic applications of a latent class model for symptoms of conduct disorders in juvenile boys. *Behavior Genetics, 23*, 5–20.

Fabsitz, R., Carmelli, D., & Hewitt, J. K. (1992). Evidence of independent genetic effects on obesity in middle age. *International Journal of Obesity, 16*, 657–666.

Falconer, D. S. (1989). *Introduction to quantitative genetics* (3rd ed.). Harlow, UK: Longman.

Fulker, D. W., Cardon, L. R., DeFries, J. C., Kimberling, W. J., Pennington, B. F., & Smith, S. D. (1991). Multiple regression analysis of sib-pair data on reading to detect quantitative trait loci. *Reading and Writing: An Interdisciplinary Journal, 3*, 299–313.

Heath, A. C., Jardine, R., & Martin, N. G. (1989). Interactive effects of genotype and social environment on alcohol consumption in female twins. *Journal of Studies on Alcohol, 50*, 38–48.

Heath, A. C., Kessler, R. C., Neale, M. C., Hewitt, J. K., Eaves, L. J., & Kendler, K. S. (1993). Testing hypotheses about direction of causation using cross-sectional family data. *Behavior Genetics, 23*, 29–50.

Hewitt, J. K., Carroll, D., Sims, J., & Eaves, L. J. (1987). A developmental hypothesis for adult blood pressure. *Acta Geneticae Medicae et Gemellologiae, 36*, 475–483.

Hewitt, J. K., Eaves, L. J., Neale, M. C., & Meyer, J. (1988). Resolving cause of developmental continuity or "tracking": Longitudinal twin studies during growth. *Behavior Genetics, 18*, 133–151.

Hewitt, J. K., Silberg, J. L., Neale, M. C., Eaves, L. J., & Erickson, M. (1992). The analysis of parental ratings of children's behavior using LISREL. *Behavior Genetics, 22*, 293–317.

Kendler, K. S., Heath, A. C., Martin, N. G., & Eaves, L. J. (1987). Symptoms of anxiety and depression: Same genes, different environments? *Archives of General Psychiatry, 44*, 451–457.

Kendler, K. S., Neale, M. C., Kessler, R. C., Heath, A. C., & Eaves, L. J. (1992). Major depression and generalized anxiety disorder: Same genes, (partly) different environments? *Archives of General Psychiatry, 49*, 716–722.

Loehlin, J. C. (1987). *Latent variable models.* Hillsdale, NJ: Erlbaum.

Loehlin, J. C. (1992). *Genes and environment in personality development.* Newbury Park, CA: Sage.

Martin, N. G., & Eaves, L. J. (1977). The genetical analysis of covariance structure. *Heredity, 38*, 79–95.

Neale, M. C., & Cardon, L. R. (1992). *Methodology for genetic studies of twins and families.* Norwell, MA: Kluwer Academic.

Rowe, D. C., & Britt, C. L. (1991). Developmental explanations of delinquent behavior among siblings: Common factor vs. transmission mechanisms. *Journal of Quantitative Criminology, 7*, 315–332.

CHAPTER 22

Genes and the Environment: Siblings and Synthesis

David Reiss

As is typical in science, a fresh synthesis in areas of dispute is made possible by important advances in methods, theory, and findings. The opportunity for synthesis, at this juncture, comes from a surprising source: siblings. Recent trends in genetic research have depended heavily on the use of research designs involving siblings—an extension of a long tradition in this field—but have now focused on the importance of sibling differences. At the same time, environmental studies are positioned to make great strides in understanding sibling differences as well. Thus, the story of a new synthesis is a story of siblings. I try to summarize some of that in this chapter.

The scientific story line has four major chapters. First, there have been improvements in the methods of behavioral genetics. Second, these improved methods have yielded results that are now conspicuously relevant for environmental theories about individual differences in psychological development. Third, there have been corresponding advances in studies of child and adolescent development and family process. These advances permit a clearer picture of sibling differences because these are becoming a central link between genes and environment. These research methods also meet the challenges of forming genuine links between social systems and genetic analysis. Fourth, several integrative projects are currently underway that are emblematic of the new opportunity for syntheses.

Improvements in Genetic Methods

The first step in this sequence involves important advances in behavioral genetics. Many of these advances are reviewed in other chapters in this book. Most conspicuous have been the opportunities for understanding individual differences in behavior that have been opened up by the major advances in molecular genetics involving recombinant DNA techniques and linkage analysis. The current yield from these methods in the area of human behavior is still debatable, but the promise remains exciting. Perhaps of greater importance are the equally impressive gains in quantitative genetics, for which the search is for genetic factors that influence individual differences in normal development as well as the genetic basis of some forms of psychopathology. The fundamental technique in quantitative genetics compares covariances among relatives: monozygotic (MZ) versus dizygotic (DZ) twins, adopting parent–child pairs versus biological parent–child pairs, and full siblings versus half siblings versus unrelated siblings (as in step families). However, these familiar techniques (many of which involve siblings) have been improved by better attention to sampling, improved statistical models, and better measurement of behavior.

However, an important but less well known advance in methods from the perspective of environmental studies is the attention given by behavioral geneticists to the potential artifact of labeling. Strictly speaking, this is not so much an advance in methods as it is a clarification of the validity of methods by systematic evaluation of a confounding factor that is difficult to isolate. Environmentalists often dismiss the covariance comparisons because the individuals studied know their genetic relatedness. For example, parents know whether their children are adopted or not and twins know whether they are identical or not. Environmentalists contend that it is this knowledge, not an underlying genetic mechanism, that may account for findings in these studies. Because this has been a neglected but fundamental objection to genetic data by nongeneticists, it is worth examining in some detail as an indicator of the current potential for synthesis.

In the past, family studies have hopelessly confused genetic and environmental sources of resemblance. Twin studies were an improvement, but there was an ongoing concern that parents treated identical twins more similarly than they did fraternal twins. Thus, higher concordance rates for any variable in MZ twins might reflect environmental and not genetic similarity. Careful investigation of this possible confound have shown that parents of MZ twins do treat them more similarly but that these differences do not account for differences in concordance rates of outcomes between MZ and DZ twins (Loehlin & Nichols, 1976).

Adoption studies were introduced to circumvent, in part, this objection to the twin method. However, it soon became clear that the adoption method had its own problems with confounding environmental and genetic effects. A vexing but empirical issue is selective placement (a correlation between a characteristic of the biological and adopting parent). Also, a potential problem in these designs is whether adopting parents treat their adoptees in ways that are systematically different from the way they treat their biological children. This might lead to greater discrepancies between adopted siblings than between biological siblings for environmental reasons alone.

Two strategies, using more stringent controls, have recently been introduced into quantitative behavioral genetics designs. Both examine the confounding effect of labeling or expectations arising from the parents' and children's knowledge of how parents, children, and siblings are, in fact, genetically related. In twin designs, MZ and DZ twins are labeled as such by an obstetrician, subsequent professionals, lay people, and the parents themselves. How might this labeling influence parental expectations and behavior toward these children, which in turn might influence concordance rates? Similarly, in adoption studies, how much does the label of "adoptee" influence parental behavior?

One approach takes advantage of "natural mislabeling." For example, in a study of 342 same-sex twins who were 10–16 years old, 41 pairs agreed on their zygosity but were wrong (e.g., DZ twins both believing they were MZ), and 93 pairs disagreed on their zygosity (Scarr & Cater-Saltzman, 1979). The results of this study suggested that perceived or labeled zygosity may have some influence on similarity between twins, particularly among DZ twins, on personality measures but not on cognitive measures. That is, DZ twins who believed they were MZ twins were more similar on two personality measures. However, the data also suggested that this effect was probably accounted for by actual variation in genetic similarity among the DZ twins as determined by blood typing; that is, the DZ twins who showed the most similar blood types were most likely to have mislabeled themselves as MZ.

In examining the role of labeling in adoption designs, the investigator cannot depend on mislabeling; it is rare that parents are confused about a child's adoptive status, although some children may have their adoption status hidden from them through childhood, adolescence, and even adult life. To address the problem of labeling in adoption designs, one study used measures of parent–child similarities, as perceived by both parents and children, as another approach to study the effects of labeling and expectations (Scarr, Scarf, & Weinberg, 1980). The investigators reasoned that as a consequence of knowing they are biologically related, parents and children in biologically linked relationships ought to be more likely to perceive themselves as similar on a range of variables. If the perception of biological relatedness, as expressed through perceived similarities, influences actual similarity, then perceived and actual similarity should be correlated in both biologically and adoptively linked parent–child relationships. However, virtually no relationships of this kind were observed.

Another strategy for circumventing this effect of labeling, and the parental expectations and attitudes that arise from it, is to study MZ and DZ twins who have been reared apart. When separation occurs shortly after birth, an investigator can be certain that differences in concordance rates between the two types of twins cannot be due to environmental labeling or any of its measurable and unmeasured sequelae.

An increasing number of recent studies that used reared-apart twins, for whom labeling can have little or no influence, has helped to clarify this issue (Tellegen et al., 1988). These designs can be carried out with particular effectiveness in Scandinavia because investigators can use uniform public records of twin births (Pedersen, Plomin, McClearn, & Friberg, 1988; Rose, Kaprio, Williams, Viken, & Obremski, 1990). In these studies, response rates of eligible twins have been high; and as a consequence, these studies are not plagued by problems of volunteer bias. This bias is especially relevant for genetic studies in that MZ twins tend to volunteer more readily than DZ twins. As a consequence, variability is greater among the former, which thereby increases covariance and overestimates genetic effects (Lykken, McGue, & Tellegen, 1987). The logic for reared-apart twin studies is straightforward and compelling. If twins are reared separately, and if the parents of one twin are in no contact with the parents of the other twin, then a knowledge of zygosity by the rearing parents cannot influence covariance between the twins. The same is true, of course, of knowledge of zygosity on the part of the twins themselves. Although there are some exceptions, for the most part inferences drawn from studies of reared-together twins have been confirmed in studies of reared-apart twins.

No experiment, of course, is perfect—particularly ones arranged by nature. The simple fact of twins being reared apart does not prevent them, as adults, from seeking each other out. In fact, there is a notable variation among reared-apart twins in their interest and success in seeking one another out, and lively debate has filled recent pages of the behavioral genetics literature as to whether this adult experience may modify any conclusions drawn from twin studies (Rose et al., 1990). There is some data to suggest that intrapair similarity in twins may lead to variation in frequency of adult contact and also some evidence that the reverse is true. In either case, most of the major conclusions drawn from studies of reared-together twins has held up, even when investigators have corrected for this variation in contact (Plomin, Chipuer, & Neiderhiser, in press).

The problem of labeling has been a stumbling block for many environmental researchers; and now that it has in large measure been cleared away, the full implications of genetic findings can be more readily integrated into a general science of individual psychological differences.

Recent Findings From Quantitative Behavioral Genetics

Two major recent findings of quantitative behavioral genetics are particularly noteworthy and are critical components of the new integration.

The first of these new findings can be summarized very briefly. In a large number of studies, two patterns of results stand out in sharp relief. For many traits or characteristics, MZ twins show correlations of .50 or less. Because MZ twins are genetic copies of one another, only environmental factors, plus measurement error, could account for these relatively low correlations. Furthermore, these environmental factors must be influencing one twin and not the other. If measurement is highly reliable, then such low correlations suggest that about half of the variation among MZ twins is due to these twin-specific effects. A second important pattern concerns siblings who are genetically unrelated, such as siblings who are adopted from different families. In almost every study so far reported, correlations on measures of many different kinds of developmental outcomes show zero or near-zero correlations for these sibling pairs, even if they have been reared together since birth. Again, if measures are very reliable, then this finding suggests that environmental influences that are common to siblings in the same family have very little influence on individual differences in development. If factors that are common to siblings (such as social class, maternal depression, or neighborhood decay) were important, then one ought to see greater-than-zero correlations for developmental outcomes for genetically unrelated siblings. For those traits that show genetic influences, we would also expect higher correlations for MZ twins. These surprising findings from twins and unrelated siblings have led to a recent flurry of excitement for studying the special form of environmental influence that has been awkwardly termed *nonshared environment* (Plomin et al., in press; Plomin & Daniels, 1987).

There is a second, more recent set of findings from quantitative behavioral genetics that is important to environmentalists. Genetic analyses have been applied not only to outcome measures, such as cognitive and social abilities and psychopathology, but to the traditional psychosocial independent variables such as parenting and sibling and peer relationships. By using the same twin and adoption designs that have been traditional for other topics of genetic inquiry, investigators have discovered that these environmental variables are influenced significantly by genetic factors. In the case of parenting, there are two possible mechanisms. First, heritable characteristics of children elicit certain responses, such as maternal warmth, from the environment. Second, genetic factors in parents may directly influence their response to their children. There are, as yet, few data on the latter; however, information on the former is accumulating (Plomin & Bergeman, 1991).

The most important single implication of these genetic influences on environmental variables is that they help researchers to understand more fully the associations that are observed between environmental variables and outcome variables. Because genes may influence both sets of variables, they may account for a significant component of any observed association between the two. For example, one frequently replicated finding is a strong correlation between a measure of parenting and measure of child outcome. The parents' measure reflects their expectations and encouragement of cognitive competence in children, whereas the children's measure reflects their actual cognitive competence. For example, Braungart, Plomin, DeFries, and Fulker (1992) found that the correlations between a measure of intellectual stimulation in the home environment, the Home Observation for Measurement of the Environment (HOME) Scale (Caldwell & Bradley, 1978), and the Bayley Index of Mental Development was higher when the children were biological offspring than when they were adopted children. A reasonable inference is that part of the correlation between parental expectations and offspring competence is due to genes shared by parent and biological offspring. The same genes that influence parenting, perhaps mediated by their influence on intellectual abilities, also influence cognitive competence in young children.

In sum, behavioral geneticists have been improving their methods and providing data that are not only credible to method-conscious environmentalists but highly relevant to their models.

Corresponding Advances in Measurements of the Psychosocial Environment

The behavioral genetics data do not, of course, indicate what aspects of the environment are critical influences on development. They provide data on form, so to speak, but not on content. However, it is reasonable to assume that familiar psychosocial variables play a major role along with other environmental variables such as toxins, infectious agents, and nutrition. In the area of measurement, there have been corresponding advances in the assessment of psychosocial variables, which now make possible research designs that rise to the challenge of these new genetic data. What are these challenges and what advances permit a positive response? The advances can be grouped into two very different areas. The first is a very general one and concerns the requirements for large and specialized samples, which are necessary to explore the questions raised by behavioral genetics data. The second concerns progress in conceptualization and measures social systems that make it possible to delineate environmental factors in the social world, which are of specific relevance for each sibling in the family.

Sampling

There are two components to the challenge of sampling. The first component is the sheer size of the samples that are required for studies of genetic factors. The second is the need for highly specialized samples of siblings. I consider each in turn.

Behavioral geneticists require, by ordinary standards of behavioral science, very large samples. Environmentalists who now seek to explore the implications of behavioral genetics findings will inherit these same requirements. Large samples are required because statistically powerful models depend on detecting rather small differences in covariances or correlations between two or more groups. For example, a twin design that is powerful enough to detect heritabilities of 20% or more must reliably detect a difference in correlations between MZ and DZ twins of just 0.10. Furthermore, sophisticated genetic designs no longer depend on just one comparison of this kind but may include twins and adoptees in the same sample or even twins with full siblings and half siblings. If six or seven comparison groups are used, this often means the entire sample will consist of at least 600–700 pairs of siblings and others in their environment who must serve as informants. Assuming just one informant other than the siblings, this means that data must be collected from at least 2,000 individuals.

Sample sizes that are this large fall ordinarily into the domain of sociologists, who in turn depend on survey research techniques to reach this many individuals. Survey researchers traditionally use highly standardized, structured interviews, which are useful for measuring many areas of concern to social science including attitudes, economic status, and occupational experience. However, these interviews are not suitable for fine-grained assessments of such variables as family process or the development of social and cognitive competence in children. It is only in the last 5 years that major strides have been made in adapting survey research methodology. It is now possible to use state-of-the-art survey research methods to assess social processes in normal and pathological development.

For example, developmental psychologists have recently taken advantage of a large, unique (but representative), multipanel sample that is referred to as the Children of the National Longitudinal Survey of Youth (Chase-Lansdale, Mott, Brooks-Gunn, & Phillips, 1991). This sample of about 13,000 young people currently has over 5,800 women, and over 2,900 of these women have borne over 4,900 children. These children are a representative sample of American children who were borne to young mothers (21–29 in 1986), with an important oversampling of minorities. Survey researchers have been trained to administer a broad range of measures that have never been applied to a sample of this kind. These measures include assessment of the parents, including their intelligence and the intellectual challenges they provide to their children as measured through home observation using a standard rating instrument, the HOME Scale (Caldwell & Bradley, 1978). Measures of child development include such sophisticated and widely used assessments as the Peabody Picture Vocabulary Test, the McCarthy Scale of Children's Abilities, and the Harter Perceived Competence Scale.

Survey research methodology has more recently been pushed much further. It is now possible to obtain finer grained assessments of child competence and pathology; but even more important, specially trained survey research teams can be used to collect high-quality videotapes of family interaction from a large, national sample of families (Reiss et al., in press).

A second component of the sampling challenge is to develop samples of siblings with known genetic relatedness; these samples should be as unbiased as possible. Twins remain as important components of research designed to be sensitive to genetic effects. But significant strides have been made to sample other types of siblings. Of increasing importance has been the use of genetically unrelated siblings, for example, as can be found in many adopting families. Most adoption studies have focused on comparisons of parents and single children. Thus, in a typical adoption design, covariances between adopting parents and their adopted children are compared with covariances between biological parents and their biological offspring. Stronger designs will provide data on the biological parents of the adopted children. Although these designs are useful for estimating genetic influences, they do not respond to the challenge of studying sibling-specific environments along with estimating genetic effects. Thus, several recent studies have used adoptive siblings, who are genetically unrelated children who are reared in the same family. For example, among the 490 families enrolled in the Colorado Adoption Project are 67 adoptive and 82 nonadoptive sibling pairs (Plomin, DeFries, & Fulker, 1988).

Another approach is to combine twins and adoptive siblings in the same study. Currently, with my colleagues Mavis Hetherington and Robert Plomin, I am conducting an ongoing study of 720 families; the sample includes families of MZ twins, DZ twins, full siblings in nondivorced families, and full siblings and unrelated siblings in step families (Reiss et al., in press). As an additional bonus, this unique sample also includes half siblings in step families. Although half siblings have been used in studies that focused exclusively on genetic factors (Cook & Goethe, 1990; Schukit, Goodwin, & Winokur, 1972), they have not been used in studies that also encompass environmental variables. The use of several different types of families (in this case, those of twins, nondivorced parents, and step families) clearly allows stronger inferences to be drawn in genetic and environmental inquiries.

The current challenge is to obtain these large, highly specialized samples in ways that avoid volunteer bias and, at the same time, represent systematically larger populations. A return to the population registries in the Scandinavian countries will be important to this cause. At the same time, improved methods of case finding and sampling are required in countries like the United States that do not have these registries. Our project has taken an important step in that direction by using market panels whose demographic characteristics are carefully matched to the norms of the United States. These panels will provide family-structure and date-of-birth data on hundreds of thousands of households, which will thereby make possible efficient case findings and systematic sampling.

Sibling-Specific Environments

The second major challenge of the new genetic data is to measure sibling-specific environments. These data challenge environmentalists to define how the social world is different or unique for each sibling in the family and how these differences influence development. This challenge is apt for many areas of studies of the environment. Consider, for example, the family. Although the myth of the "schizophrenogenic mother" has now been debunked, one may still tend to think of good families versus bad families. For example, families in which there is sustained, high marital conflict are thought to put children at risk for a variety of developmental problems. But the behavioral genetics data argue strongly that marital conflict itself cannot be a major influence on psychological development. Rather, it is how this conflict is refracted uniquely for each child in the family that matters: Some may be drawn into it, whereas others may be protected from its impact. From the perspective of family research, there are two major advances that are important for assessing these sibling-specific environments.

The first advance involves concepts. Primarily as a result of the family therapy movement, there have been major advances in thinking about the family as a system. These ideas include concepts of subsystems and their interrelationships as in the case of marital distress and its differential effects on children. Existing data support a concept of family conflict that ties together many subsystems within the overall family system (Cowan, Cowan, Heming, & Miller, 1991; Gilbert, Christensen, & Margolin, 1984; Hetherington & Clingempeel, 1992; Markman & Jones-Leonard, 1985). This concept rests on the notion that the resolution of conflict is a fundamental task of all enduring marriages. Marital partners who cannot resolve conflicts by other means use their relationships with their children as an ongoing strategy in the resolution of this dilemma: Some children are brought closer into the marriage, whereas others are distanced from it. This differential distancing forms a core of the nonshared family environment for the siblings (Gilbert et al., 1984). In an analogous fashion, siblings may resolve their own conflicts by differentiating themselves from each other and their relationship with each parent (Schachter, 1982; Schachter, Gilutz, Shore, & Adler, 1978; Schachter, Shore, Feldman-Rotman, Marquis, & Campbell, 1976). Furthermore, when one sibling is developing a stigmatized deviance, such as alcoholism, the other sibling may make special efforts to avoid developing the same condition (Cook & Goethe, 1990). It is interesting that, at least in some circumstances, there are clear relationships between marital and sibling conflicts (Hetherington & Clingempeel, 1992; MacKinnon, 1988) such that nonshared environments may arise as a family-level strategy for dealing with ongoing conflict within the system. The serious impact of enduring conflict on child development, through its impairment of parenting, has now been fully documented in three major longitudinal studies of family development (Cowan et al., 1991; Hetherington & Clingempeel, 1992; Markman & Jones-Leonard, 1985).

Ideas of this kind can now be assessed with considerable accuracy because of advances in methods of measurement and analysis. Improved methods of self-report and direct observations, which can be performed in the family's own home, are now widely available. Moreover, in the last 2 years, researchers have shown that many of these methods can be accurately administered by appropriately trained survey researchers, and thus, data can be obtained from large samples (Reiss et al., in press).

Equally important, new methods of analysis are now resolving long-standing dilemmas of how to integrate self-report and direct observational measures of family process (Bank, Dishion, Skinner, & Patterson, 1990).

The Current Emphasis on Integration

These two parallel lines of development offer an opportunity for one of the most exciting confluences in behavioral science today. Indeed, there are currently several major studies underway that combine sophisticated analyses of the psychosocial environment with research designs that are capable of detecting genetic influences. One example is the current 9-year-long study of the nonshared environment currently being conducted by Mavis Hetherington, Robert Plomin, and me (Reiss et al., in press). This study, to summarize briefly, focuses on family and peer influences on the development of both competence and psychopathology in adolescents. It capitalizes in two ways on the current advances just noted.

First, it uses a large sample of families, 720 to be exact, which has been divided into six groups to detect even moderate genetic influences. These groups are families of MZ twins, families of DZ twins, families of ordinary siblings for which there has been no divorce, and step families of three kinds: those with full siblings, those with half siblings, and those with genetically unrelated siblings. Second, it uses state-of-the-art methods for characterizing family subsystems: It weds the techniques of self-report and direct observation through videotape to traditional survey research methods. Thus, 39 interview teams have been trained to collect data from urban, suburban, and rural families in all 48 contiguous states. Findings from this study have only recently become available for publication, but two trends are very clear. First, there are sizable genetic influences on measures of parenting, peer relationships, and even sibling relationships. Second, there are also strong effects of the sibling-specific or nonshared family environment.

Conclusion

Taking the long view, I believe there will be two immediate gains and a longer term, more integrative outcome. First, in the next 3–5 years, I think there will be a rapid increase in knowledge about the environmental mechanisms that are required for the expression of genetic influence. For example, several studies (e.g., Plomin & Bergeman, 1991) have shown substantial effects on genetic influence on the affection parents show toward their children. One possible role that genes may play in affective disorders may operate by this route: not directly from genes to protein synthesis to synaptic function to affect regulation but from genes (with the necessary protein and neural mediation) through social processes such as the elicitation of affection and social support that, in turn, protect against affective disorders.

The second short-term gain is the specification, for the first time, of environmental effects on development that do not reflect genetic processes. It is in this area that researchers need to focus psychosocial model building, and the odds are that these models will focus on siblings, specifically, the causes and consequences of their differential exposure to and experience of the environment. The dynamics of sibling relationships will become the fulcrum of new psychosocial models of development, a prospect that was unthinkable as recently as 5 years ago. Over the long haul, I believe that models will emerge that will integrate the ideas of genetic influences on environmental process and differential sibling experience; these models may also integrate the new findings from molecular genetics.

References

Bank, L., Dishion, T., Skinner, M., & Patterson, G. R. (1990). Method variance in structural equation modeling: Living with "Glop." In G. R. Patterson (Ed.), *Depression and aggression in family interaction* (pp. 247–280). Hillsdale, NJ: Erlbaum.

Braungart, J. M., Plomin, R., DeFries, J. C., & Fulker, D. W. (1992). Genetic influence on tester-rated infant temperament: Nonadoptive and adoptive siblings and twins. *Developmental Psychology, 28*, 40–47.

Caldwell, B. M., & Bradley, R. H. (1978). *Home observation for measurement of the environment.* Little Rock: University of Arkansas Press.

Chase-Lansdale, P. L., Mott, F. L., Brooks-Gunn, J., & Phillips, D. A. (1991). Children of the national longitudinal survey of youth: A unique research opportunity. *Developmental Psychology, 27*, 918–931.

Cook, W. L., & Goethe, J. W. (1990). The effect of being reared with an alcoholic half-sibling: A classic study reanalyzed. *Family Process, 29*, 87–93.

Cowan, C. P., Cowan, P. A., Heming, G., & Miller, N. B. (1991). Becoming a family: Marriage, parenting and child development. In P. A. Cowan & M. Hetherington (Eds.), *Family transitions* (pp. 79–109). Hillsdale, NJ: Erlbaum.

Gilbert, R., Christensen, A., & Margolin, G. (1984). Patterns of alliances in nondistressed and multiproblem families. *Family Process, 23*, 75–87.

Hetherington, E. M., & Clingempeel, W. G. (1992). Coping with marital transitions: A family systems perspective. *Monographs of the Society for Research in Child Development, 57.*

Loehlin, J. C., & Nichols, R. C. (1976). *Heredity, environment and personality.* Austin: University of Texas Press.

Lykken, D. T., McGue, M., & Tellegen, A. (1987). Recruitment bias in twin research. *Behavior Genetics, 17*, 343–362.

MacKinnon, C. E. (1988). Influences on sibling relations in families with married and divorced parents: Family form or family quality? *Journal of Family Issues, 9*, 469–477.

Markman, H. J., & Jones-Leonard, D. (1985). Marital discord and children at risk: Implications for research and prevention. In W. Frankenberg & R. Emde (Eds.), *Early identification of children at risk* (pp. 59–77). New York: Plenum.

Pedersen, N. L., Plomin, R., McClearn, G. E., & Friberg, L. (1988). Neuroticism, extraversion, and related traits in adult twins reared apart and reared together. *Journal of Personality and Social Psychology, 55*, 950–957.

Plomin, R., & Bergeman, C. S. (1991). The nature of nurture: Genetic influence on "environmental" measures. *Brain and Behavioral Sciences, 14*, 373–427.

Plomin, R., Chipuer, H. H., & Neiderhiser, J. (in press). Behavioral genetic evidence for the importance of the nonshared environment. In E. M. Hetherington, D. Reiss, & R. Plomin (Eds.), *Separate social worlds of siblings: Impact of the nonshared environment on development.* Hillsdale, NJ: Erlbaum.

Plomin, R., & Daniels, D. (1987). Why are children in the same family so different from one another? *Behavioral and Brain Sciences, 10*, 1–16.

Plomin, R., DeFries, J. C., & Fulker, D. W. (1988). *Nature and nurture during infancy and early childhood.* Cambridge, UK: Cambridge University Press.

Reiss, D., Plomin, R., Hetherington, E. M., Hower, G., Rovine, M., & Tryon, A. (in press). The separate social worlds of teenage siblings. In E. M. Hetherington, D. Reiss, & R. Plomin (Eds.), *Separate social worlds of siblings: Impact of the nonshared environment of development.* Hillsdale, NJ: Erlbaum.

Rose, R. J., Kaprio, J., Williams, C. J., Viken, R., & Obremski, K. (1990). Social contact and sibling similarity: Facts, issues and red herrings. *Behavior Genetics, 20*, 766–778.

Scarr, S., & Carter-Saltzman, L. (1979). Twin method: Defense of a critical assumption. *Behavior Genetics, 9,* 527–542.

Scarr, S., Scarf, E., & Weinberg, R. A. (1980). Perceived and actual similarities in biological and adoptive families: Does perceived similarity bias genetic inferences? *Behavior Genetics, 10,* 445–457.

Schachter, F. F. (1982). Sibling deidentification and split-parent identification: A family tetrad. In M. E. Lamb & B. Sutton-Smith (Eds.), *Sibling relationships: Their nature and significance across the life span* (pp. 123–151). Hillsdale, NJ: Erlbaum.

Schachter, F. F., Gilutz, G., Shore, E., & Adler, M. (1978). Sibling deidentification judged by mothers: Cross-validation and developmental studies. *Child Development, 49,* 543–546.

Schachter, F. F., Shore, E., Feldman-Rotman, S., Marquis, R. E., & Campbell, S. (1976). Siblings deidentification. *Developmental Psychology, 12,* 418–427.

Schukit, M. A., Goodwin, D. A., & Winokur, G. (1972). A study of alcoholism in half siblings. *American Journal of Psychiatry, 128,* 1132–1136.

Tellegen, A., Bouchard, T. J., Wilcox, K. J., Segal, N. L., Lykken, D. T., & Rich, S. (1988). Personality similarity in twins reared apart and together. *Journal of Personality and Social Psychology, 54,* 1031–1039.

CHAPTER 23

Whither Behavioral Genetics?—A Developmental Psychopathological Perspective

Michael Rutter, Judy Silberg, and Emily Simonoff

Statistical model fitting has provided behavioral geneticists with a powerful tool for examining the interplay between genetic and environmental factors in determining individual differences with respect to a wide range of human characteristics (see Loehlin, 1992, for a very readable account). Great progress has come about through a better appreciation of the concepts involved and the ways in which these may be tested through quantitative techniques and because major improvements in computer technology have meant that it has been possible to deal with the complex algebra that is entailed. Methodological advances of various kinds have provided improved means of comparing competing genetic models, examining gene–environment interactions, testing diagnostic distinctions, and determining whether extremes of psychopathology are ge-

We are grateful to the Medical Research Council of the United Kingdom, the Wellcome Trust, and the National Institute of Mental Health (grant no. MH45268 to John K. Hewitt) for support in the preparation of this chapter.

netically separate from, or continuous with, variations within the normal distribution. Yet, four rather different groups have sometimes seemed to suggest that, for all its accomplishments, behavioral genetics is something for the history of genetics and not for its future.

Is Behavioral Genetics Research Still Worthwhile?

Critique From the Perspective of Molecular Genetics

Within the field of genetics, the center of the field has been taken over by molecular genetics (Weatherall, 1991). There is massive international investment in the research enterprise of mapping the whole of the human genome (Bodmer, 1990). Potentially, genetics need no longer be a "black box" subject. Instead, individual genes can be localized and identified. Already, in the field of medicine, there has been the successful localization of genes for many different diseases; and for a smaller but growing number of conditions, the gene product has actually been identified. Initially, the focus was on diseases that were known to follow a classical Mendelian pattern, which indicates the operation of a single major gene. However, molecular genetic approaches are being increasingly applied to multifactorial disorders, such as coronary artery disease or diabetes, in which one or several major genes play only a contributory role along with a range of environmental factors. Even polygenic transmission may be studied through molecular genetics techniques if the tools developed in plant genetics can be successfully applied to quantitative trait loci in humans (Plomin, McClearn, Gora-Maslak, & Neiderhiser, 1991). Advances in molecular cytogenetics have also opened up new vistas through the demonstration that the genes themselves may actually be changed through transmission across the generations. Thus, our understanding of the fragile X anomaly has been greatly increased as the result of the discovery that the disease phenotype is associated with the presence of a large methylated insert that has become magnified during intergenerational transmission (see Jacobs, 1991; Webb, 1991). Similarly, it has been found that quite different clinical disorders are associated with deletion of Chromosome 15 according to whether or not it comes from the father or the mother (Pembrey, 1991). With all the well-merited excitement generated by these rapid advances in the so-called "new genetics," one may reasonably ask whether there is still any need for the "old genetics."

There can be no doubt that an important part of the future of genetics must lie in molecular genetics, and clearly there is a need to bring molecular genetics and behavioral genetics together. Nevertheless, at least so far as psychology and psychiatry are concerned, it is clear that not only does behavioral genetics continue to have much to offer but also that there are many questions for which the traditional methods are still required (Plomin, Rende, & Rutter, 1991; Rutter, Bolton, Harrington, Le Couteur, Macdonald, & Simonoff, 1990; Rutter, Macdonald, Le Couteur, Harrington, Bolton, & Bailey, 1990). Psychiatric molecular genetics got off to a bad start with premature claims that had to be withdrawn as the result of repeated failures to replicate the original findings (Kelsoe et al., 1989; McGuffin & Murray, 1991; Watt & Edwards, 1991). There are few psychiatric conditions that follow a Mendelian pattern; many disorders are likely to involve the operation of several genes rather than just one, and many varieties of psychopathology are likely to represent extremes of normal variation rather than disease entities as such. Moreover, there are considerable problems in the application of genetic techniques when there is continuing uncertainty about the definition of the phenotype, as is the case with many psychiatric conditions. As discussed more fully later, there are still many ways in which behavioral genetics has a great deal to offer.

Critique From the Perspective of Environmentalists

An entirely different set of criticisms comes from those who espouse an extreme environmentalist position (e.g., Kamin, 1974; Schiff & Lewontin, 1986). Much of their fire has been directed at the supposed weaknesses and inconsistencies in genetic methods as they have been applied in the past. Although it must be accepted that some geneticists have made unwarranted excessive claims (as have some environmentalists), most of this fire is misdirected. With respect to, say, schizophrenia and intelligence (two traits for which critics have wished to deny the importance of genetic factors), the most impressive feature of the research findings is the extent to which a range of different genetic strategies, on quite varied populations, have come up with much the same answer (Gottesman & Shields, 1982; McGuffin, 1988; Plomin, 1990).

However, there is rather more substance to some of their criticisms of the concept of heritability, which dominated most genetic reports until relatively recently. It has been noted that the concept of heritability is a population statistic that applies strictly to the sample being studied. If environments change, estimates of the strength of genetic effects will necessarily also change (see Rutter, 1991b). A corollary of that point is that environmental influences may affect the average population level of a trait without necessarily altering genetic effects on individual differences within it (Scarr, 1992). This point was nicely illustrated by Tizard (1975) with respect to the increase in the height of London boys over the first half of this century, presumably as a result of improvements in diet. The same considerations mean that the notion that genes provide a limit to potential is also seriously misleading. If environments change, then so will the potential.

The height example illustrates that effect but so do the contrasted environments adoption study findings on intelligence (Locurto, 1990). Two major studies were undertaken in France by Schiff and Lewontin (1986) and Capron and Duyme (1989, 1991). Both showed that there was a mean IQ advantage of some dozen points as a result of rearing in a socially advantaged, as compared with a socially disadvantaged, home. Genetic factors do affect an individual's "reaction range" (Gottesman, 1963) in any given environment but they do not limit potential in any absolute sense. For all these reasons, a high level of heritability by no means excludes major environmental effects that may result from a major change in environment. We may well agree with the environmentalists' criticism that knowledge of the level of heritability of any given trait is of very little interest with respect to policy and practice because even quite a high heritability does not rule out effective environmental interventions. Nevertheless, it is a serious mistake to suppose that behavioral genetics is mainly involved with quantifying heritability. There is indeed very little interest in calculating the precise level of heritability as such, but it is a most useful tool in the overall research process used to tackle a whole range of important questions (Rutter, Simonoff, & Silberg, in press), as discussed later.

A somewhat different criticism of heritability is that research findings show that heritability levels do not provide any very clear or useful differentiation between many psychological attributes (Plomin, 1986). At one time, it had been hoped that a strong genetic component would serve to differentiate temperamental characteristics from other psychological features. However, in practice, this expectation has not been borne out. As any biologist might have expected, all human behavior involves a substantial genetic component, and this is so even for attitudes and beliefs, as well as behavior. Most psychological traits show a heritability somewhere in the range between 20% and 60%. That is, there is a substantial genetic component, but so also is there a strong environmental one.

It should be noted that the generalization that heritability provides a very poor differentiation between psychological attributes involves a number of important exceptions. For example, it is clear from several studies that the genetic component in autism accounts for more than 90% of the overall variability in liability (see chapter 14). Also, there is a substantial difference between the heritability of serious unipolar and bipolar affective disorders (for which the genetic component is strong) and that of the much commoner milder depressions that affect some one third of the population at one time or other during their lives, in which environmental influences predominate (Kendler, Neale, Kessler, Heath, & Eaves, 1992; McGuffin & Katz, 1986; McGuffin et al., in press). In this case, the difference in the heritability findings has been important in pointing to the need to distinguish between different varieties of depression.

Somewhat similarly, twin data (admittedly of a rather mixed quality) indicate that the heritability of antisocial behavior in adulthood involves a much stronger genetic component than does apparently similar behavior occurring in childhood (DiLalla & Gottesman, 1989; McGuffin & Gottesman, 1985). Interestingly, too, the genetic data for juvenile delinquency are important in showing that there is a rather modest genetic component in spite of a very strong tendency for delinquency to run in families. It is clear that genetic findings have drawn attention to some very important phenomena, and behavioral genetics techniques also provide the means of investigating these further (see later discussion).

Critique From the Perspective of Psychopathologists

The third source of criticism comes from the field of psychopathology. In this case, the main complaint is not about the genetic methods as such but, rather, about the extremely crude measures to which they often have been applied. Certainly, it has to be admitted that many genetic findings concern the heritability of people's responses to questionnaires they have received through the mail. These can provide only a very limited understanding of human behavior, especially in the field of psychopathology. However, the problem does not end there. Numerous studies in the field of psychology have shown the need to use multimethod, multioccasion methods of assessment (Rutter & Pickles, 1990). Moreover, statistical modeling techniques to tap the latent construct are needed if distortions resulting from both systematic and random error are to be avoided. Similarly, most genetic studies have tended to ignore the need to examine constellations of behavior (Magnusson & Bergman, 1988) and to take into account variations in age of onset (Lahey, Loeber, Quay, Frick, & Grimm, 1992). Thus, for example, within the broad and heterogeneous field of conduct disorders, those that are most likely to persist into adult life seem to be the ones that begin earliest in childhood and that are associated with hyperactivity, inattention, and poor peer relationships (Farrington, Loeber, & Van Kammen, 1990; Robins, 1991). The need is to apply modern behavioral genetics methods to these more differentiated psychopathological concepts and to do so with the appropriate mathematical methods to tap the latent construct. A start has been made in that direction (Eaves et al., 1993), and there is every reason to suppose that much more can be done. It should be added, of course, that much of psychiatric genetics has been concerned with clearly articulated diagnostic concepts and that systematic attempts have been made to explore the effects on genetic findings of differences in diagnostic definition (cf. Kendler et al., 1992, in the field of depression; Farmer, McGuffin, & Bebbington, 1988, in the field of schizophrenia).

It is important, too, to consider the methods and approaches that are applied in behavioral genetics according to the state of knowledge at the time. It may seem absurd now to suppose that, say, schizophrenia and intelligence might be thought to have a negligible genetic contribution, but it is not that long ago that such claims were indeed being made (cf. Jackson, 1960; Kamin, 1974). Even today, there are surprising areas of ignorance. For example, so far there have been no systematic twin studies of mild mental retardation. Following Lewis's (1933) differentiation of so-called "sociocultural retardation," many people have assumed that mild retardation is largely environmentally determined as a result of social disadvantage. However, it remains a very open question whether or not this is so. It is equally plausible that genetic factors play a major role. Of course, it would not be enough to examine the heritability of mild retardation as a broad grouping. Rather, it would be important to examine the ways in which the genetic contribution varied according to other characteristics, such as obstetric complications, social circumstances, and neurodevelopmental impairment. Such research has yet to be undertaken, and until it has been, it is not clear what direction more focused research ought to take.

Critique From the Perspective of Medicine

The fourth set of criticisms comes from the perspective of medicine, for which there has been a concern that there has been rather a neglect of the biological differences between twins and singletons and between monozygotic (MZ) and dizygotic (DZ) pairs. For a variety of historical, as well as practical, reasons, behavioral genetics has tended to place most emphasis on the twin research strategy. As Galton (1876) observed many years ago, twins constitute an extremely important "experiment of nature" that allows the effects of nature and nurture to be separated. That is because MZ and DZ twin pairs are broadly similar in the extent to which they share the same environment, but they differ in the extent to which they share segregating genes. The method is indeed a powerful one, and it has well justified its place as a key genetic design (Plomin, 1986). A key assumption of the design is, of course, that the environment is no more similar for MZ twins than it is for DZ twins and also that the psychological development of twins is broadly comparable with that of singletons. Most critics of the twin study have focused on the first assumption and have noted that the social environment of MZ twins is more similar within pairs than is that of DZ twins. Thus, for example, they are more likely to be dressed alike. However, this is not such a flaw as it might appear at first sight. That is because such evidence as is available suggests that, for the most part, the similarity with which twins are treated is to a considerable extent the result (and not the cause) of their genetically influenced behavioral similarity and because those environmental features that do vary by zygosity seem not to be the ones that have much effect on the degree of concordance for psychopathology. It seems unlikely that this zygosity difference in environmental similarity creates a serious bias in the use of twin designs for the study of most psychological features.

On the other hand, it has to be accepted that there has been remarkably little investigation into the psychological development and parental treatment of twins (Rutter & Redshaw, 1991). It is important that there be more systematic research into the development of twin–twin relationships and, more generally, of the ways in which the psychological experiences and development of twins may differ somewhat from that of singletons. For example, Goodman (1991) has drawn attention to the neglect of the possible influence of both assimilation and deidentification effects (i.e., the tendency for siblings to identify with each other and share activities and the reverse tendency for them to accentuate their differences and emphasize their individuality).

There has also been a relative neglect of investigations of the possible influence of biological differences between MZ and DZ twins (Macdonald, in press; Rutter, Simonoff, & Silberg, in press). Thus, it is well known that there tends to be a greater discrepancy in birth weight within MZ pairs than within DZ pairs, largely because of the fetofetal transfusion syndrome, and that congenital anomalies are more common in MZ twins than in DZ twins. However, placentation differences have been almost entirely ignored in twin studies, and there is a need that they be taken into account in the future. Similarly, the possible importance of minor congenital anomalies (not as causal factors in themselves but rather as an index of something having gone wrong in the biological developmental process) has also been neglected.

Two further features also require attention. First, over the last 30 years or so, there has been a revolution in the quality of prenatal and neonatal care. As a consequence, many very low-birth-weight babies who would have died in a previous era are now surviving. This means that the research into the effects of obstetric complications that was based on research that took place before this revolution may well be largely irrelevant in relation to present-day circumstances. There is a need for further longitudinal studies extending at least into middle and later childhood if researchers are to understand the psychological sequelae associated with very low birth weight and, more specifically, for understanding the risk factors within this group (Casaer, De Vries, & Marlow, 1991). Of course, neurodevelopmental sequelae are likely to be important only with respect to certain sorts of psychopathology, but they may well be relevant, for example, in relation to schizophrenia (Jones & Murray, 1991) and both specific and general types of cognitive impairments (Casaer et al., 1991).

The second consideration with respect to twin studies is that modern methods for the treatment of infertility have had a dramatic effect on the frequency of multiple births. This has several consequences (Macdonald, in press). Thus, it is well known that there are greater environmental influences on the occurrence of DZ twinning than on MZ twinning; and changes in, for example, the age at which women have children have led to alterations in the expected ratio of MZ to DZ twin pairs. It is necessary that this be taken into account in using this ratio to check on the adequacy of sampling in twin studies. Also, however, because people who seek infertility treatment tend to be older than average parents, this may well result in families of twins being less similar to those of singletons than used to be the case; and it may well also mean that there are now more differences between families of MZ and DZ twins than there used to be.

One may conclude that it is important that genetic twin designs in the future pay more attention to the biology and psychology of twinning than was the case in earlier studies. For example, it would be helpful to include the study of singleton siblings as well as twins. In addition, it is important that the twin design be complemented with other genetic strategies. For all its evident power, the twin design does have certain limitations, and it is important that genetic researchers also use adoptee and family genetic designs. Each of these designs has inherent limitations, but from the viewpoint of genetic research, it is crucially important that the limitations of each design are different. Accordingly, there is great strength in tackling the same question with a combination of designs (Rutter, Bolton et al., 1990).

Some Priorities for Future Research

Although it has to be accepted that there is substance in some of the criticisms of behavioral genetics, it is also apparent that there are effective ways of dealing with the problems raised by critics and, indeed, that much of modern behavioral genetics is already using such methods. It remains to consider some of the priorities for future research in behavioral genetics as they particularly apply in the field of developmental psychopathology. The focus here is on the issues to be tackled rather than on the details of the genetic methods. It is obvious that there is much to be gained from a bringing together of quantitative behavioral genetics and molecular genetics (Plomin & Neiderhiser, 1991), but the question of the additional leverage provided by the latter is outside the scope of this chapter.

Testing for Environmental Effects

One of the major mistaken stereotypes held by many nongeneticists is that genetic strategies are of no value for studying environmental influences. To the contrary, they are crucially important just because only genetic strategies can allow determination of which effects are truly environmental (Rutter, 1991b). The fact that a particular variable "looks" as if it refers to environmental influences does not necessarily mean that its effects are environmentally mediated. Genetic studies have indeed shown that some supposed environmental effects to an important extent reflect genetically mediated influences. Thus, how parents bring up their children is going to be affected by their own qualities as individuals, and those qualities are going to include a substantial genetic component (Plomin & Bergeman, 1991). However, this issue is not just one of determining whether supposedly environmental effects are truly environmental. Rather, the main interest and importance lies in the potential power of the genetic design for sorting out which aspects of a person's environment are having environmental influences on psychological development or psychopathology. However, for that use to be made of genetic designs, it is absolutely crucial that genetic studies include high-quality, specific, discriminating measures of the environment. It is very striking how few have done this so far. All too often, the environment is simply treated as an unmeasured "black box" variable that is defined in terms of that which is not genetic. Kendler et al. (1992) have shown, with respect to parental loss, the big difference in conclusions when specific environmental measures are incorporated. Partitioning the variance in the usual ways used in behavioral genetics provided no evidence of any effect at all of shared environmental influences. However, when parental loss was included as a specific measured variable, it was found to have a significant shared environmental effect.

One of the important contributions of behavioral genetics up to now has been the drawing of investigators' attention to the relatively greater importance of nonshared environmental influences, as compared with shared ones (Plomin & Daniels, 1987). It is important to appreciate that the partitioning of environmental effects into those that are shared and those that are nonshared is an abstraction and not something that can be directly measured (Rutter, Simonoff, & Silberg, in press). Sometimes nongeneticists have assumed that this must mean that variables such as family discord or parental neglect cannot be important because they are obviously family-wide variables; that is a false inference. The point is not that family-wide variables cannot be influential but rather that such apparently family-wide influences tend to impinge differentially on different children in the same family. Thus, when there is parental quarreling and hostility, some children tend to get drawn into the conflict, whereas others are able to remain outside it. The really important message from the behavioral genetics findings is that there is a need to study differences between siblings in their experiences within, as well as outside, the home;

and this clearly does have an influence on the ways in which environmental effects are conceptualized and measured (Dunn & Plomin, 1990).

Although this message is certainly an important one that researchers would do well to heed, it is also necessary that researchers do not overgeneralize the findings. To begin with, there are methodological hazards involved in the calculation of shared and nonshared effects (see Goodman, 1991). Also, there are exceptions to the general inference that nonshared influences have the major effect. Thus, it seems likely that this does not hold when studying extremes of environments. For example, the cross-fostering design used by Capron and Duyme (1989, 1991) showed that the social circumstances of adoptive parents had an important effect on the children's IQ. Also, shared environmental influences seem quite important with respect to conduct disorders and delinquency. Unlike the situation with many psychological characteristics, there is quite a strong tendency for siblings to be relatively alike in their propensity to antisocial behavior (DiLalla & Gottesman, 1989; McGuffin & Gottesman, 1985) and possibly also fearfulness (Stevenson, Batten, & Cherner, 1992).

Another important message from behavioral genetics for the study of environmental effects is that we need to investigate the reasons why there is such huge individual variation in the extent to which people are exposed to psychologically risky environments. Environments are not randomly distributed, and if researchers are to use research findings on environmental effects to develop effective methods of prevention and intervention, then they need to understand how these individual differences arise (Rutter & Rutter, 1993). Clearly, part of the answer will lie in society-wide influences such as racial discrimination, economic policies that trap people in poverty, and the distribution of public housing. However, part of the explanation will also lie in the ways in which people select and shape the environments they experience (Scarr, 1992; Scarr & McCartney, 1983). Thus, longitudinal studies have shown that antisocial boys experience a greatly increased rate of stressful environments in adult life (e.g., as reflected in unemployment, marital breakdown, lack of social support, and rebuffs from friends) (Robins, 1966). Similarly, Quinton and Rutter's follow-up of institution-reared girls (Quinton & Rutter, 1988) showed they had a markedly increased tendency to marry and have children in their teens, as well as a much-increased tendency to marry behaviorally deviant men. Behavioral geneticists have emphasized the possible role of genetic factors in the processes involved in the selecting and shaping of environments (Plomin & Bergeman, 1991), but the issues are much broader than that. The point is that it is apparent that people do behave in ways that shape their life circumstances, and that is so for reasons that stem from both genetic and environmental influences on their own behavior. Genetic designs are important for studying this phenomenon because they can be used to differentiate genetic and environmental mechanisms and not just because they test genetic effects. There is a need for longitudinal twin studies, therefore, in which people's later environments are studied as if they were part of a behavioral phenotype.

It is also important that there be further study of possible person–environment interactions (including, but not restricted to, gene–environment interactions). It is clear from a range of studies in biology and medicine that people do vary considerably in their susceptibility to various environmental influences (Rutter & Pickles, 1991). There seems to be a paradox, therefore, between the pervasiveness of such interactions in biology and the extreme difficulty of demonstrating them in behavioral genetics studies (Plomin, DeFries, & Fulker, 1988). Part of the resolution of that paradox lies in the differences in methods that are used to test for interactions, but a much greater part of the explanation lies in the fact that most demonstrated interactions apply to subsections of the population and to highly specific environmental effects (Rutter & Pickles, 1991). It is not particularly likely that gene–environment interactions will be found for general environmental effects on continuously distributed characteristics such as IQ or temperament. On the other hand, there is a reason to suppose that interactions may well play a part in people's responses to specific environmental hazards and that these interactions may play a part in the development of psychopathology. The extent to which this is the case is not known; but, provided that there is a focus on specific genetic and specific environmental effects, the means are available to study such processes more effectively than has been the case in the past. However, rather than search for interaction effects through examination of the interaction term in statistical analyses, it may well often be preferable to use more focused research strategies that are specifically designed to test for particular types of interactions. Kendler and Eaves (1986) have made some very useful suggestions in this connection.

Definition of Phenotypes

Another mistaken stereotype is that genetic studies will not help to identify diseases (Rutter, 1991b). It is of course true that genetic research requires a prior specification of the disorder to be investigated, but there are also many examples in medicine of genetic findings that have led to a reconceptualization of the condition. Sometimes what has been thought to be one disease turns out to include several different genetic conditions (as is the case with both "gargoylism" and retinitis pigmentosa); conversely what had been thought to be separate disorders sometimes prove to be genetically the same. For example, within the field of psychopathology, adoptee data have been useful in suggesting that the phenotype for schizophrenia includes certain sorts of paranoid conditions and personality disorders (perhaps especially of a schizotypal type) but does not include a range of other psychiatric conditions (Kendler & Gruenberg, 1984). Similarly, although clinical studies suggest that anorexia nervosa and bulimia are often associated at the individual level, twin studies suggest that the genetic component is probably much stronger in the former than in the latter (Treasure & Holland, 1991). Family studies have suggested that Tourette's syndrome may include some types of obsessive–compulsive disorder, as well as multiple tics, in its phenotype (Paul, Towbin, Leckman, Zahner, & Cohen, 1986). Also, both twin and family studies of autism indicate that the phenotype is likely to include a combination of cognitive and social deficits because they occur in individuals of normal intelligence and not just the more severely handicapping syndrome of autism as traditionally diagnosed (see chapter 14).

Clearly, there is a very considerable potential for further use of behavioral genetics data for the purposes of providing better definitions of psychiatric phenotypes. Of course, this needs to be an iterative process in which clinical concepts provide the starting point for genetic research, with genetic findings then leading to a reconceptualization of the diagnostic category; further genetic studies will then be needed to validate or invalidate the hypothesized phenotypic definition. Two features of this iterative process require particular emphasis. First, it is not enough to show that different phenotypic definitions lead to similar heritability figures; rather, what is required is direct testing of whether they share the same genetic component. For example, it seems that autism shows high heritability regardless of whether it is accompanied by a normal level of intelligence or very severe mental retardation (see chapter 14). However, known medical conditions (often due to a single major gene) are much more a feature of autism accompanied by severe mental retardation (Rutter, Bailey, Bolton, & Le Couteur, in press). For example, Steffenburg (1991) found that known medical conditions occurred in 43% of cases of autism in severely mentally retarded individuals but in just 18% of those who were mildly retarded or of normal intelligence. Similarly, both the major studies of tuberous sclerosis (Hunt & Shepherd, in press; Smalley, Tanguay, Smith, & Gutierrez, 1992) found that, with but one exception, autism occurred only in those who were also mentally retarded. The implication is that the genetic mechanisms may not be quite the same in cases of autism accompanied by severe, and especially profound, mental retardation.

Second, the need is not just to replicate findings using the same approach. It is also necessary to test the implications that follow from altering the phenotype. For example, if it is truly the case that the autism phenotype includes cognitive and social deficits in individuals of normal intelligence, then it should follow that comparable findings will be evident if proband status is defined in that way. There is a need, therefore, for twin and family studies based on this conceptualization of the phenotype.

Comorbidity

Comorbidity constitutes a closely related research issue. Numerous studies, both epidemiological and clinical, have shown the very high levels of comorbidity among child psychiatric disorders (Caron & Rutter, 1991). However, this is not a single phenomenon—it may reflect a diverse range of mechanisms (Caron & Rutter, 1991; Klein & Riso, in press). Thus, for example, the concurrence of two apparently different disorders may arise simply because the disorders are wrongly specified. Comorbidity may just mean that a single disorder manifests itself by a mixed picture of symptomatology. It is possible that this may explain the overlap between some of the many different subvarieties of anxiety disorder. Alternatively, the concurrence may arise because the two disorders share an overlapping set of risk factors in terms of, say, temperamental variables, cognitive deficits, or family adversity. This may be the case with respect to depression and conduct disorder, in which both are related to parental depression but possibly through different mechanisms—with genetic factors being more important for depression, and family discord as an environmental variable and risk factor for conduct disorder (Downey & Coyne, 1990). Yet again, one disorder may represent an earlier manifestation of the second. This may well be the case, for example, with oppositional defiant disorder and conduct disorder. Yet another alternative is that one disorder constitutes a risk factor for the other. Thus, the adoptee study undertaken by Cadoret, Troughton, Merchant, and Whitters (1991) showed that antisocial personality disorder and depressive problems in adult life were genetically distinct but that the former predisposed to the latter through environmental mechanisms (perhaps through the role of antisocial behavior in creating psychosocial stress situations). Genetic designs (particularly if they include a longitudinal component with multiple data points over time) provide a powerful means of testing these alternative hypotheses on mediating mechanisms. Longitudinal twin studies represent a particularly important research strategy—one that has been rather underused up to now.

Developmental Change and Heterotypic Continuity

Developmental change and heterotypic continuity constitute two key features of necessary interest to developmental psychopathologists. Thus, depressive disorders and suicidal acts show a very marked rise in frequency over the adolescent age period (Angold & Rutter, 1992; Rutter, 1991a). A range of hypotheses has been put forward to explain this phenomenon, including the possibility that there is an increase in psychosocial stresses and a loss of support during the teenage years and the possibility that the rise represents a "switching on" of genetic influences. Twin studies may be informative in determining whether depressive disorders in this age period have a stronger or weaker genetic component compared with those arising in adult life. Both twin and family designs can be helpful in testing whether or not there is genetic continuity between childhood-onset depression and the affective disorders of adult life.

Twin findings have suggested that genetic influences on cognitive functioning increase in their effects over the early and middle years of childhood (Plomin, 1986). However, the apparent rise with age in the heritability of intelligence may reflect discontinuities in the measures of cognitive functioning at different ages. Thus, it is well known that developmental quotients in the preschool years have a very low correlation with IQ scores obtained in later childhood and adolescence. Research over the last decade has been important in showing that cognitive functioning in infancy may be much better indexed by measures of attention and habituation than by the timing of developmental milestones (Bornstein & Sigman, 1986; Rose, Feldman, Wallace, & Cohen, 1991; Slater, Cooper, Rose, & Morison, 1989). Methodological problems remain in the use of these infant measures, but genetic findings are beginning to show that they may have genetic continuity with later measures of intelligence (DiLalla et al., 1990; Plomin & Neiderhiser, 1991). Longitudinal studies have been consistent in showing the high frequency with which conduct disorder in childhood leads to personality disorder in adult life (Robins, 1978; Zoccolillo, Pickles, Quinton, & Rutter, 1992). However, the mechanisms involved in this continuity remain largely unknown. Genetic designs are needed to determine whether continuity reflects the operation of environmental risk mechanisms or rather age-related variations in phenotypic manifestations of the same underlying genotype.

Continuities and Discontinuities Between Normality and Disorder

Child and adult psychiatry include numerous examples of psychiatric conditions that seem to have behavioral parallels within the range of normal variation (Rutter & Sandberg, 1985). Thus, for example, it is necessary to consider whether "ordinary" feelings of misery represent lesser varieties of major affective disorder; whether the eating problems shown by so many adolescent girls constitute a milder variety of the rarer severe disorder of anorexia nervosa; whether heavy drinking is on the same continuum as alcoholism; whether the minor delinquencies shown by most boys in inner city areas are milder varieties of antisocial personality disorder; and whether severe disorders of language development and severe reading retardation represent extremes of normal variations in language and reading acquisition. The issues are very important because of their implications for the understanding of both normal development and psychiatric disorder. Again, genetic designs provide an invaluable means of testing whether the same genetic and environmental factors account for individual variations in disorder as for individual variation within the normal range. However, if the issue of continuities and discontinuities between normality and disorder is to be tackled in an effective manner, then it is crucial to recognize that it will be rare for disorders to be conceptualized only in terms of extremes on a single behavioral dimension. Most diagnostic concepts involve constellations of symptomatology and sometimes considerations of age of onset. It is necessary that these features shape both the data gathering and the methods of data analysis in genetic research. Discriminating standardized interview methods, rather than questionnaires, will almost always be required; and the sampling will need to ensure that there is an adequate number of high-risk subjects so that there will be enough cases of disorder for there to be adequate power in the testing of continuities and discontinuities between normality and disorder.

Conclusions

There has been space in this chapter to consider only a few of the ways in which behavioral genetics may provide powerful research leverage for gaining an increased understanding of normal developmental processes and psychopathology. So far, behavioral genetics has scarcely begun to tackle the key research questions for which it is particularly well adapted, and further research is likely to have major implications for psychiatric concepts and practice. Genetics has come a long way from the early studies that were mainly preoccupied with the calculation of heritabilities for a range of psychological traits. There is, of course, value in the gaining of an understanding of genetic mechanisms per se. However, as this chapter has sought to illustrate, the main strength of genetic research strategies lies in their power to provide leverage on a wide range of questions in developmental psychopathology that at first sight do not seem to have much to do with genetics as such. Progress in the future is likely to be aided by effective collaborations between clinical researchers, developmental investigators, and behavioral geneticists who have a sufficient understanding of the contributions of each field of knowledge for there to be an effective, well-integrated collaboration.

References

Angold, A., & Rutter, M. (1992). Age and pubertal status in a large clinical sample. *Development and Psychopathology, 4*, 5–28.

Bodmer, W. F. (1990). Genetic sequences. *Proceedings of the Royal Society of London. Series B. Biological Sciences, 24*, 85–92.

Bornstein, M. H., & Sigman, M. D. (1986). Continuity in mental development from infancy. *Child Development, 57*, 251–274.

Cadoret, R. J., Troughton, E. , Merchant, L. M., & Whitters, A. (1991). Early life psychosocial events and adult affective symptoms. In L. Robins & M. Rutter (Eds.), *Straight and devious pathways from childhood to adulthood* (pp. 300–313). Cambridge, UK: Cambridge University Press.

Capron, C., & Duyme, M. (1989). Assessment of effects of socio-economic status on IQ in a full cross-fostering study. *Nature, 340*, 552–554.

Capron, C., & Duyme, M. (1991). Children's IQs and SES of biological and adoptive parents in a balanced cross-fostering study. *European Bulletin of Cognitive Psychology, 11*, 323–348.

Caron, C., & Rutter, M. (1991). Comorbidity in child psychopathology: Concepts, issues and research strategies. *Journal of Child Psychology and Psychiatry, 32*, 1063–1080.

Casaer, P., De Vries, L., & Marlow, N. (1991). Prenatal and perinatal risk factors for psychosocial development. In M. Rutter & P. Casaer (Eds.), *Biological risk factors for psychosocial disorders* (pp. 139–174). Cambridge, UK: Cambridge University Press.

DiLalla, L. J., & Gottesman, I. I. (1989). Heterogeneity of causes for delinquency and criminality: Lifespan perspective. *Development and Psychopathology, 1*, 339–349.

DiLalla, L. F., Thompson, L. A., Plomin, R. , Phillips, K., Fagan, J. F., Haith, M. M., Cyphers, L. H., & Fulker, D. W. (1990). Infant predictors of preschool and adult IQ: A study of infant twins and their parents. *Developmental Psychology, 26*, 759–769.

Downey, G., & Coyne, J. C. (1990). Children of depressed parents: Integrative review. *Psychological Bulletin, 108*, 50–76.

Dunn, H., & Plomin, R. (1990). *Separate lives: Why siblings are so different*. New York: Basic Books.

Eaves, L. J., Silberg, J. L., Hewitt, J. K., Rutter, M., Meyer, J. M., Neale, M. C., & Pickles, A. (1993). Analysing twin resemblance in multi-symptom data: Genetic applications of a latent class model for symptoms of conduct disorder in juvenile boys. *Behavior Genetics, 23*, 5–19.

Farmer, A. E., McGuffin, P., & Bebbington, P. (1988). The phenomena of schizophrenia. In P. Bebbington & P. McGuffin (Eds.), *Schizophrenia: The major issues*. Oxford, UK: Heinemann Medical.

Farrington, D., Loeber, R., & Van Kammen, W. B. (1990). Long-term criminal outcomes of hyperactivity–impulsivity–attention deficit and conduct problems in childhood. In L. Robins & M. Rutter (Eds.), *Straight and devious pathways from childhood to adulthood* (pp. 62–81). Cambridge, UK: Cambridge University Press.

Galton, F. (1876). The history of twins as a criterion of the relative powers of nature and nurture. *Royal Anthropological Institute of Great Britain and Ireland Journal, 6*, 391–406.

Goodman, R. (1991). Growing together and growing apart: The non-genetic forces on children in the same family. In P. McGuffin & R. Murray (Eds.), *The new genetics of mental illness* (pp. 212–224). Oxford, UK: Heinemann Medical.

Gottesman, I. I. (1963). Genetic aspects of intelligent behaviour. In N. Ellis (Ed.), *Handbook of mental deficiency: Psychological theory and research* (pp. 253–296). New York: McGraw-Hill.

Gottesman, I. I., & Shields, J. (1982). *Schizophrenia, the epigenetic puzzle*. Cambridge, UK: Cambridge University Press.

Hunt, A., & Shepherd, C. (in press). A prevalence study of autism in tuberous sclerosis. *Journal of Autism and Developmental Disorders*.

Jackson, D. D. (1960). *The etiology of schizophrenia*. New York: Basic Books.

Jacobs, P. A. (1991). The fragile X syndrome [Editorial]. *Journal of Human Genetics, 28*, 809–810.

Jones, P. B., & Murray, R. M. (1991). Aberrant neurodevelopment as the expression of schizophrenia genotype. In P. McGuffin & R. Murray (Eds.), *The new genetics of mental illness* (pp. 112–129). Oxford, UK: Butterworth-Heinemann.

Kamin, L. J. (1974). *The science and politics of IQ*. Hillsdale, NJ: Erlbaum.

Kelsoe, J. R., Ginns, E. I., Egeland, J. A., Gerhard, D. S., Goldstein, A. M., Bale, S. J., Pauls, D. L., Long, R. T., Kidd, K. K., Coute, G., Housman, D. E., & Paul, S. M. (1989). Re-evaluation of the linkage relationship between chromosome 11p loci and the gene for bipolar affective disorder in the Old Order Amish. *Nature, 342*, 238–243.

Kendler, K. S., & Eaves, L. J. (1986). Models for the joint effect of genotype and environment disorder in women: A population-based twin study. *Archives of General Psychiatry, 143*, 279–287.

Kendler, K. S., & Gruenberg, A. M. (1984). An independent analysis of the Danish adoption study of schizophrenia: VI. The relationship between psychiatric disorders as defined by *DSM-III* in the relatives and adoptees. *Archives of General Psychiatry, 41*, 555–564.

Kendler, K. S., Neale, M. C., Kessler, R. C. Heath, A. C., & Eaves, L. J. (1992). Childhood parental loss and adult psychopathology in women: A twin study perspective. *Archives of General Psychiatry, 49*, 109–116.

Klein, D. N., & Riso, L. P. (in press). Psychiatric disorders: Problems of boundaries and comorbidity. In C. G. Costello (Ed.), *Basic issues in psychopathology*. New York: Guilford Press.

Lahey, B. B., Loeber, R., Quay, H. C., Frick, P. J., & Grimm, J. (1992). Oppositional defiant and conduct disorders: Issues to be resolved for *DSM-IV*. *Journal of the American Academy of Child and Adolescent Psychiatry, 31*, 539–546.

Lewis, E. O. (1933). Types of mental deficiency and their social significance. *Journal of Mental Science, 79*, 298–304.

Locurto, C. (1990). The malleability of IQ as judged from adoption studies. *Intelligence, 14*, 75–290.

Loehlin, J. C. (1992). *Genes and environment in personality development*. Newbury Park, CA: Sage.

Macdonald, A. (in press). What can twin studies contribute to the understanding of childhood behavioral disorders? In T. J. Bouchard & P. Propping (Eds.), *Twins as a tool of behaviour genetics*. New York: Wiley.

Magnusson, D., & Bergman, L. R. (1988). A pattern approach to the study of pathways from childhood to adulthood. In L. Robins & M. Rutter (Eds.), *Straight and devious pathways from childhood to adulthood* (pp. 101–115). Cambridge, UK: Cambridge University Press.

McGuffin, P. (1988). Genetics of schizophrenia. In P. Bebbington & P. McGuffin (Eds.), *Schizophrenia: The major issues* (pp. 107–126). Oxford, UK: Heinemann Medical.

McGuffin, P., & Gottesman, I. I. (1985). Genetic influences on normal and abnormal development. In M. Rutter & L. Hersov (Eds.), *Child and adolescent psychiatry: Modern approaches* (2nd ed., pp. 17–33). Oxford, UK: Blackwell Scientific.

McGuffin, P., & Katz, R. (1986). Nature, nurture and affective disorder. In J. W. K. Deakin (Ed.), *The biology of depression* (pp. 26–52). London: Gaskell Press.

McGuffin, P., Katz, R., Rutherford, J., Watkins, S., Farmer, A. E., & Gottesman, I. I. (in press). Twin studies as vital indicators of phenotypes in molecular genetic research. In T. J. Bouchard & P. Propping (Eds.), *Twins as a tool of behaviour genetics*. New York: Wiley.

McGuffin, P., & Murray, R. (Eds.). (1991). *The new genetics of mental illness*. Oxford, UK: Heinemann Medical.

Paul, D. L., Towbin, K. E., Leckman, J. F., Zahner, G. E. P., & Cohen, D. J. (1986). Evidence supporting an etiological relationship between Gilles de la Tourette syndrome and obsessive compulsive disorder. *Archives of General Psychiatry, 43*, 1180–1182.

Pembrey, M. (1991). Chromosomal abnormalities. In M. Rutter & P. Casaer (Eds.), *Biological risk factors for psychosocial disorders* (pp. 67–99). Cambridge, UK: Cambridge University Press.

Plomin, R. (1986). *Development, genetics and psychology*. Hillsdale, NJ: Erlbaum.

Plomin, R. (1990). The role of inheritance in behavior. *Science, 248*, 183–188.

Plomin, R., & Bergeman, C. S. (1991). The nature of nurture: Genetic influence on "environmental" measures. *Behavioral and Brain Sciences, 14*, 373–386.

Plomin, R., & Daniels. (1987). Why are children in the same family so different from one another? *Behavioral and Brain Sciences, 10*, 1–15.

Plomin, R., DeFries, J. C., & Fulker, D. W. (1988). *Nature and nurture during infancy and early childhood*. Cambridge, UK: Cambridge University Press.

Plomin, R., McClearn, G. E., Gora-Maslak, G., & Neiderhiser, J. M. (1991). Use of recombinant inbred strains to detect quantitative trait loci associated with behavior. *Behavior Genetics, 21*, 99–116.

Plomin, R., & Neiderhiser, J. (1991). Quantitative genetics, molecular genetics, and intelligence. *Intelligence, 15*, 369–387.

Plomin, R., Rende, R., & Rutter, M. (1991). Quantitative genetics and developmental psychopathology. In D. Cicchetti & S. L. Toth (Eds.), *Internalizing and externalizing expressions of dysfunction: Rochester Symposium on Developmental Psychopathology* (Vol. 2, pp. 155–202). Hillsdale, NJ: Erlbaum.

Quinton, D., & Rutter, M. (1988). *Parenting breakdown: The making and breaking of intergenerational links*. Aldershot, UK: Avebury.

Robins, L. (1966). *Deviant children grown up*. Baltimore: Williams & Wilkins.

Robins, L. (1978). Sturdy childhood predictors of adult antisocial behaviour: Replications from longitudinal studies. *Psychological Medicine, 8*, 611–622.

Robins, L. N. (1991). Conduct disorder. *Journal of Child Psychology and Psychiatry, 32*, 193–212.

Rose, S. A., Feldman, J. F., Wallace, I. F., & Cohen, P. (1991). Language: A partial link between infant attention and later intelligence. *Developmental Psychology, 27*, 798–805.

Rutter, M. (1991a). Age changes and depressive disorders: Some developmental considerations. In J. Garber & K. A. Dodge (Eds.), *The development of emotion regulation and dysregulation* (pp. 273–300). Cambridge, UK: Cambridge University Press.

Rutter, M. (1991b). Nature, nurture and psychopathology: A new look at an old topic. *Development and Psychopathology, 3*, 125–136.

Rutter, M., Bailey, A., Bolton, P., & Le Couteur, A. (in press). Autism and known medical conditions. *Journal Child Psychology and Psychiatry.*

Rutter, M., Bolton, P., Harrington, R., Le Couteur, A., Macdonald, H., & Simonoff, E. (1990). Genetic factors in child psychiatric disorders: I. A review of research strategies. *Journal of Child Psychology and Psychiatry, 1*, 3–37.

Rutter, M., Macdonald, H., Le Couteur, A., Harrington, R., Bolton, P., & Bailey, A. (1990). Genetic factors in child psychiatric disorders: II. Empirical findings. *Journal of Child Psychology and Psychiatry, 1*, 39–83.

Rutter, M., & Pickles, A. (1990). Improving the quality of psychiatric data: Classification, cause and course. In D. Magnusson & L. R. Bergman (Eds.), *Data quality in longitudinal research* (pp. 32–57). Cambridge, UK: Cambridge University Press.

Rutter, M., & Pickles, A. (1991). Person–environment interactions: Concepts, mechanisms and implications for data analysis. In T. Wachs & R. Plomin (Eds.), *Conceptualization and measurement of organism–environment interaction* (pp. 105–141). Washington, DC: American Psychological Association.

Rutter, M., & Redshaw, J. (1991). Annotation: Growing up as a twin. Twin–singleton differences in psychological development. *Journal of Child Psychology and Psychiatry, 32*, 885–895.

Rutter, M., & Rutter, M. (1993). *Developing minds: Challenge and continuity across the lifespan*. Harmondsworth, UK: Penguin; New York: Basic Books.

Rutter, M., & Sandberg, S. (1985). Epidemiology of child psychiatric disorder: Methodological issues and some substantive findings. *Child Psychiatry and Human Development, 15*, 209–233.

Rutter, M., Simonoff, E., & Silberg, J. (in press). How informative are twin studies of child psychopathology? In T. J. Bouchard & P. Propping (Eds.), *Twins as a tool of behaviour genetics*. New York: Wiley.

Scarr, S. (1992). Developmental theories for the 1990s: Development and individual differences. *Child Development, 63*, 1–19.

Scarr, S., & McCartney, K. (1983). How people make their own environments: A theory of genotype → environmental effects. *Child Development, 54*, 424–435.

Schiff, M., & Lewontin, R. (1986). *Education and class: The irrelevance of IQ genetic studies*. Oxford, UK: Clarendon.

Slater, A., Cooper, R., Rose, D., & Morison, V. (1989). Prediction of cognitive performance from infancy to early childhood. *Human Development, 32*, 137–147.

Smalley, S. L., Tanguay, P. E., Smith, M., & Gutierrez, G. (1992). Autism and tuberous sclerosis. *Journal of Autism and Developmental Disorders, 22*, 339–355.

Steffenburg, S. (1991). Neuropsychiatric assessment of children with autism: A population-based study. *Developmental Medicine and Child Neurology, 33*, 495–511.

Stevenson, J., Batten, J., & Cherner, M. (1992). Fears and fearfulness in children and adolescents: A genetic analysis of twin data. *Journal of Child Psychology and Psychiatry, 33*, 997–985.

Tizard, J. (1975). Race and IQ: The limits of probability. *New Behaviour, 1*, 6–9.

Treasure, J. L., & Holland, A. J. (1991). Genes and the aetiology of eating disorders. In P. McGuffin & R. Murray (Eds.), *The new genetics of mental illness* (pp. 198–211). Oxford, UK: Heinemann Medical.

Watt, D. C., & Edwards, J. H. (1991). Doubt about evidence for a schizophrenia gene. *Psychological Medicine, 21*, 279–285.

Weatherall, D. J. (1991). *The new genetics and clinical practice.* London: Oxford University Press.

Webb, T. (1991). Molecular genetics of fragile X: A cytogenetics viewpoint. Report of the Fifth International Symposium on X-Linked Mental Retardation, Strasbourg, France, 12–16 August 1991 (organizer Dr. J.-L. Mandel). *Journal of Human Genetics, 28*, 814–817.

Zoccolillo, M., Pickles, A., Quinton, D., & Rutter, M. (1992). The outcome of conduct disorder: Implications for defining adult personality disorder. *Psychological Medicine, 22*, 971–986.

PART SEVEN

Summary

CHAPTER 24

Nature and Nurture: Perspective and Prospective

Robert Plomin

The chapters in this book provide a good overview of the field of behavioral genetics and its relationship with psychology as psychologists approach the 21st century. They also point the way to the future. The purpose of this concluding chapter is to highlight some of these themes concerning the past, present, and future of genetics research in psychology.

Concerning the past, in the first two chapters, Kimble and McClearn, respectively, describe the field of behavioral genetics as one of the oldest in psychology. The theoretical blueprint of quantitative genetics has guided construction of the field for more than 75 years. A special source of satisfaction to me is that, as much as any area in psychology, behavioral

Preparation of this chapter and some of the quantitative genetics research it describes were supported, in part, by grants from the National Institute of Aging (AG-04563), National Institute of Child Health and Human Development (HD-10333 and HD-18426), National Institute of Mental Health (MH-43373 and MH-43899), National Science Foundation (BNS-91-08744), and the John D. and Catherine T. MacArthur Foundation.

genetics research is cumulative. This contributes to a sense of building a solid edifice, still far from completion, the construction of which stretches back over many generations of researchers.

The accelerating pace of the present research will propel the field far into the next century. This momentum comes from the findings themselves, from new methods that make it possible to broach evermore interesting issues, from the many large-scale ongoing projects, from the psychologists who have begun to incorporate genetic strategies into their research, and from the promise of molecular genetics. In this concluding chapter, I consider these five contributions to the field's momentum: findings, methods, projects, people, and molecular genetics.

Findings

The chapters in this book indicate the progress that has been made in behavioral genetics research in the traditional domains of cognitive abilities and disabilities, psychopathology, and personality. These findings in the domains of psychology that have traditionally considered individual differences will eventually lead to similar research in other domains. In addition, a new topic mentioned in several chapters involves the use of environmental measures in genetic designs. The fifth topic in this section looks to the future more than the present.

Cognitive Abilities

The chapters in Part Two, which concerns cognitive abilities, reflect the fact that much more is known about this psychological domain than any other. For example, cognitive abilities is one domain in which a case can be made for differential heritability. That is, some cognitive abilities, such as verbal and spatial abilities, appear to be more heritable than others, such as memory (Plomin, 1988). Moreover, although most verbal tests are moderately heritable, spatial tests show a greater range of heritability. The most difficult spatial tests, such as those that require three-dimensional rotations of objects pictured in two dimensions, appear to show the greatest heritability.

Multivariate genetic analyses of the type described by Cardon and Fulker in chapter 5 indicate that specific tests and group factors show some genetic effects unique to each test and factor. Nonetheless, many of the genetic effects are shared in common across diverse tests and factors. In other words, Spearman's *g*, which refers to general cognitive ability, is attributable substantially to genetic effects shared by diverse cognitive tests. Another recent finding makes a related point: The heritabilities of cognitive tests are strongly correlated with their *g* loadings, which are their factor loadings on an unrotated first principal component (Jensen, 1987). That is, the more a test measures *g*, the more heritable it is. For example, researchers who used the powerful designs of reared-apart twins and matched reared-together twins found that the correlation between heritabilities and *g* loadings was 0.77 after differential reliabilities of the two tests were controlled (Pedersen, Plomin, Nesselroade, & McClearn, 1992).

Another example of multivariate genetic analysis concerns the relationship between cognitive abilities and measures of school achievement. Recent research indicates that the substantial overlap between these domains is largely mediated by genetic factors (Cardon, DiLalla, Plomin, DeFries, & Fulker, 1990; Thompson, Detterman, & Plomin, 1991). Differences in performance in the two domains is largely environmental in origin.

Developmental genetic analyses of cognitive ability are also producing some interesting findings. As indicated by McGue, Bouchard, Sacono, and Lykken in chapter 3, heritability of *g* appears to increase throughout development, reaching what may be the highest heritability (80%) reported in the behavioral sciences, in the first study of older adults (Pedersen et al., 1992). In addition to addressing developmental changes in heritability, an intriguing story is emerging from longitudinal analyses of age-to-age change and continuity. As indicated by Fulker, Cherny, and Cardon in chapter 4, genetic effects on *g* contribute to stability during childhood, but what is more surprising is the extent to which genetic effects appear to contribute to change from age to age. Particularly interesting is the suggestion of substantial new genetic variation during the transition from early to middle childhood.

Important developmental findings have also emerged from the environmental side of behavioral genetics analyses. Cognitive ability is the only domain that has shown solid evidence of shared environmental influence. However, research during the past decade indicates that this finding is limited to childhood. By adolescence, shared environmental influence on cognitive ability diminishes to negligible levels, which suggests that environmental influences that have an effect in the long run are of the nonshared variety (see chapter 3; Plomin, 1988). As usual, new findings lead to new questions. What are these shared environmental influences that decline in importance by the adolescent years? What are the nonshared environmental factors that constitute the environmental contribution to individual differences in cognitive ability after childhood? The longitudinal genetic analyses described by Fulker et al. in chapter 4 provide some hints. During childhood, shared environmental influence is monolithic and continuous (which perhaps suggests the influence of a static factor such as socioeconomic status), whereas nonshared environmental effects contribute to change from year to year (which perhaps suggests the influence of idiosyncratic experiences).

A key question for all domains of behavior is the etiological relationship between the normal and abnormal. A new technique called *DF analysis* (named after its creators, DeFries and Fulker) can begin to address this issue at the level of etiology rather than symptomatology. The technique is described by DeFries and Gillis in chapter 6 in relation to research suggesting that the genetic etiology of the disorder of reading disability may differ from that of the full dimension of reading ability. In other words, reading disability may be something other than the extreme of a continuum of reading ability. Although severe mental retardation is etiologically distinct from the normal distribution of IQ, there is some evidence from sibling studies that mild mental retardation may represent the lower end of the normal distribution, but as yet no twin or adoption studies have been reported for mild mental retardation (Plomin, 1991). The technique of latent class analysis, described in chapter 15 by Eaves et al., is also promising for understanding the links between the normal and abnormal.

Despite these advances in understanding the origins of individual differences in cognitive abilities, researchers may be closer to the beginning than to the end of the behavioral genetics story (Plomin & Neiderhiser, 1992).

Personality

In the introduction to Part Three, which is a brief overview of genetic research in personality, Goldsmith indicates that behavioral genetics is playing a role in the general renaissance of personality research. Chapter 10 by Kagan, Arcus, and Snidman represents a case study of growing interest in temperament. The chapter by Brody (chapter 8) provides an interesting contrast between what is known about the genetics of intelligence and the genetics of personality. This strategy allows Brody to review much current behavioral genetics research in personality and to provide hypotheses for future research. In chapter 9, Rowe reviews new developments such as multivariate genetic analysis and investigations of methodological assumptions of behavioral genetics research in personality (see also chapter 22 in this book). He also considers evidence and implications concerning environmental influences in personality and discusses links with evolutionary psychology.

Psychopathology

There is more genetic research on psychopathology than all other areas of psychology combined. The section on psychopathology begins with a broad historical overview of the area by Irving I. Gottesman, who also deftly summarizes in chapter 12 the scores of genetic studies on schizophrenia that converge on the conclusion of significant genetic influence. Particularly noteworthy is his multifactorial "ecumenical model," which allocates most of the genetic influence on schizophrenia to multiple-gene risks that interact during development with several "toxic" environmental factors.

For depression, many more questions than answers are available, as indicated by Peter McGuffin and Randy Katz in chapter 11. Bipolar disorder appears to be more strongly familial than unipolar depression, although McGuffin's recent research indicates that narrowly defined unipolar depression shows a genetic influence. In their chapter, McGuffin and Katz also present pioneering research that incorporates environmental measures of stress in the context of family and twin studies of depression, research that yields several surprising results.

Alcoholism is one area in the behavioral sciences in which acceptance of genetic influence might have outstripped the data (Searles, 1988). In his chapter on alcoholism (chapter 13), McGue judiciously reviews the data and concludes that alcoholism is moderately heritable in men, especially men with early onset, but that women show modest heritability at most. McGue offers interesting ideas for investigating psychological processes underlying genetic influences on alcoholism in men, and he concludes that behavioral genetics offers an integrative framework for comprehensive analyses from the molecular to the molar.

The chapter on autism by Rutter, Bailey, Bolton, and Le Couteur (chapter 14) is particularly interesting in historical perspective. Just 15 years ago, autism was thought to be entirely environmental in origin. Research by Rutter and others during this time has indicated that autism may be one of the most heritable disorders. I used to think that autism was one example of a disorder for which there would be no etiological links with normal dimensions of behavior. However, Rutter and colleagues' recent research suggests that autism is broader than current diagnoses would suggest, including minor cognitive and social problems but not mental retardation.

Environment

The most novel direction for research lies at the interface between nature and nurture. Although the issues involved can be complicated, the specific suggestion with which everyone agrees is the usefulness of incorporating environmental measures in genetic designs. This is a major theme in the chapters by Bronfenbrenner and Ceci; McGuffin and Katz; Reiss; Rowe and Waldman; Rutter, Silburg, and Simonoff; and Wachs. This will facilitate identification of specific sources of nonshared environment and will continue the investigation of genetic contributions to measures of the environment. It will also encourage more research on the neglected issues of genotype–environment interaction and correlation, as emphasized by Bronfenbrenner et al. in chapter 16 and by Wachs in chapter 20. The results presented by Kagan et al. in chapter 10 can be viewed as examples of interactions in which maternal behavior relates to fearfulness only for highly reactive infants and of correlation in which temperament results in a tendency to select compatible environments, even perhaps basic philosophical positions. In chapter 11, McGuffin and Katz illustrate the surprises that may lie ahead when environmental hypotheses are tested using specific measures of the environment in genetically sensitive designs.

A special feature of this book is the inclusion of chapters by three eminent environmental researchers: Urie Bronfenbrenner, Frances Degen Horowitz, and Theodore D. Wachs. Although these researchers are by no means in complete agreement with the theory and methods of behavioral genetics, nor do they use behavioral genetics strategies in their research, each proposes an environmental theory that attempts to encompass genetic influence. The common theme in the "bioecological model" of Bronfenbrenner et al., Horowitz's "comprehensive new environmentalism," and Wach's "multidetermined probabilistic systems framework" is the need to address Anastasi's (1958) question of how genotypes and environments interact in development. These authors also discuss barriers to communication and collaboration and are separated by chapters addressing similar issues by behavioral geneticists (Goldsmith [chapter 17]; Rowe & Waldman [chapter 19]; Rutter et al. [chapter 23]). The contrasts between these chapters indicate that there is still a long way to go in bridging the gaps between environmentalists and geneticists.

My experience is that abstract arguments are unlikely to resolve these complicated issues. I agree with Wachs that what is needed is research collaboration "to construct actual empirical bridges between nature and nurture" in which environmentalists and geneticists are full and equal partners. A pioneering effort of this type is described by Reiss in chapter 22.

New Areas

What is not known can be as stimulating to future research as what is known. In contrast to the fields of cognitive abilities, psychopathology, and personality, for some major domains of psychology, nature–nurture questions have scarcely been considered. Some of the oldest areas of psychology (e.g., perception and language), as well as some of the newest (e.g., neuroscience and social cognition), primarily describe species-typical themes rather than individual-differences variations on those themes. Until the spotlight falls on the description of individual differences, questions about the genetic and environmental etiology of these differences are unlikely to be asked. The relative disregard of individual differences is unfortunate because these areas have developed especially sensitive and process-oriented measures that could be applied profitably to the study of individual differences.

An example of the excitement engendered in the shift to individual differences comes from the heartland of normative experimental research: perceptual development. Three relevant chapters in the *Handbook of Child Psychology* (Mussen, 1983)—those devoted to visual perception, auditory and speech perception, and attention, learning, and memory—contain a total of 189 pages of text but not one single page on individual differences. However, one of the most important advances in the field is the discovery during the past decade of the long-term predictiveness of individual differences in infant novelty preferences. After using the novelty preference technique for several years to describe normative perceptual development, Joseph Fagan began to consider individual differences (Fagan & McGrath, 1981; Fagan & Singer, 1983). He obtained vocabulary scores from children whom he had tested for novelty preferences several years earlier in the first 6 months of life and found significant correlations (Fagan, 1985). To my knowledge, this is the first evidence for any behavioral measure in the first year of life that predicts later cognitive ability. Once a researcher's focus shifts to individual differences, descriptive and predictive questions invariably lead to questions about the origins of individual differences. Such questions put a researcher at risk for behavioral genetics. Fagan's interest in individual differences in infant novelty preferences led to a twin study that showed a genetic influence (DiLalla et al., 1990).

In addition, the "whether" and "how much" questions are only beginning to be asked about some areas of psychology that focus on individual differences. (In contrast to several authors in this book, it seems reasonable to me to ask not only whether genetic effects are significant but also to assess the effect size, the "how much" question of heritability.) For example, possible genetic contributions to health psychology variables have hardly been considered. There is next to no research on the genetic and environmental provenances of such favorite health psychology variables as stress; mechanisms for coping; life-styles; attributions of self-efficacy and sense of control in relation to health and illness; and nonadherence to regimens of medical treatment, exercise, and nutrition. Another example involves behaviors important in the context of life-course transitions, in contrast to traditional research on traits relevant throughout the life span. For example, fertile areas for genetic research include the stresses of beginning school; friends and peers; the physical and social transitions of adolescence; entrance into the adult world of work, marriage, and child rearing; and adjustment to the changes of later life.

Even within the domains most often investigated using genetic strategies—cognitive abilities, psychopathology, and personality—the basic nature–nurture questions need to be asked as different measures are used. For example, in the cognitive realm, attempts are being made to move beyond paper-and-pencil tests toward tests of elemental cognitive processes, often referred to as information-processing measures. So far, such measures appear to show a wider range of heritabilities than traditional paper-and-pencil measures of cognitive abilities (Ho, Baker, & Decker, 1988; McGue & Bouchard, 1989; Vernon, 1989). Similarly, the "whether" and "how much" questions will need to be asked again in the areas of psychopathology and personality as new measures are used. For psychopathology, Rutter, Bailey, Bolton, and Le Couteur (chapter 14) recommend the use of more sophisticated asssessments that take into account constellations of behavior and age of onset. For personality, Brody (chapter 8) suggests that measures other than questionnaires need to be used.

Methods

Also fueling the current momentum of the field is the development of new research tools. Especially important developments during the past decade include model fitting, multivariate analysis, and longitudinal analysis, which are discussed in several chapters in this book. Also represented in this book are two new techniques of far-reaching significance: (a) the DF analysis of extremes that can examine links between the normal and abnormal described in chapter 6 by DeFries and Gillis; and (b) latent class analysis, which is the focus of chapter 15 by Eaves et al. An emerging development that will revolutionize behavioral genetics is the application of molecular genetic techniques. This development is more like a paradigm shift than a methodological advance, and, for this reason, I highlight the topic in a separate section at the end of this chapter.

Model Fitting

Sometimes called *causal*, *structural*, *biometrical*, or *path* modeling, model fitting gains its name because it tests the fit between a model and observed data. Behavioral genetic research is now usually reported in terms of model-fitting analyses. For example, the twin method can be considered to be a model consisting of two equations. One equation states that the identical twin correlation is equal to all genetic variance plus resemblance caused by shared environment. The second equation equates the fraternal twin correlation to half the genetic variance plus resemblance caused by shared environment. Solving these two simultaneous equations is nothing more than doubling the difference between the correlations to estimate heritability and then attributing twin resemblance not explained by heredity to shared environmental influence. Researchers cannot test the fit of this model because the number of unknowns (two) equals the number of equations (two). Unless the model is overdetermined (i.e., unless there are more equations than unknowns), researchers can estimate parameters but cannot test the fit of a model. Model fitting is especially useful when combination designs are used that yield many different familial correlations. With multiple groups, model fitting is especially valuable because it analyzes all of the information simultaneously, weights each piece of information according to its sample size, tests the adequacy of the model and its assumptions, yields parameter estimates and standard errors that best fit the model, and compares alternative models (Plomin, DeFries, & McClearn, 1990).

The state of the art of model fitting is represented in the chapters by Fulker et al. and Cardon and Fulker. An introduction to model fitting is available (Loehlin, 1987), an issue of the journal *Behavior Genetics* is devoted to model-fitting analyses of twin data (Boomsma, Martin, & Neale, 1989), and a new book provides a comprehensive treatment of model fitting in behavioral genetics (Neale & Cardon, 1992).

Multivariate Genetic Analysis

Other methodological advances are tested by model-fitting techniques but represent important concepts in their own right. One of the most important advances is the extension of univariate analyses of the variance of a single trait to multivariate analysis of the covariance between traits. Multivariate genetic analysis allows researchers to estimate the overlap (pleiotropy) of genetic effects across traits. For example, to what extent do genetic effects on verbal ability also contribute to spatial ability? From a genetic perspective, a multivariate approach is important because it is highly unlikely that completely different sets of genes affect the various behaviors researchers examine. In chapter 5, Cardon and Fulker use a hierarchical model of multivariate genetic analysis to investigate specific cognitive abilities and their relationship to general cognitive ability. As emphasized by Hewitt in chapter 21 and by Rutter, Silberg, and Simonoff in chapter 23, multivariate genetic analysis can be used to address the fundamental issues of heterogeneity and comorbidity in psychopathology at the level of etiology rather than just symptomatology.

Longitudinal Genetic Analysis

Another major methodological advance involves the application of multivariate analysis to longitudinal genetic data. This permits analysis of the etiology of age-to-age change as well as continuity (i.e., to what extent do genetic effects at one age overlap with genetic effects at another age?). Genetic change means that genetic effects at one age differ from genetic effects at another age. Even though heritability is substantial for a trait in childhood and in adolescence, different genetic effects may operate at the two ages. In chapter 4, Fulker et al. present a model-fitting longitudinal analysis of general cognitive ability. Developmental genetic analysis is also discussed in chapters by Hewitt; McClearn; and Rutter, Silberg, and Simonoff.

Analyses of Dimensions Versus Disorders

Two more recent methodological advances are described in this book. The technique described in the chapter by DeFries and Gills can address the fundamental question of the etiological association between the normal and abnormal. The approach—DF analysis (DeFries & Fulker, 1985, 1988)—requires that data be obtained using a quantitative measure of a disorder-relevant dimension. DeFries and Gillis present analyses of this type for reading disability as related to a quantitative discriminant function score. The latent class approach described by Eaves et al. (chapter 15) can also address questions about the etiological links between dimensions and disorders. As emphasized by Rutter, Bailey, Bolton, and Le Couteur (chapter 14), this is a key issue for psychopathology.

Projects

The third reason why the present momentum will carry the field far into the future is the large number of major ongoing research projects. A partial list of large-scale behavioral genetic projects of normal development includes the Colorado Adoption Project (Plomin, DeFries, & Fulker, 1988), the Louisville Twin Study (Matheny, 1989), the MacArthur Longitudinal Twin Study (Emde et al., in press), the Nonshared Environment and Adolescent Development Project (Reiss et al., in press), the Virginia Study of Adolescent Behavioral Development (Hewitt, Silberg, Neale, Eaves, & Erikson, 1992), the Minnesota Study of Twins Reared Apart (Bouchard, Lykken, McGue, Segal, & Tellegen, 1990), the Minnesota Twin Registry (Lykken, Bouchard, McGue, & Tellegen, 1990), the Texas Adoption Project (Loehlin, Horn, & Willerman, 1989), and the Swedish Adoption/Twin Study of Aging (Pedersen et al., 1991). The list of large-scale projects in psychopathology would be much longer.

People

Now that behavioral genetics has flowed into the mainstream of psychology, a fourth reason for optimism about its future is that the field is successfully being given away. I believe that the best behavioral genetics research will be done by scientists who are not primarily behavioral geneticists. Experts in substantive domains will ask the theory-driven questions and interpret their research findings in a way that will make the most sense to other researchers in their field.

Some of the leading researchers in the behavioral sciences are beginning to incorporate genetic strategies in their research. For example, this book profits from contributions by three scientists who have recently come to use behavioral genetics strategies in their research: Jerome Kagan, David Reiss, and Michael Rutter. Many other well-known psychologists or psychiatrists whose research is not represented in this book have made a similar transition. For example, Kagan's twin study of behavioral inhibition (Robinson, Kagan, Reznick, & Corley, in press) is part of a collaboration between developmental psychologists and behavioral geneticists known as the MacArthur Longitudinal Twin Study. Other collaborators include the project's leader, Robert N. Emde (Emde et al., in press); Carolyn Zahn-Waxler, who has reported the first twin analysis of empathy (Zahn-Waxler, Robinson, & Emde, in press); and Joseph Campos, who was also involved in an earlier twin study on temperament (Goldsmith & Campos, 1986). David Reiss's longitudinal twin/stepfamily study, the Nonshared Environment in Adolescent Development Project, is a collaborative project with E. Mavis Hetherington, who has also edited a book on nonshared environment (Hetherington, Reiss, & Plomin, in press). Judy Dunn has brought her expertise in siblings to bear on issues of nonshared environment using genetic designs (Dunn & Plomin, 1990). In the area of personality, the ranks include Warren Eaton (Saudino & Eaton, 1991), Auke Tellegen (Tellegen et al., 1988), and Marvin Zuckerman (1991). In developmental psychopathology, Craig Edelbrock has begun a midcareer shift to incorporate genetic strategies in his research on dimensional and diagnostic issues (e.g., Edelbrock, Rende, Plomin, & Thompson, in press). In adult psychiatric research, genetics research is being conducted by Phillip Holzman and Myrna Weissman, to name just two of the many psychiatric researchers who have incorporated genetic strategies into their research. Examples in the area of cognitive abilities include Douglas Detterman, who is a coinvestigator in the Western Reserve Twin Project, which includes extensive information-processing measures (e.g., Detterman, Thompson, & Plomin, 1990). Joseph Fagan and Marshall Haith are involved in twin research on infant information processing (DiLalla et al. 1990). K. Warner Schaie has incorporated a familial concept to his decades-long Seattle Longitudinal Study of cognitive abilities (Schaie et al., 1993). Methodologists such as Peter Molenaar (Boomsma, Martin, & Molenaar, 1989) and John Nesselroade (Plomin & Nesselroade, 1990) are applying their expertise in model fitting and the analysis of intraindividual change to behavioral genetics, respectively.

Other psychologists may be eager to use genetic strategies in their research but do not know how to begin. Most of the scientists mentioned earlier began to use genetic strategies as part of collaborative "big science" projects, but this is not the only route to genetics. My suggestion is to begin by adding siblings in one's research. More than 80% of families have more than one child, and it is relatively easy to recruit a sibling of a subject. After analyzing the topic of interest, the data can be examined from a new perspective that considers sibling similarities and differences. How similar are siblings in the same family for this phenomenon? For most traits, siblings are not very similar, which leads to the following nonshared question: Why are siblings in the same family so different? Multivariate questions can also be asked: Do the familial or nonfamilial influences on one aspect of the phenomenon overlap with effects on another aspect? Developmental questions can be asked about age differences and age changes in sibling resemblance. Although sibling analyses are familial rather than genetic, such analyses represent an important first step in explaining the etiology of individual differences. In addition, twins are not nearly as difficult to find as one may think. Approximately 1% of all births are twins. Moreover, twins, especially parents of young twins, are particularly willing to participate in research because twins are so obviously special. Even adoption designs are not impossible. During the 1960s and early 1970s, approximately 1% of all births involved nonfamilial adoptions and about one third of adoptive parents adopted a second child (Mech, 1973). With contraception and abortion, the numbers of such adoptions declined dramatically during the 1970s. Nonetheless, an adoption study in young adulthood that compares adoptive siblings with nonadoptive siblings is possible. Moreover, as pointed out by Reiss (chapter 22), little use has been made in behavioral genetics of the large numbers of half-siblings that can be found in stepfamilies.

Molecular Genetics

A final reason for optimism about the future of genetic research in psychology is that the field will be the beneficiary of the incredible advances currently being made in molecular genetics. This is surely the first book on genetics and psychology in which so many authors mention molecular genetics. It seems clear that researchers are at the dawn of a new era in which molecular genetic techniques will revolutionize behavioral genetics by identifying specific genes that contribute to genetic variance in behavior (e.g., Aldhous, 1992; McGuffin, Owen, & Gill, 1992; Plomin, 1990). The purpose of this final section is to introduce psychologists to some of the terms and techniques of this revolution.

It was only 10 years ago that the now-standard techniques of the "new genetics" were first used to identify genes responsible for disorders. These techniques began in the 1970s with the discovery of restriction enzymes isolated from various bacteria that cut DNA in specific sites. These restriction enzymes made it possible to recombine a gene from the human species with the DNA of bacteria and thus to clone the gene when the bacteria reproduces. Such recombinant DNA could be used to produce human gene products in bacteria. It also led to the ability to sequence the 3 billion nucleotide base sequences of DNA, which is the ultimate goal of the Human Genome Project (U.S. Department of Energy, 1992).

DNA Markers

For psychologists, the most important outcome of the new genetics is the development of thousands of new DNA markers, genetic differences among people that involve the DNA itself rather than gene products such as the blood groups. Because these markers are the stuff of the molecular genetic revolution in behavioral genetics, I provide an overview of these markers to introduce them to psychologists.

Before 1980, only a few score classical single-gene markers from enzymes in blood, urine, and saliva were available, such as the blood groups. They were limited to the relatively small amount of DNA that is transcribed and expressed as polypeptides in such peripheral systems. In 1980, the first anonymous (function unknown) DNA marker was found using restriction enzymes (Wyman & White, 1980). This type of DNA marker is called a "restriction fragment length polymorphism" (RFLP) because it detects the presence or absence of a restriction enzyme cutting site. If two individuals have any different nucleotide bases for a site recognized by a particular restriction enzyme, the restriction enzyme will cut only the DNA at that site for the individual who has the restriction site. This results in a DNA fragment that is longer for the individual without the restriction site. In other words, the DNA marker is a fragment length difference (polymorphism) caused by the presence or absence of a restriction site recognized by a particular restriction enzyme. The DNA fragment lengths can be detected by a technique called Southern blotting, named after the person who developed the technique (Southern, 1975). This technique begins by making the fragment lengths single stranded and spreading them out in a gel according to length using electrophoresis, which applies an electrical current to the gel. A single-stranded DNA probe (target DNA that has been cloned in bacteria) is made radioactive and washed over the fragments to allow the probe to seek its complement. When the radioactively labeled probe finds its match, it hybridizes to it and shows up as a bright band in autoradiography, which uses an x-ray plate to detect radiation. For example, when cut with a particular restriction enzyme, the β-globin gene yields two fragments: one 2.7 thousand base pairs (kilobases [kb]) and the other 7.2 kb. Both bands are lit up by autoradiography because the probe for β-globin gene encompasses both fragments. A particular allele of this gene is responsible for sickle-cell anemia. Individuals with sickle-cell anemia have a substitution in the gene that is at this restriction site. Thus, their DNA is not cut at this site and, for this reason, they do not show 2.7-kb and 7.2-kb fragments but a single 9.9-kb fragment.

RFLP markers typically have two alleles that indicate the presence or absence of a restriction site. A special type of RFLP, known as a "minisatellite repeat" or "variable number tandem repeat," yields multiple alleles. For unknown reasons, as much as one third of the human genome consists of repetitive DNA sequences. For example, some sequences several hundred base pairs in length repeat hundreds of times, and the number of repeats differs for individuals. When a restriction enzyme cuts out a fragment with a repeat element in it, the resulting fragment will differ in length among individuals as detected by a radioactive probe for the repeat element. Thus, minisatellite repeat markers are not restriction site markers but repeat length markers. A special adaptation of this approach led to DNA fingerprinting, in which probes are used that detect several such minisatellite loci simultaneously (Jeffreys, Wilson, & Thein, 1985). The resulting "bar code" of bands is unique for each individual and thus has practical utility for identifying individuals (e.g., in forensics).

In the past few years, attention has turned to a new type of DNA marker called *microsatellite repeat markers*. Microsatellite repeat markers involve two to four nucleotide base pairs that, again for unknown reasons, repeat dozens of times. Similar to minisatellite repeat markers, microsatellite repeat markers are highly polymorphic (i.e., have many alleles) because the number of repeats differs for different individuals. Hundreds of microsatellite repeat markers have been found, and thousands are potentially available (Weber & May, 1989).

These and other new types of markers rely on a technique that has revolutionized molecular biology: polymerase chain reaction (PCR). PCR amplifies minute amounts of DNA, even a single copy, to make millions of copies in a few hours (Saiki et al., 1988). Two DNA fragments called *primers* (sequences of approximately 20 nucleotide bases) are found that flank the target DNA. PCR finds and copies the DNA sequence between the two primers because the *P* in PCR is a specialized polymerase enzyme that copies each strand of DNA from one primer site to the other site of the other primer. A machine that costs only a few thousand dollars automatically repeats this cycle and thus multiplies the target DNA exponentially. In about 1 hr, 20 PCR cycles can amplify the target by 1 million. PCR makes genotyping much more efficient and capable of working with minuscule amounts of DNA. For example, when the amplified DNA is cut with a restriction enzyme, an RFLP can be detected without hybridization to a radioactively labeled probe because so many copies of the DNA fragments are produced that they show up as a dark band during electrophoresis and differences can be observed directly. PCR has also led to new types of DNA markers. For example, single-strand conformation polymorphisms involve detection of slight differences in the migration of single-stranded DNA in a gel that depends on its conformation, which in turn depends on its nucleotide sequence (Orita, Suzuki, Sekiya, & Hayashi, 1989). In this way, polymorphisms, even single base pair differences, can be detected directly without using restriction enzymes or radioactive probes.

PCR led to a new class of markers of the genome called *sequence tagged sites* (STS) for which pairs of primers are available. Particularly useful are expressed sequence tagged sites (ESTS), pairs of primers for the 10% of DNA that is transcribed and translated into polypeptide products. ESTS that are especially interesting for psychologists are the tens of thousands of genes expressed in the brain. Several thousand brain ESTS have recently been identified (Adams et al., 1991).

The new DNA markers have led to a comprehensive linkage map of the human genome, which incorporates 1,676 DNA markers and spans more than 90% of the human genome (NIH/CEPH Collaborative Mapping Group, 1992). Even more impressive is a new map of the genome that consists of 813 newly developed microsatellite repeat markers that are highly informative for linkage analyses (Weissenbach et al., 1992). Detailed physical maps—an array of overlapping clones—are now available for the two smallest human chromosomes, 21 (Chumakov et al., 1992) and Y (Foote, Vollrath, Hilton, & Page, 1992). A physical map is also being built for other chromosomes (e.g., approximately 40% of the moderately large X chromosome; Mandel, Monaco, Nelson, Schlessinger, & Willard, 1992). These physical maps will greatly facilitate the sequencing of the genome.

DNA Markers and Behavior

DNA markers have been primarily used to identify a chromosomal region and, eventually, to isolate a gene for single-gene traits, most notably, cystic fibrosis and Duchenne muscular dystrophy. These are dichotomous traits, such as Mendel's smooth-versus-wrinkled seeds, in which one gene is necessary and sufficient to explain the observed difference. Although several thousand single-gene traits have been reported, behavior is much more complex. Behavior reflects the functioning of the whole organism, and it is dynamic, changing in response to the environment. Genes that affect behavioral traits are transmitted hereditarily according to Mendel's laws in the same way as genes that affect any other phenotype, but behavior is special in three ways. First, unlike Mendel's smooth-versus-wrinkled seeds, most behavioral dimensions and disorders are not distributed in simple either–or dichotomies, although in psychopathology psychologists often pretend that a line exists that sharply separates the normal from the abnormal. Second, behavioral traits are substantially influenced by nongenetic factors: Heritabilities rarely exceed 50%. Third, behavioral dimensions and disorders are likely to be influenced by many genes, each causing small effects. The challenge is to use DNA markers to find genes in the complex system of behavior that is influenced by multiple genes as well as multiple nongenetic factors.

For a single-gene trait, linkage is a method guaranteed to find the chromosomal location of the gene, even when nothing is known about the gene product. Linkage traces the cotransmission of a marker and a disorder within a family pedigree (Ott, 1985). The exemplar is Huntington's disease, which was the first disorder mapped to a chromosome using the new RFLP markers (Gusella et al., 1983). Huntington's disease has long been known to be a single dominant gene that is lethal later in life regardless of a person's other genes or environment. Other single-gene disorders are quickly being put on the genome map through the use of linkage.

The problem is that behavioral dimensions and behavioral disorders are different. They do not show simple single-gene inheritance. Although linkage can be used for more complexly determined traits (Lander & Botstein, 1989), it is limited to finding a major gene that is largely responsible for a disorder. Moreover, linkage is difficult to use to analyze quantitative traits unless a dichotomy is imposed, and its utility is lessened considerably when the mode of inheritance is unknown for the putative major gene, which is the typical case for behavioral disorders and dimensions (Risch, 1990).

Several chapters in this book mention the failures to replicate early reports of linkage for behavioral disorders. Reliance on linkage techniques that can detect only major gene effects seems like an example of losing one's wallet in a dark alley but looking for it in the street because the light is better there. It is now generally recognized that no major gene for behavioral dimensions or disorders is likely to be found in the population. However, current linkage research assumes that a major gene can be found in certain families. For this reason, linkage studies focus on large pedigrees with many affected individuals in the hope of finding a major gene responsible for the disorder in a particular pedigree. In this view, multiple-gene influence is seen at the population level because different major genes are responsible for the disorder in different families.

The alternative view espoused here is that major genes will not be found for behavior either in the population or in the family. Rather, for each individual, many genes make small contributions to variability and vulnerability. In this view, the genetic quest is to find not *the* gene for a behavioral trait but the many genes that affect the trait in a probabilistic rather than predetermined manner. Although some sledgehammer effects of major genes may be found, it seems more likely that many other alleles nudge development up as well as down for many individuals and do not show a dramatic disruption of development as in the classical single-gene disorders.

The point is not that behavior is too complex to take advantage of the new DNA markers but that researchers need to bring the light of molecular biology into the dark alley. New strategies are needed to identify genes that affect behavioral traits, even when the genes account for only a small amount of variance, when nongenetic factors are important, and when the traits are quantitatively distributed. That is, researchers need to use molecular genetic techniques in a quantitative genetic framework. The sibling-pair quantitative trait loci approach to linkage mentioned at the end of the chapters by DeFries and Gillis and Fulker et al. is a promising step in this direction.

Another strategy is called *allelic association* (Edwards, 1991). *Linkage* refers to loci rather than alleles: Linked traits such as hemophilia and color-blindness do not occur together in the population. By contrast, allelic association occurs when a DNA marker is so close to a gene (or it is part of the gene) that affects the trait that a marker allele is correlated with the trait in unrelated individuals in the population. The particular combination of the marker allele and the effective gene allele that happen to be on the same chromosome is rarely separated by recombination (meiotic crossing over of chromosomes) if their loci are close together on the chromosome.

Allelic associations have been found between disease states and candidate genes such as the human leukocyte antigen (HLA) histocompatibility complex (Tiwari & Terasaki, 1985). For normal variation, the best example is serum cholesterol levels, for which about one quarter of the variance can be explained by four apolipoprotein gene markers (Sing & Boerwinkle, 1987). In psychiatry, an RFLP allele of the D2 dopamine receptor has been reported in several studies to be associated with alcoholism (Cloninger, 1991). That is, the frequency of this allele appears to be greater in severe alcoholics than in controls, although failures to replicate have been reported (see chapter 13 in this book). In seven of nine studies, an association has been found between an HLA allele and paranoid schizophrenia that accounted for about 1% of the liability to the disorder (Sturt & McGuffin, 1985).

A major advantage of allelic association analysis is that it can use samples of unrelated individuals, whereas linkage requires pedigrees of related individuals. In addition, allelic association is just as applicable to quantitative traits as to disorders. Finally, by increasing the sample size of relatively easy-to-obtain unrelated subjects, association analysis can be made sufficiently powerful to detect small genetic effects.

A major problem is that there are so many DNA markers. The allelic association approach is like a myopic search for a few needles in a haystack. In contrast to linkage, which can detect a major gene far away from a marker, allelic associations can be detected only when a marker is very close to a gene that affects the trait. For behavioral traits influenced by many genes as well as nongenetic factors, a near-sighted strategy such as association may be needed to see fine details of the landscape near a marker even if it has to sacrifice the ability to see distant mountains. This is no sacrifice because there are no mountains to be seen. Nonetheless, there are so many markers that randomly drawing straws from the haystack is unlikely to pay off. The odds can be stacked in researchers' favor by beginning the search using the markers in or near genes of neurological relevance (Boerwinkle, Chakraborty, & Sing, 1986). The odds can also be improved by using large samples and well-measured extreme groups to increase the power to detect small effects. The goal is to identify some, certainly not all, genes that contribute to the ubiquitous genetic variance found for behavioral traits.

Another strategy, mentioned by McClearn in chapter 2, is the use of the much more powerful methods available in research on nonhuman animals, especially recombinant inbred strains of mice, in a search for candidate genes for human analysis. Mouse genes are generally similar enough to human genes that they can be used as probes for the human genes.

Given the speed of technological advances in molecular genetics (e.g., Landegren, Kaiser, Caskey, & Hood, 1988), the safest bet is that at the turn of the century, researchers will be investigating multiple-gene influences for complex dimensions and disorders using completely different techniques from those in use today. The bottom line message is that DNA markers will be found that are associated with behavioral traits. When this happens, psychologists will be able to use these markers in their research by collecting saliva and sending it to a commercial firm that will genotype the samples for the relevant markers.

Molecular genetics will provide powerful tools that can be used by psychologists to identify DNA differences among individuals without relying on familial resemblance. In addition to providing indisputable evidence of genetic influence, it will revolutionize behavioral genetics by providing a measured genotype for investigating multivariate and longitudinal genetic issues, the links between the normal and abnormal, and interactions and correlations between genotype and environment. In the larger scheme of things, it will help to integrate genetics research on human and nonhuman animals at the universal level of DNA. It will also help to integrate the increasingly fractionated biological and behavioral sciences. The much-used phrase *paradigm shift* seems no exaggeration for advances of this magnitude. As is the case with most important advances, it will raise new ethical issues as well (Wright, 1990).

Epilogue

In summary, the mounting momentum of behavioral genetics is guaranteed to propel the field far into the next century, especially as it joins the mainstream of psychological research. In this chapter, I have explored the sources of this momentum. These findings, methods, projects, people, and the promise of molecular genetics converge to make this an especially exciting time to study nature and nurture in psychology.

References

Adams, M. D., Kelley, J. M., Gocayne, J. D., Dubnick, M., Polymeropoulos, M. H., Xiao, H., Merril, C. R., Wu, A., Olde, B., Moreno, R. F., Kerlavage, A. R., McCombie, W. R., & Venter, J. C. (1991). Complementary DNA sequencing: Expressed sequence tags and human genome project. *Science, 252*, 1651–1656.

Aldhous, P. (1992). The promise and pitfalls of molecular genetics. *Science, 257*, 164–165.

Anastasi, A. (1958). Heredity, environment, and the question "How?" *Psychological Review, 65*, 197–208.

Boerwinkle, E., Chakraborty, R., & Sing, C. F. (1986). The use of measured genotype information in the analysis of quantitative phenotypes in man. *Annals of Human Genetics, 50*, 181–194.

Boomsma, D. I., Martin, N. G., & Molenaar, P. C. M. (1989). Factor and simplex models for repeated measures: Application to two psychomotor measures of alcohol sensitivity in twins. *Behavior Genetics, 19*, 79–96.

Boomsma, D. I., Martin, N. G., & Neale, M. C. (1989). Structural modeling in the analysis of twin data. *Behavior Genetics, 19*, 5–8.

Bouchard, T. J., Jr., Lykken, D. T., McGue, M., Segal, N. L., & Tellegen, A. (1990). Sources of human psychological differences: The Minnesota Study of Twins Reared Apart. *Science, 250*, 223–228.

Cardon, L. R., DiLalla, L. F., Plomin, R., DeFries, J. C., & Fulker, D. W. (1990). Genetic correlations between reading performance and IQ in the Colorado Adoption Project. *Intelligence, 14*, 245–257.

Chumakov, I., et al. (36 authors). (1992). Continuum of overlapping clones spanning the entire human chromosome 21q. *Nature, 359*, 380–386.

Cloninger, C. R. (1991). D2 dopamine receptor gene is associated but not linked with alcoholism. *Journal of the American Medical Association, 266,* 1833–1834.

DeFries, J. C., & Fulker, D. W. (1985). Multiple regression analysis of twin data. *Behavior Genetics, 15,* 467–473.

DeFries, J. C., & Fulker, D. W. (1988). Multiple regression analysis of twin data: Etiology of deviant scores versus individual differences. *Acta Geneticae Medicae et Gemmellologiae, 37,* 205–216.

Detterman, D. K., Thompson, L. A., & Plomin, R. (1990). Differences in heritability across groups differing in ability. *Behavior Genetics, 20,* 369–384.

DiLalla, L. F., Thompson, L. A., Plomin, R., Phillips, K., Fagan, J. F., Haith, M. M., Cyphers, L. H., & Fulker, D. W. (1990). Infant predictors of preschool and adult IQ: A study of infant twins and their parents. *Developmental Psychology, 26,* 759–769.

Dunn, J., & Plomin, R. (1990). *Separate lives: Why children in the same family are so different.* New York: Basic Books.

Edelbrock, C., Rende, R., Plomin, R., & Thompson, L. A. (in press). Genetic and environmental effects on competence and problem behavior in childhood and early adolescence. *Journal of Child Psychology and Psychiatry.*

Edwards, J. H. (1991). The formal problems of linkage. In P. McGuffin & R. Murray (Eds.), *The new genetics of mental illness* (pp. 58–70). London: Butterworth-Heinemann.

Emde, R. N., Plomin, R., Robinson, J., Reznick, J. S., Campos, J., Corley, R., DeFries, J. C., Fulker, D. W., Kagan, J., & Zahn-Waxler, C. (in press). Temperament, emotion, and cognition at 14 months: The MacArthur Longitudinal Twin Study. *Child Development.*

Fagan, J. F. (1985). A new look at infant intelligence. In D. K. Detterman (Ed.), *Current topics in human intelligence* (pp. 223–246). Norwood, NJ: Ablex.

Fagan, J. F., & McGrath, S. K. (1981). Infant recognition memory and later intelligence. *Intelligence, 5,* 121–130.

Fagan, J. F., & Singer, L. T. (1983). Infant recognition memory as a measure of intelligence. In L. P. Lipsitt (Ed.), *Advances in infancy research* (pp. 31–79). Norwood, NJ: Ablex.

Foote, S., Vollrath, D., Hilton, A., & Page, D. C. (1992). The human Y chromosome: Overlapping DNA clones spanning the euchromatic region. *Science, 258,* 60–66.

Goldsmith, H. H., & Campos, J. J. (1986). Fundamental issues in the study of early temperament: The Denver Twin Temperament Study. In M. E. Lamb, A. L. Brown, & B. Rogoff (Eds.), *Advances in developmental psychology* (pp. 231–283). Hillsdale, NJ: Erlbaum.

Gusella, J. F., Wexler, N. S., Conneally, P. M., Naylor, S. L., Anderson, M. A., Tanzi, R. E., Watkins, P. C., & Ottina, K. (1983). A polymorphic DNA marker genetically linked to Huntington's disease. *Nature, 306,* 234–238.

Hetherington, E. M., Reiss, D., & Plomin, R. (Eds.). (in press). *Separate social worlds of siblings: Impact of nonshared environment on development.* Hillsdale, NJ: Erlbaum.

Hewitt, J. K., Silberg, J. L., Neale, M. C., Eaves, L. J., & Erikson, M. (1992). The analysis of parental ratings of children's behavior using LISREL. *Behavior Genetics, 22,* 293–317.

Ho, H.-Z., Baker, L. A., & Decker, S. N. (1988). Covariation between intelligence and speed of cognitive processing: Genetic and environmental influences. *Behavior Genetics, 18,* 247–261.

Jeffreys, A. J., Wilson, V., & Thein, S. L. (1985). Individual-specific "fingerprints" of human DNA. *Nature, 316,* 76–79.

Jensen, A. R. (1987). The *g* beyond factor analysis. In R. R. Ronning, J. A. Glover, J. C. Conoley, & J. C. Witt (Eds.), *The influence of cognitive psychology on testing* (pp. 87–142). Hillsdale, NJ: Erlbaum.

Landegren, U., Kaiser, R., Caskey, C. T., & Hood, L. (1988). DNA diagnostics: Molecular techniques and automation. *Science, 242,* 229–237.

Lander, E. S., & Botstein, D. (1989). Mapping Mendelian factors underlying quantitative traits using RFLP linkage maps. *Genetics, 121,* 185–199.

Loehlin, J. C. (1987). *Latent variable models: An introduction to factor, path, and structural analysis.* Hillsdale, NJ: Erlbaum.

Loehlin, J. C., Horn, J. M., & Willerman, L. (1989). Modeling IQ change: Evidence from the Texas Adoption Project. *Child Development, 60,* 993–1004.

Lykken, D. T., Bouchard, T. J., McGue, M., & Tellegen, A. (1990). The Minnesota Twin Registry: Some initial findings. *Acta Geneticae Medicae et Gemmellologiae, 39,* 35–70.

Mandel, J.-L., Monaco, A. P., Nelson, D. L., Schlessinger, D., & Willard, H. (1992). Genome analysis and the human X chromosome. *Science, 258,* 103–109.

Matheny, A. P., Jr. (1989). Children's behavioral inhibition over age and across situations: Genetic similarity for a trait during change. *Journal of Personality, 57,* 1–21.

McGue, M., & Bouchard, T. J., Jr. (1989). Genetic and environmental determinants of information processing and special mental abilities: A twin analysis. In R. J. Sternberg (Ed.), *Advances in the psychology of human intelligence* (Vol. 5, pp. 7–45). Hillsdale, NJ: Erlbaum.

McGuffin, P., Owen, M., & Gill, M. (1992). Molecular genetics of schizophrenia. In J. Mendlewicz & H. Hippius (Eds.), *Genetic research in psychiatry* (pp. 25–48). New York: Springer-Verlag.

Mech, E. V. (1973). Adoption: A policy perspective. In B. M. Caldwell & H. N. Ricciuti (Eds.), *Reviews of child development research: Vol. 3. Child development and social policy* (pp. 467–508). Chicago: University of Chicago Press.

Mussen, P. H. (Ed.). (1983). *Handbook of child psychology.* New York: Wiley.

Neale, M. C., & Cardon, L. R. (1992). *Methodology for genetic studies of twins and families.* Norwell, MA: Kluwer Academic.

NIH/CEPH Collaborative Mapping Group. (1992). A comprehensive genetic linkage map of the human genome. *Science, 258*, 67–86.

Orita, M., Suzuki, Y., Sekiya, T., & Hayashi, K. (1989). Rapid and sensitive detection of point mutations and DNA polymorphisms using the polymerase chain reaction. *Genomics, 5*, 874–879.

Ott, J. (1985). *Analysis of human genetic linkage*. Baltimore: Johns Hopkins University Press.

Pedersen, N. L., McClearn, G. E., Plomin, R., Nesselroade, J. R., Berg, S., & DeFaire, U. (1991). The Swedish Adoption/Twin Study of Aging: An update. *Acta Geneticae Medicae et Gemellologiae, 40*, 7–20.

Pedersen, N. L., Plomin, R., Nesselroade, J. R., & McClearn, G. E. (1992). A quantitative genetic analysis of cognitive abilities during the second half of the life span. *Psychological Science, 3*, 346–353.

Plomin, R. (1988). The nature and nurture of cognitive abilities. In R. Sternberg (Ed.), *Advances in the psychology of human intelligence* (Vol. 4, pp. 1–33). Hillsdale, NJ: Erlbaum.

Plomin, R. (1990). The role of inheritance in behavior. *Science, 248*, 183–188.

Plomin, R. (1991). Genetic risk and psychosocial disorders: Links between the normal and abnormal. In M. Rutter & P. Caesar (Eds.), *Biological risk factors for psychosocial disorders* (pp. 101–138). Cambridge, UK: Cambridge University Press.

Plomin, R., DeFries, J. C., & Fulker, D. W. (1988). *Nature and nurture during infancy and early childhood*. New York: Academic Press.

Plomin, R., DeFries, J. C., & McClearn, G. E. (1990). *Behavioral genetics: A primer* (2nd ed.). New York: Freeman.

Plomin, R., & Neiderhiser, J. M. (1992). Quantitative genetics, molecular genetics, and intelligence. *Intelligence, 15*, 369–387.

Plomin, R., & Nesselroade, J. R. (1990). Behavioral genetics and personality change. *Journal of Personality, 58*, 191–220.

Reiss, D., Plomin, R., Hetherington, E. M., Howe, G., Rovine, M., Tryon, A., & Stanley, M. (in press). The separate worlds of teenage siblings: An introduction to the study of the nonshared environment and adolescent development. In E. M. Hetherington, D. Reiss, & R. Plomin (Eds.), *Separate social worlds of siblings: Impact of nonshared environment on development*. Hillsdale, NJ: Erlbaum.

Risch, N. (1990). Genetic linkage and complex diseases, with special reference to psychiatric disorders. *Genetic Epidemiology, 7*, 3–16.

Robinson, J. L., Kagan, J., Reznick, J. S., & Corley, R. (in press). The heritability of inhibited and uninhibited behavior: A twin study. *Developmental Psychology*.

Saiki, R. K., Gelfand, D. H., Stoffel, S., Sharf, S. J., Higuchi, R., Horn, G. T., Mullis, K. B., & Erlich, H. A. (1988). Primer-directed enzymatic amplification of DNA with a thermostable DNA polymerase. *Science, 239*, 487–491.

Saudino, K. J., & Eaton, W. O. (1991). Infant temperament and genetics: An objective twin study of motor activity level. *Child Development, 62*, 1167–1174.

Schaie, K. W., Plomin, R., Willis, S. L., Gruber-Baldini, A., Dutta, R., & Bayen, U. (1993). Longitudinal studies of family similarity in intellectual abilities. In J. J. F. Schroots & J. E. Birren (Eds.), *The next generation of longitudinal studies of health and aging* (pp. 183–198). Amsterdam: Elsevier.

Searles, J. S. (1988). The role of genetics in the pathogenesis of alcoholism. *Journal of Abnormal Psychology, 97*, 153–167.

Sing, C. F., & Boerwinkle, E. A. (1987). Genetic architecture of inter-individual variability in apolipoprotein, lipoprotein and lipid phenotypes. In G. Bock & G. M. Collins (Eds.), *Molecular approaches to human polygenic disease* (pp. 99–122). New York: Wiley.

Southern, E. M. (1975). Detection of specific sequences among DNA fragments separated by gel electrophoresis. *Journal of Molecular Biology, 98*, 503–517.

Sturt, E., & McGuffin, P. (1985). Can linkage and marker association resolve the genetic aetiology of psychiatric disorders: Review and argument. *Psychological Medicine, 15*, 455–462.

Tellegen, A., Lykken, D. T., Bouchard, T. J., Wilcox, K., Segal, N., & Rich, S. (1988). Personality similarity in twins reared apart and together. *Journal of Personality and Social Psychology, 54*, 1031–1039.

Thompson, L. A., Detterman, D. K., & Plomin, R. (1991). Associations between cognitive abilities and scholastic achievement: Genetic overlap but environmental differences. *Psychological Science, 2*, 158–165.

Tiwari, J., & Terasaki, P. I. (1985). *HLA and disease associations*. New York: Springer.

U.S. Department of Energy. (1992). *Human genome: 1991–1992 program report*. U.S. Department of Commerce: National Technical Information Service.

Vernon, P. A. (1989). The heritability of measures of speed of information processing. *Personality and Individual Differences, 10*, 573–576.

Weber, J. L., & May, P. E. (1989). Abundant class of DNA polymorphisms that can be typed using the polymerase chain reaction. *American Journal of Human Genetics, 44*, 388–396.

Weissenbach, J., Gyapay, G., Dib, C., Vignal, A., Morissette, J., Millasseau, P., Vaysseix, G., & Lathrop, M. (1992). A second-generation linkage map of the human genome project. *Nature, 359*, 794–801.

Wright, R. (1990, July 9). Achilles' helix. *The New Republic*, pp. 21–31.

Wyman, A. R., & White, R. L. (1980). A highly polymorphic locus in human DNA. *Proceedings of the National Academy of Sciences, 77*, 6754–6758.

Zahn-Waxler, C., Robinson, J., & Emde, R. (in press). The development of empathy in twins. *Developmental Psychology*.

Zuckerman, M. (1991). *Psychobiology of personality*. New York: Cambridge University Press.

Index

About the Editors

Robert Plomin is a professor of human development at the Department of Human Development and Family Studies, Pennsylvania State University, and a visiting professor at the Department of Psychological Medicine, University of Wales, College of Medicine, Cardiff, Wales. Dr. Plomin's recent honors include the Fulbright Scholar Award and Fogarty Senior International Fellowship, and he has been named Distinguished Scientist Lecturer by the American Psychological Association. He previously served as president of the Behavior Genetics Association and has written widely on the topic.

Gerald E. McClearn is Dean, College of Health and Human Development, Pennsylvania State University. His honors include the Dobzhansky Memorial Award for Eminent Research in Behavioral Genetics from the Behavior Genetics Association. Dr. McClearn has served on numerous national panels for the National Institutes of Health and the National Academy of Sciences, and he has written widely on behavioral genetics issues.